PHILOSOPHY AND THE CHALLENGE OF THE FUTURE

PHILOSOPHY AND THE CHALLENGE OF THE FUTURE

JOHN LANGE

ISBN: 978-1-5040-2470-9

Distributed in 2015 by Open Road Distribution
345 Hudson Street
New York, NY 10014
www.openroadmedia.com

CONTENTS

Introductory Considerations

Part One: Anticipations

Part Two: Conceptualizations

Part Three: Prescriptions

INTRODUCTORY
CONSIDERATIONS

PHILOSOPHY HAS NEW THINGS TO DO, AND NEW places to go.

I am not sure she is aware of that.

But revolutions are afoot.

Perceptions are limited, and generalizations are dangerous, but one cannot help but have perceptions, and it is difficult to think without generalizations, particularly when one finds them widely shared and significantly supported. To be sure, as one would expect, strictly considered, most of them are false. But most of them, too, are "true enough," or "pretty much true," like "dogmatism leads to intellectual stagnation," "fanaticism is dangerous," and so on.

Most people, including philosophers, have rather ambivalent feelings toward philosophy. Morale, in the discipline, is not high, particularly amongst older philosophers. Indeed, it seems contests are underway to see who can belittle and insult philosophy most cleverly. It is hard to know what to do with philosophy today. Some of it flourishes, famously and spectacularly, on the cutting edge of triviality, applying itself more and more to less and less; some of it keeps its eye on the rear view mirror, what might Kant have meant by that; and some of it is more interested in changing a world than understanding it, more interested in politics than thought, more interested in enforcing conformity than freeing the mind, less interested in educating than indoctrinating.

Philosophy become propaganda is philosophy betrayed.

Disciplines, of course, tend to mutate. And there are ascents, and plateaus; stagnations and revolutions. Sometimes they march; sometimes they mark time. Today, philosophy is on a plateau, many not noticing her position at the foot of an ascent.

I think the current doldrums in philosophy, her peregrinations into minutiae, apparently the best that can be managed by many, at least currently, her antiquarianism, and her occasional political experiments in Pavlovian intellection and behavioral modification, are symptoms less the results of decadence, misdirection, and treachery, or suicidal self-betrayal, than "plateau phenomena," the phenomena of a discipline in dormancy, confusion, boredom, the compulsion to tidy things up, the temptation to substitute one ideal for another, one activity for another, if only to have something to do, which seems important.

Yet, what could be more important than philosophy?

In one useful sense of "philosophy," an ancient and profound sense, it has to do with the love of, and the seeking of, wisdom. And that, obviously, has no essential relationship to academies, administrations, divisions, departments, and such. That is available to all human beings, to the physicist, the chemist, the physician, the architect, the journalist, the accountant, the chef, the electrician, the plumber, the truck driver, the musician, the artist, the student, even the professional philosopher. And wisdom transcends understandings and knowledge; it also involves what might be done, and perhaps well done, in the light of understandings and knowledge.

That is important, not only what might come about, but how might we relate to it.

The sciences, as opposed to politics and religion, have their roots in philosophy. Philosophy has been spoken of as the mother of the sciences, although she is, in many cases, more of a grandmother or great grandmother. Scientists were originally regarded as "natural philosophers," namely, philosophers interested in nature. There has often been an intimate relationship between science and philosophy, exceeding the facts of intellectual genealogy. The mansions of philosophy are best built, after all, on the foundations of fact, and she is deeply indebted to her "natural philosophers," whom, in my view, she is entitled to claim and see as her own. What wondrous contributions to a human world view, or world vision, are owed to Copernicus and Kepler, Galileo, Newton, and Einstein, Darwin, and Freud, and thousands of others. Philosophy divorced from nature is malnourished, if not starved, if not barren. The telescope and microscope contributed more to philosophy than Hegel and Aquinas.

A tension is generated, in eras of cognitive advance, between a status-quo, but revisable, "common sense" and new realities, new discoveries. One of the places where philosophy spends its time is the border territory between a new science and an old common sense. What sense can we make of new truths, new possibilities? How shall we think about them? Should they alter our view of the world, not merely additively, as in learning a new telephone number, but radically, as in taking seriously the hypothesis that wind is a meteorological phenomenon and not the breath of a god, that the earth moves, that man may be transformed and find himself become an alien onto himself.

Philosophy has been on the plateau long enough. I do not think she will stay there much longer.

Probably no one is more aware of the limitations, lacks, weaknesses, and faults of this small book than I am. Certainly I have considered never writing it, which leaves a nice opening for critics, because, it seems to me, such a book, at least now, cannot be well written. On the other hand, it also seems to me that this sort of book, or something much like it, should be written, indeed, must be written, by someone.

So, I have tried.

Philosophy has new things to do, and new places to go.

I am not sure she is aware of that.

But revolutions are afoot.

PART ONE

ANTICIPATIONS

CHAPTER ONE:
PROCREATIVE LIBERTY

THE TWO GREAT ENEMIES OF PHILOSOPHY ARE politics and religion, or power and superstition. Whereas there is a bromide to the effect that might does not make right, it is reasonably clear someone's view of right may be inflicted on others easily, by means of convenient, at-hand tools, ostracization, isolation, exclusion, avoidance, marginalization, intimidation, insult, scorn, ridicule, disemployment, fines, imprisonment, torture, and death. Thrasymachus was wiser in the ways of the world than naive Socrates. Those in power, with weapons and men at their disposal, will define right and justice as they wish, in a self-serving manner, subject only to practicalities, their capacity to resist opposition, this largely a function of the disorganization and sessility of the ruled, if not oppressed, and the loyalty of minions. One may rule by the spear or rule by the mind. And ruling by the mind is most effective, economizing on resources. An individual self-imprisoned is well imprisoned, and must pay for his own keep. So the democracy sentences Socrates to death, and he does not fight or flee, but goes like a sheep to the slaughter, to lap the hemlock, obedient to the very laws by which he was condemned. And Plato expects us to applaud. But perhaps Socrates was merely weary, or seriously expected to trade in an old life for a new one, somewhere, somehow. Plato, of the Athenian aristocratic party, indeed, even related to the notorious and murderous Critias, of the Thirty Tyrants, viewed democracy as a beast, or monster, of many heads. Surely justice and right could not depend on the outcome of a poll, or vote, shifting about, depending on the weathers of an electorate. But on what could it depend? There is one advantage to a beast with many heads, which is that the heads may hiss, strike, spit, and bite at one another, and serve thusly to balance tyrants against tyrants.

In any event, might may not make right, despite appearances to the contrary, but it is not so obvious what would make right. Indeed, it is not obvious what right is, other than a self-serving posit, or invention, by interested parties, and there are many such posits, or inventions, incompatible with one another. And people hate, and suffer, and die, using right or justice as an excuse, each dogmatic in his zealotry. Who cannot see that, that having one's way, that competition for power, or resources, is the smithy within which warring rights are forged?

These dismal remarks regrettably emplaced, one turns to a related question, the natural as opposed to the artificial, or unnatural. The notion of right and wrong surely occurs in this context. For example, is what is natural right, is what is different, surprising, unnatural, unusual, or artificial wrong?

One supposes that hominids in the wild, say, chimpanzees, gorillas, baboons, and such, behave in a natural manner. It does not seem to make much sense to say that the way they behave is right or wrong, though it might be prudential or not. One supposes that humans, or protohumans, once lived similarly. On the other hand, is it natural or unnatural for, say, a clever ape to grasp a stone and strike another ape, thus utilizing a tool, or weapon, or, say, is it natural that an ape, or apelike thing, might invent a word, or wrap itself in a hide against the cold, or make a fire? Humans no longer drink from puddles or race about naked, or hide at night in trees. We do not regard such revisions, abstinences, and departures as unnatural. It seems plausible to consider superorganic evolution, rather like organic, or biological, evolution, as something natural to an occasionally rational species. If not, then most of us would suppose that a number of "unnatural things" are unobjectionable, even right. Surely they increase comfort, security, and power. Many things once castigated as unnatural, and thus wrong, are now accepted without demur, say, inoculations against disease, anesthesia to ease the pangs of childbirth, and such. Are spectacles natural? Certainly they are generally considered acceptable. It is nice to be able to see.

Philosophy, here and there, begins to dip into subjects such as reprogenetics, i.e., reproductive biology and genetics, cloning, and genetic engineering. This is healthy, as adventure, exploration, and heresy often are. Such things, of course, are the first hints that the philosophical world may be shaken once more into motion, and discover itself once more dislodged from the center of one universe, with which it was familiar, to find itself adrift anew in larger spaces, not yet explored. It would not be the first time that our "natural philosophers," with their discoveries, inventions, proposals, theories, and practices have opened new vistas and pioneered new world views. These new fields, precipitated by burgeoning sciences, may, for a time, be ignored, and sealed off from philosophical respectability, but, eventually, they will have their inevitable impact on philosophy, as they will on law and common sense. Surely they will lure a discipline, or force a discipline, into new paths.

Happily, most moral issues, and their appropriate resolutions, given particular cultures, and the systemic processes that produce workable moralities, are reasonably clear. This is to be expected, given the likely recurrence of similar sorts of situations, over generations, amongst relatively homogeneous populations, under reasonably stable conditions. Problematicities are most likely to be generated in virtue of cultural diffusion, cultural differences, intergroup contact, peaceful or otherwise, migration, immigration, and social and economic change, for example, populational increase, a shift from farm to city, from agriculture to industry, and such. Such shifts necessitate adaptation, as much so as a change in climate might, over a thousand years, require an arboreal species to adapt to the ground, or die. Similarly, a rapidly evolving technology is likely to produce situations to which traditional guidelines are either irrelevant, or remote. The old measuring rods and scales, so wonderfully useful in their time, and invaluable yet, may not be clearly applicable to novel, unanticipated realities. For new game new nets might need be woven.

Much is involved here.

I will, shortly, largely for illustrative purposes, call attention to four possibilities connected with reprogenetics. On the other hand, to give a sense, more generally, of what is going on here, it may be helpful to review, or allude to, briefly, some of the concepts associated with, or consequences attendant on, developments in this remarkable field.

Traditionally, for example, one distinguished between the social mother and the biological mother, the social mother being the woman who raises the child, and the biological mother being the woman who conceives and delivers the child. Today, given reprogenetics, the concept of the biological mother is more complex, being divided between the gene mother, whose egg is fertilized, and the gestational mother, or host mother, who, the gene mother's fertilized egg implanted in her womb, brings the child to term. Classically, it was common for all three roles to be fulfilled by the same woman. With artificial wombs under development, it is possible that, eventually, a gestational mother will not be required.

Controversy has surrounded research utilizing stem cells, which might be natural or bred. Natural stem cells are totipotent, pluripotent, or multipotent. The totipotent stem cell is capable of developing into a complete individual; the pluripotent stem cell can develop into any cells in a human body; the multipotent stem cells are found in mature tissue, and have only a limited ability to grow into different types of cell. In bred stem cells molecular signals are adjusted to retain potentiality. Other molecular signals could presumably be used to trigger the growth of particular organs, or such. Cells from mature tissue restored to potency by molecular signals might allow less controversial research, as embryos would not be required. Religious individuals, or many, commonly believe that the embryo is ensouled at, or shortly after, conception. Accordingly, understandably, they are reluctant to sanction an act which, in effect, is prenatal infanticide. Even nonreligious individuals may have reservations about terminating what is clearly human life, though at an early stage of development. The justification for utilizing embryos is commonly utilitarian, indexed to their value in connection with medical research. Genetic therapies, perhaps better discussed under genetic engineering, are germ-line therapies or somatic-cell therapies. Germ-line changes will be transmitted to offspring. Chromosomes may be either natural or constructed. Constructed chromosomes might be the recipients of "gene-packs," their insertion into the cell nuclei tending to avoid genetic disruption.

A large variety of techniques and possibilities are, or may eventually be, involved in reprogenetics. Some of these are donor insemination; laboratory fertilization; egg donation; embryo donation; embryo multiplication ("twinning"); embryo selection; sperm concentration; cryopreservation, say, of embryos; embryo transfer, from one animal or person to another; nuclear transfer, which is cloning, to be discussed later; testicular transfer; artificial wombs and nonhuman surrogate mothers.

Most of the above techniques are familiar to reprogenetics and commonly applied. For example, it is estimated that in the United Stated alone there are something like 200 million DI-conceived (donor-insemination-conceived) individuals. Given the screenings involved in donor selection, the preferential criteria utilized, and such, the genomes in this group are likely to differ from those more common in the population at large. Further, in time, given interbreeding, such individuals may have their effect on the gene pool as a whole. This is analogous to selective, improvement breeding, or eugenics, reminiscent of plant and animal husbandry. There is no essential connection between this sort of thing and racism, or such, though quality, conceived in one way or another, is commonly involved.

Now, to give a better idea of reprogenetic possibilities, present and future, one might consider, briefly, four possibilities.

1. Fetal Mating

Both oocytes and spermatogonia exist in fetuses. For example, millions of immature eggs, oocytes, are stored in a woman's ovaries, which are all she will ever produce, and they are all there some six and a half months before she is born. These oocytes and comparable spermatogonia, from a male fetus, may be retrieved from aborted fetuses. The spermatogonia may be matured within the testes of an appropriate animal species, and the oocytes could be matured into fertilizable eggs, in a laboratory incubator. Later, given IVF techniques (in-vitro-fertilization techniques), the matured sperm could be used to fertilize the matured eggs, producing normal, viable embryos which, in turn, could be implanted in a host mother, presumably human, or an artificial womb, should such then be available. Thus, children might be born to parents who themselves were never born. They would be orphans, of parents who died before they were born. Biologically, such children would have relatives, for example, grandparents. Indeed, they might turn out to be the next of kin to such grandparents, be entitled to their name, entitled to an estate, an inheritance, and such. Additionally, they would be citizens of a state, and subject to the rights and privileges of citizenship, and so on. Interestingly, such matings, as forced, nonconsensual matings, might be construed as violations of civil rights, or even crimes, such as kidnapping, abuse, rape, or such.

2. The Twenty-Thousand-Year-Old Father

Spermatogonia cells produce sperm. These cells may be frozen and transferred amongst animals. In this fashion, such cells being self-renewing, the host animal can produce the sperm in question indefinitely. Such cells may then be refrozen, and transferred, sequentially, to a line of animals. As these transfers may be continued indefinitely, it would be empirically possible, given IVF techniques, for a given individual not only to produce thousands of offspring but to continue to do so, indefinitely, for thousands of years. A rich individual might endow a foundation with this sort of thing in mind. It would be easy to conceive of cults being interested in this sort of thing, as well, perhaps each year honoring a sacred virgin, or such, chosen from amongst their flock, to bear the child of the leader, savior, prophet, or such, in question, who may have died thousands of years before.

3. Pluriparents

By fusing embryos one may produce offspring with genetic endowments from several individuals. This would produce, in a strict sense, a chimera, namely, a mixed organism. The children resulting from such techniques seem to be healthy and normal. One could conceive of friends resorting to this technique in order to produce a common offspring. Most typically, it might be favored by lesbian couples. In any event, given this technique, it is possible for a child to have several parents. Here, as in many cases, one has the question as to how best to understand "natural." Obviously this is unnatural in the sense of being an unusual way in which to produce a child, but, on the other hand, the technique utilizes natural processes to produce a normal, say, natural, child.

4. Embryo Design

In theory it is possible to rig the genetic lottery. Any given couple has its several thousands of genes. One supposes they wish the child to be theirs, to be, on the whole, at least, the product of their genes, but of which genes? Rather than hope for the best, they might prefer to arrange the best, according to their lights. If, as most couples, they carry some less favored, undesirable, or even deleterious genes, they may not want those in the embryo. Too, they might have some preferences, with respect, say, to appearance, intelligence, aptitudes, and such. It would be nice to include those in the embryo. Too, of course, it might be nice to add in some genes from the outside, as gifts to the offspring, so to speak.

Let us suppose several artificial, or "constructed," chromosomes are available on the market. Each of these contains a "gene packet," of one type or another, which may be inserted into the cell nucleus. Inserting a chromosome as a whole is a way, as suggested earlier, to minimize the risk of genetic disruption. There is nothing random, or unanticipated, about these gene packets, unless the client would like to be surprised. Certainly there is nothing in them likely to be disappointing. The couple might then peruse the colorful brochures available from the various genetic supply houses. Such brochures will list the contents of the various packets, which are diversely priced, depending on the offerings, the prices, of course, subject to change without notice. Presumably this would be something like selecting amongst cable services, and deciding what one might most like from amongst the selections available on such a service. It would be nice to have that channel, so let's add it in. Software might even be provided, by means of which one might get a sense of what the child might look like, at various ages, and into early adulthood, and such. As one is concerned with one's own child, one, as a typical parent, concerned and loving, will wish to do the best for the child that lies within one's power, with respect to advantages, opportunities, and such. The typical parent is likely to be willing to sacrifice a good deal for their offspring, economically and otherwise. It is not only that it would be nice to have that gene, so let's add it in, but rather, how could one possibly bring oneself to deny that gene to the child? What sort of parent is one? How selfish can one be? This gene packet or that may be expensive, but, remember, one is dealing with one's own child. As a parent one does what one can for one's child, even if it means considerable sacrifice, even if it means going heavily into debt. Nothing is too good for the child.

Whereas the scenario suggested in this fourth possibility is not practical given the state of current reprogenetics it is in theory practical at some future date, perhaps within the current century, and the sort of thing it envisages, namely, embryo production and modification, to one extent of another, is already a matter of reprogenetic fact.

It is generally, and plausibly, supposed that embryo design, when it becomes available, would be extremely expensive, ranking perhaps with apartment buildings and yachts. Accordingly, one supposes it would be primarily available to the rich, even the very rich, or to those in a position of power, with the capacity to divert the resources of others to their own ends.

It is difficult to know what might be the social and economic consequences of the availability of such procedures and techniques. It is quite possible that nothing much would come of it. Perhaps the engineered children would, on the whole, from the point of view of the population at large, if not from that of their parents, not be all that different from other children. They might be popular, fit in nicely, and ruffle few feathers. Perhaps, too, the whole movement might be nipped in the bud, by public disinterest, disapproval, scorn, or ridicule, or perhaps it might be extinguished politically. Perhaps an outspoken, well-organized minority of technological Luddites, so to speak, would raise such a righteous, scandalized hue and cry that legislators, fearful for their tenure in office, would appoint loaded committees to look into the matter, following the reports of which such techniques, predictably, would be outlawed, or regulated into practical extinction. To be sure, in such situations one would expect the technology to go underground or emigrate. But, in either case, it would presumably continue to be rare and expensive, and would most likely have little impact on the world as a whole. Too, if it were illegal one supposes many individuals would be less likely to avail themselves of such opportunities, if not from moral or prudential reservations, perhaps because of skepticism concerning the quality of such services, provided under uncertain or dubious conditions. And certainly one would expect that horror stories of procedures gone wrong would be rampant. But that would be true in any case.

There are, of course, darker possibilities. Envy, hatred, and violence, are common amongst human beings. Human beings, many of them, tend to be good at such things. For example, it is very common for many human beings to literally hate the wealthy and successful. It is not so much that Smith wants Jones' money, as that Smith does not want Jones to have it; it is not so much that Smith wants to be successful, which usually requires a considerable expenditure of time and effort, which may not pay off, as that he does not want Jones to be successful. "At least the rich people no longer have their jewels." This sort of thing is endemic in many cultures, and, socially, tends to reduce enterprise and initiative, for fear of the social penalties, even hazards, attached to such antisocial behaviors. In freer cultures it seems the concept of "luck" is often used to, and may have been invented to, divert envy, as though success was not really the fault of the succeeder, who thusly should not be blamed for it, but is to be attributed rather to luck, the unpredictable and capricious smiles of Fortuna, its fickle goddess. Today, one hears much of dark rumblings, having to do with social and economic "polarization." It is not that our rumblers are badly off, so much as they do not approve of others being better off than they, particularly if they are much better off, which is not to deny, of course, that some people are not well off, who are usually getting on with their lives and not complaining. No one, of course, at least as far as I know, favors destitution. I suppose that needed to be said, which is in itself a pathetic commentary on our times. In any event, given the nature of human beings, let us suppose that our social and economic "polarization," such as it is, was exponentially aggravated by a new polarization, to further outrage those interested in such things, "biological polarization."

Contrary to all likelihood, let us suppose that, over time, say seven or eight generations, careful not to "marry out," our engineered children thrive, and become recognizable in the society at large. We shall suppose, again contrary to all likelihood, that they are all extremely intelligent, and are, on the whole, larger, stronger, better looking, and more talented, than their less favored brethren in the population as a whole. It is natural to suppose that, in an unrigged system, a free system, one not practicing systematic discrimination for or against one group or another, the results of the designed embryos, their descendants, and such, would come to positions of authority, prestige, and power in a society. This is not necessary, of course, as they might, in the classical Epicurean tradition, prefer to chat in the garden, making free with the barley bread and water. Or, following Aristotle, given their high intelligence, and such, they might prefer a contemplative life, leaving the hurly-burly, tumult, and competition of the vulgar life and the market to folks to whom it was still important.

On the other hand, if the "different types," a more carefully chosen word than "higher types," despite their coaching in techniques of envy avoidance, say, attending baseball games, affecting a homely dialect, wearing suspenders, uttering populist rhetoric, and such, were generally recognizable, one might expect them to identified and selected out, under one pretext or another, perhaps racial or ethnic, or such, or perhaps simply singled out as threats to society, "enemies of the people," social contaminants, plague carriers, tainted monsters, moral inferiors, heretics, or such. Something would be worked out.

One speculates, of course, herein, based on historical precedents, relative to the dangers of difference, the sensitivity of establishments to peril, their resolve to preserve and enhance their position and power, and such, at any cost. But what if the future should be much different, culturally, psychologically, and morally, from the present. Perhaps the "different types" would be prized in society, and welcomed as its saviors and redeemers. Might they not be accepted with jubilation and gratitude, and worshipped as gods amongst men? One need not suppose that demagogues, from either class, would emerge to rouse and inflame "normals," utilizing them as a power base, and a militant weapon for the transformation of society. Caesars have always, whatever their class of origin, had purposes to which to put the people. Many are the means by which people may be inveigled, bribed, and swayed, amongst which "bread and circuses" are doubtless the least objectionable, doubtless the most benign. But, let us, per hypothesis, set aside the hints, if not the lessons, of history, and suppose a peaceful, stable, two-tiered society emerges. Indeed, perhaps the reprogenetic techniques might become more generally affordable, as nowadays calculators and computers have become more affordable. Or, if not, perhaps resentment might be defused through lotteries, where access to such reprogenetic techniques might be won by occasional couples, and other couples might continue to hope, to buy tickets, and such. Surely such lotteries have a similar role, if only as a byproduct, in contemporary society. One would not wish to do away with a society in which personal, fabulous wealth is a genuine, if improbable, prospect. Too, much may be done though conditioning, education, religion, indoctrination, and other forms of social engineering, and the "different types" may be in charge of such matters, and would be likely to use them in their own best interest, an interest which would, if wise, be not only acceptable to "normals," but designed to enhance their lives and self-image. Historical precedents certainly suggest themselves. For example, one might suppose that one or more gods are particularly fond of "normals," holding them more dear than "differents." Each "normal" would have, perhaps, an infinitely precious "x" or something, which is as valuable, or more valuable, than the "x" of a "different," who might not even have an 'x." Perhaps the god or gods were even "normals" to begin with, before becoming divine, or perhaps they spent some time being "normals," familiarizing themselves with the difficulties of the condition, before resuming their godship. Similarly, whatever might be the case with the "differents," the "normals" are assured they are immortal and, if they behave nicely, will go to a nice place after they die, or, if they have not behaved so well, perhaps to a less nice place. These beliefs, with their attendant hopes, and more importantly, terrors, are stamped into the "normals" during early childhood, and may remain indefinitely influential, affecting behavior into adulthood, and old age, drawing on strings from behind the curtain of consciousness, particularly, one supposes, in the psyche of a "normal." Society's arrangements, thus, are endowed with divine sanction, which is all right with the "differents." Certainly the "normals" may regard themselves as being morally, if not socially or economically, superior to the "differents," which is a viewpoint easily come by, and one not likely to be challenged by a rational "different." Indeed, such a "different" would be well advised to encourage such a view, in public. Too, nothing prohibits the "normal" from regarding himself not only as morally superior to the "different," but intellectually superior, as well, though, of course, in the light of a somewhat different concept of intelligence, that of "true intelligence," perhaps "gut thinking," "blood thinking," or such, which would also be acceptable to a rational "different," publicly.

In closing, one might briefly consider this general possibility from the point of view of the beneficiaries, or victims, of embryo design. It is difficult to know what the psychological effects of this sort of thing would be on the offspring, if they were apprised of their origins. Probably the effects would tend to be a function of the view of society as a whole. Would they be accepted and welcomed, envied and hated, viewed as gifts to the human race, regarded as artifacts and freaks, or what? As there are fashions in pets, and names, and such, might there not be fashions in embryo design? From time to time one packet might be more prized and popular than another. Setting aside forms of intelligence, diversities of aptitudes, and such, one might suppose a run on, say, blond-haired, blue-eyed children in one decade, and brown-haired, brown-eyed children in the next. Perhaps fashions in height and skin color, or in gender, might vary from time to time, and so on. If one has a normal child in a normal way one is usually happy to get what one gets. Certainly it would be unusual to question the roll of nature's dice. Few are the diminutive barbarians who are not jubilantly welcomed into the gates of the city. On the other hand, suppose one had literally selected and paid for a product with such-and-such properties, rather as one might select an automobile, of a given make, design, and color, with or without an automatic transmission, with or without power steering, with or without antilock brakes, and so on. It seems then that one might possibly, if disappointed, regret one's choices. Perhaps one should have gone for a standard shift, after all. Worse, perhaps one asked for a standard shift and received an automatic transmission, with the added reduction in fuel economy. How would one then view the product, and how would the product feel about how it was viewed? Would the company be sued? Was the packet not as advertised, and so on. Did one get what one paid for, and, if not, what then? Children may or may not be ensouled organisms, but, whatever they are, they do not seem well regarded as commodities.

Whereas I suspect that "biological polarization" is, for a number of reasons, a very unlikely eventuation in a human society, one already has, and has always had, something rather similar. For example, intelligent individuals tend, statistically, to mate with intelligent individuals, and, statistically, tend to produce intelligent offspring. Such people tend to associate with one another, cooperate with one another, advance one another's interests, and so on. Intelligent parents, statistically, as others, love their children, and, on the whole, earning more, are likely to provide them with social and economic opportunities and advantages to a greater degree than is the case in the population at large. This sort of thing, incidentally, tends to be the case even in totalitarian societies. One of the major differences is that in the totalitarian society, as there is little to compete for but power, those who rise to positions of power are likely to be characterized by a certain lack of scruples, useful in attaining and keeping power, and a willingness to accept the necessity of intervening in, and controlling, the lives of others, a necessity appealing to some but unlikely to be universally palatable.

CHAPTER TWO:
Cloning

1. Explanation and Observations

THE WORD 'CLON' OR 'CLONE' ENTERED ENGLISH early in the 20th Century, being a transliteration of the Greek expression for a twig, 'κλον', which might be literally transliterated into English as 'klon'. Grafting, of course, in a botanical, or horticultural, sense, has been long familiar, a procedure where a bud, shoot, or such, of one plant is inserted into another plant, say, a fruit tree, where it, the scion, will grow, sustained by, and becoming a part of, the receiving plant, the stock. Grafting may serve many purposes, for example, adapting plants to diverse conditions, facilitating pollination, and producing diversely fruited or flowered plants. For some species, interestingly, it is apparently the only method of propagation. Embryo multiplication, mentioned in the preceding chapter, is sometimes referred to as cloning, but this, given the division techniques involved, is more usually referred to as "twinning." To be sure, an indefinite number of embryos may be produced from an original embryo, so "embryo multiplication" might be a better characterization of the procedure. In the preceding chapter, we referred, in listing some aspects of reprogenetics, to "nuclear transfer," and this is what is involved in cloning, as it is usually understood today. In this more germane sense of 'cloning' an egg and a donor cell are denucleated; the nucleus of the donor cell being transplanted into the denucleated egg, the original nucleus of the egg and the denucleated donor cell being discarded. The renucleated egg, now bearing the nucleus of the donor cell, is then stimulated with the end in view of producing a new organism, in the case of an animal, an embryo. Plants and amphibians have been cloned for some time. In 1980 the first mammals, mice, were cloned. Since that time cloning appears to have been successful with several forms of mammalian life, such as goats, cows, cats, and sheep. Pigeons have also been cloned. Traditional mammalian cloning utilized embryonic tissue. In 1996, however, Dr. Ian Wilmut and his associates, at the Roslin Institute in Scotland, produced the first animal, the sheep, Dolly, to be cloned utilizing adult tissue. This breakthrough brought cloning out of the laboratories and seminars into widespread public attention. Given the tree of life, and the evolutionary continuities amongst organisms, the development of the cloning field, and such, there seems no scientific reason why humans, including adult humans, might not be cloned, and, indeed, it seems to be a practical certainty that this possibility will eventually be actualized, if it has not already taken place.

The philosophical journey begins.

Philosophical questions are seldom so pressing that one must solve them before lunch. Indeed, many individuals seem unaware that they exist, and some who are aware of them forget about them as soon as possible, and with apparent impunity. Philosophy is one of several things without which one may live, but it is also one of the several things which make life worth living. There is a sense, of course, in which everyone carries philosophy, or something like philosophy, about with them, in their wallet or purse, so to speak, primarily a world view, involving a notion as to the nature of the world and how one might live within it. In this broad sense, religion, mythology, superstition, folk lore, an age's common sense, and such, have their philosophical aspects, or share elements and interests with philosophy. The differences, on the other hand, between, say, societal givens and the philosophical enterprise are discernible and unmistakable. Most such givens, once established, are essentially inert or passive, and involve little activity, unless in their propagation or enforcement. On the whole, they are inculcated, and accepted, by most, uncritically. One takes them much as the ground on which one may walk, or the air which one breathes. Philosophy, on the other hand, is usually taken to be independent, active, and critical, more concerned to inquire than believe, to question rather than acquiesce, to examine than submit. Those who prize the freedom of the mind prize philosophy. Those who fear the freedom of the mind resent and fear her.

Cloning seems, on the whole, less problematic than a number of other techniques and possibilities likely to appear on tomorrow's doorstep. Its conceptual problems, for example, seem less provocative, or daunting, than those likely to be associated with areas such as genetic engineering or artificial life. The three conceptual questions most likely to arise in cloning, the civic conceptualization of the clone, its identity-or-difference relation to the donor, and its relation to the concept of survival after death or to empirical immortality, seem reasonably easily resolved, within the compass of a widely available, accepted conceptual structure.

First, is the clone a person?

The answer to this question seems clearly affirmative, or, in any event, best resolved as affirmative. The donor is a person, presumably, lest we wish, in virtue of his decision to be cloned, to remove him from that category, perhaps as having forfeited his right to that status, rather as an outlaw might be regarded as having removed himself from the moral community. If the donor is a person, then the clone, which, per hypothesis, is genetically identical to the donor, except in virtue of the same sort of minutiae which might distinguish identical twins, would be a person. For example, logically, if A is a person, and B is identical, or pretty much so, with A, then B would also count as a person. It is possible, of course, that one might be a citizen and the other not. For example, perhaps the donor is not a citizen, but the clone is born elsewhere, on a different side of a border, or something, and thus counts as a citizen. It might be claimed that the clone cannot be a person, as, say, given its antecedents, it lacks a soul, or something. This would seem to presuppose that a soul-giver would be unable to ensoul a clone, or would not care to do so, perhaps disapproving of clones. This would be difficult to prove, and the same sort of arguments might seem to appertain to any individual. For example, perhaps twenty percent of the human race is ensouled and the rest not. Who belongs to which category? If one cannot demonstrate the presence of a soul, or such, in the donor, it is not surprising that it would be difficult to do so in the case of the clone, as well. Presumably either the donor and the clone would both lack such an entity, or possess it. In any event, the legal concept of person does not require a proof of soul possession, which is fortunate, as such a proof seems unavailable. It might be argued in certain cases whether or not human life counts as a person, for example, if the life recently began, or if it is vegetating in a coma, or such, but in the clone case it seems clear either that the clone would be a person, or, at least, should be regarded as a person, pending proof to the contrary, given the development of a pertinent technology of soul detection. Legally, of course, a clone might be ruled a nonperson, as, for example, a member of a given race, religion, ethnicity, or such, might be ruled a nonperson, subject to extermination on sight and so on. Such social arrangements are not unprecedented. Provisionally, we would recommend that the clone be counted as a person, as would be a child, a younger sibling, or such.

Second, is the clone the same person as the donor?

Later in this study we will deal questions of sameness of person, as we will with questions of sameness of species, for there are several interesting aspects to such questions. Here, however, it seems clear, given generally accepted, systemic-process intuitions, social practice, and such, the answer to that question is clearly negative. Even were the donor and the clone genetically identical, in all respects, as they would not be, they would not be the same person, any more than identical twins would be the same person, or two automobiles of the same make and model would be the same automobile. The usual principle of individuation in such a case is matter, not form. The donor and the clone, aside from being of different sizes, shapes, ages, and such, would not be in the same place at the same time. They would occupy different localities, different spaces. The only sense in which they could be regarded as the same would be in the sense, irrelevant in this context, of possessing identical, or near identical, genetic endowments, this analogous to the sense in which a salesman might assure a customer that he need have no fear, as automobile one is the same as automobile two.

Third, does the clone assure the donor of survival after death, or would a succession of clones assure the donor of an indefinite survival after death, or, possibly, an empirical immortality?

The answer to this question, given the considerations proposed in answer to the previous question, is clearly in the negative. If the donor possesses, say, an immortal soul then he does not need the clone for immortality, and, if he lacks such an entity, then the clone is not going to be of any help in the matter. The donor might derive some comfort from the thought that something much like himself, at least genetically, will survive him, but it is not him, clearly. This is not an argument against cloning, *per se*, of course, any more than it is an argument against having children, of any sort. The parent is not the child; the child is not the parent. One identical twin is not the other, and so on. As the genes replicate themselves it is not even the same genes. An analogy might be a succession of robots, each programmed to construct an identical successor. Robot one might be identical to robot twenty thousand, but robot twenty thousand is not the same robot as robot one. Also, some might prefer a variety of robots constructing a further variety of robots, a diversity of robots, perhaps of a sort unforeseen. Also, one has no guarantee that the clone will be interested in cloning himself. If not, the immortality will be one of the shortest on record.

The most serious questions connected with cloning are not conceptual questions but moral questions. For example, if one has reservations pertaining to the utilization of embryos for scientific purposes, one may well have reservations pertaining to cloning. The embryo which produced Dolly, of Roslin fame, was one of thirteen, and these thirteen embryos were all that resulted from 277 attempts to produce such embryos. This suggests, if nothing else, that cloning technology is in an early phase of development. These are sheep embryos but, presumably, something similar might be the case if they were human embryos. Indeed, even in current reprogenetics more embryos are routinely produced than are utilized. One might see this sort of thing as a byproduct of an end sought, to be justified, if one were concerned with justification, in terms of a human child which, otherwise, would never have had life. More broadly, one might reference, where appropriate, considerations of medical research. For purposes of the following discussion, we shall suppose that cloning technology has advanced to the point where such problems are minimized, if not avoided. We now disengage ourselves from such issues , and address ourselves to cloning itself, rather than possible concomitants.

There seems to be a frequently encountered view that cloning, *per se*, is somehow immoral, for example, according to Mr. William Jefferson Clinton, at the time president of the United States of America, cloning would "violate the sacred bonds that are the very core of our ideals and our society," and "make our children objects rather than cherished individuals." This certainly seems serious, particularly as issuing from Mr. Clinton, well known as a role model and moral inspiration. Similarly, one hears that cloning is against the teachings of the Roman Catholic Church, which is surprising as the church had not heard about it until recently. Presumably what was meant was that cloning was regarded as "unnatural," and thus, favoring natural things, at least selectively, the church was ready, implicitly, to condemn the practice, should it come up. This is also the church which is opposed to birth control, except by means of, say, the notoriously inefficient rhythm method, or, better, abstinence, i.e., not having sex at all. The last technique is almost invariably effective. Some states, for example, California, have banned cloning, and, it seems, some countries here and there around the world, as well. All this seems very surprising. To be sure, the furor and hysteria have muchly subsided, and seems to have exercised political and religious leaders to a greater extent than their constituents and flocks. Perhaps the religious leaders felt threatened by an envisaged threat to the "soul hypothesis," an element important in their world view and one perhaps important to the security of their position in society, rather as they have failed, statistically, to welcome anthropoids conversing in Amslan (ASL, or American Sign Language), and such, and where religious leaders are alarmed, astute politicians are not likely to be far behind. In any event, the following discussion, whatever its flaws, will attempt to consider several of the advantages and disadvantages, the values and disvalues, of cloning, against a background of systemic-process, secular ethics, rather than in terms of compatibility or incompatibility with the views of particular authorities, institutional or otherwise, contemporary or historical. As will be usual in this study, we will be more concerned to discern and clarify moral issues than resolve them. Even within the context of a systemic-process, secular ethics moral disagreement is likely, and, indeed, without such disagreement, it is unlikely that ethical adaptation, moral evolution, and such, could take place.

Before entering into various aspects commonly associated with cloning, two questions of fact might be noted, first, are clonal generations plausible, which would be important for those who might wish to propagate an indefinite succession of identicals; and, second, is there any genuine public interest in cloning, such that the question might have practical as well as hypothetical aspects. The answer to the first question seems clearly affirmative, as several generations of mice have been cloned, the clones being cloned, and those clones cloned, and so on. In this sense, one could have several generations of "Smiths" through the ages. Had the technology been available and utilized, then, one might have had a several Caesars, Antonys, Cleopatras, Christs, Hadrians, Napoleons, and so on, through the ages, or, better, genetic identicals of such individuals. As to the second question, polls suggest that most individuals would not have much interest in being cloned, but, on the other hand, somewhere between five and ten percent of Americans would actually wish to have themselves cloned. Five percent of a population of 300 million would be fifteen million individuals, a not inconsiderable number. In the light of such a figure, it seems clear that the issues having to do with cloning are of more than a purely abstract or hypothetical interest.

In most of what follows, it is taken for granted that cloning would be likely to be undertaken only by individuals with individual aims in mind. Scenarios in which large numbers of, say, robotic workers or fearless, steadfast soldiers are cloned, to be engineered into docilely serving a smaller population, industrially or militarily, are, at least for the most part, ignored. Such things are surely empirically possible, but politically and economically unlikely. Factories and barracks may be filled, and perhaps more efficiently, by more familiar methods. A world already concerned with large populations and limited resources is not likely to increase its hazards by engaging in massive cloning. This is not to deny that a totalitarian state with access to reprogenetic techniques, cloning and otherwise, providing itself with, say, hundreds of thousands of artificial wombs, might not engineer an expanding population with the intent of eventually applying it to aggressive policy goals, but it is to suggest that this development is unlikely, if only because such goals are more easily achievable by conditioning a population into mindless, self-sacrificing zealotry and providing it with nuclear, biological, and chemical devices, and proficient systems, human and otherwise, by means of which they may be delivered. On the other hand, in the face of a global disaster, one in which, say, most men and women have been rendered radiologically sterile, states might have recourse to various reprogenetic techniques, such as cloning. In such a situation, such techniques, including cloning, might save the human race from extinction.

The possibilities involved in cloning are multiplex, and the following considerations are typical.

Cloning could be of reproductive assistance. There are sterile men and barren women, who wish to have children. Cloning would furnish a way in which this might be possible. It is not clear they should be denied such an opportunity, which is readily available to their fellows in the population at large.

There is no doubt, on the other hand, that cloning might be abused, or, at least, used for purposes likely to be more suspect. For example, an unusually attractive or desirable child might be repeatedly cloned, and the embryos sold. Objections to this sort of thing would presumably be similar to those lodged against "baby black markets." Similarly, cells might be obtained commercially, or surreptitiously, as in medical examinations, from celebrities, athletes, and such, and utilized in cloning, with the consequent embryos being marketed, presumably privately. Parents might clone a dying child, to obtain a replacement, which seems unobjectionable, if unsettling, or, more objectionably, clone a healthy child in order to have a source of spare organs on hand, should they be later needed. This would be, so to speak, a "warehouse child." More ghoulish is the possibility of producing headless clones, from which organs might be subsequently obtained. This sort of thing has been done with mice. More benignly, as most birth defects and miscarriages are the result of either an atypical number of chromosomes or the presence of deleterious genes, cloning would assure the acquisition of an embryo of likely normality, at least in the respects cited. One might note, too, that cloning, associated with other reprogenetic attentions, would permit gay couples, male or female, to have children. The embryos would be fused, recall the "chimera" possibility, and thus would contain genes from both parents. In this way, the child would be "theirs," in the same way that a typically conceived child would be the child of its parents.

The state, foundations, or such, might be interested in cloning individuals of unusual intelligence or talent. Might it not be nice to have some extra Newtons or Einsteins, Mozarts or Beethovens, Rembrandts and Michelangelos, about? One has no guarantee, of course, that, say, a clone of Einstein is going to be another Einstein. Perhaps the clone would prefer to compose recorder music or design aircraft engines. What produces a human phenomenon is a function not merely of genetic endowment but the influences brought to bear on the individual, the opportunities in his environment, and so on. What would have happened if, say, a genetic identical of Mozart had been born ten thousand years ago, among Cro-Magnon tribesmen, if Napoleon had been born in Latium in the time of Tarquin the Proud instead of in Corsica in 1769, or had been born in Corsica in 1969? Today, living, working in hardware stores or advertising agencies, there might be individuals who, in a different time, might have led migrations or founded world religions, or, were they to have been born a thousand years from now might have pioneered colonies on exoplanets, or extrasolar planets, or, amongst the ashes of extinct civilizations, on a small planet, fashioned an unusual and problematic instrument, an alphabet. It is true that if there were a thousand Mozarts Mozarts would be likely to be less esteemed, but it is also a possibility, seldom recognized, that if there had been a thousand Mozarts, there might never have been a "Mozart." Perhaps essential to certain achievements is the notion that one is different, unique, and that no one can do what one can do, and this lures and drives the individual to create and bestow what no other could so create and bestow. Without him, such things will never be done. Thus he must do them. To be sure, a thousand Michael Jordans might transform basketball, but then we might not remember Michael Jordan, and that would be a shame.

Monozygotic twins are apparently of great interest with respect to research. Science keeps track of hundreds of them. In the manipulation of variables, or the assessment of the influence of variables, such twins are of enormous value, as the genetic variables, so to speak, are constant, or nearly so. One could conceive a totalitarian state, with an interest in science, benign or otherwise, producing large sets of clones. These sets would be invaluable in experimental situations, for example, for use as test groups and control groups. The sets, then, would be analogous to laboratory animals, bred for various purposes. This possibility is likely to affect one as morally abrasive, but there are few social arrangements which, here or there, in one place or another, have not found a moral sanction of one sort or another. After all, a principle may be found to justify anything, or almost anything. In the current supposition, the individuals' very *raison d'être* is to serve as test subjects; this is the purpose for which they have been produced; without this they would not be alive. In this sense they are produced by the state, and exist for the state. They are special, and they and the society find the matter acceptable, having been so conditioned. Perhaps they are even assured a favored place in some future rainbow world, one with unlimited ice cream and cake, and no scientists. If a justification is requested it is one of a utilitarian nature, in the sense that this research, besides furthering various purposes of the state, will save many lives, increase health, and improve the quality of life in the community. There is doubtless much of scientific value which might emerge from an arrangement of this sort. On the other hand, such an arrangement is one unlikely to be viewed universally with moral equanimity.

Aside from reproductive assistance, there might be a variety of personal reasons why an individual might wish to have himself cloned. These need not all reduce to reasons of personal vanity, although that might be involved in many cases. Whereas the vanity of others may be annoying, amusing, endearing, or such, it seldom seems morally opprobrious. Too, it is not difficult to recognize the pervasiveness of this property, and, indeed, upon occasion, its social utility. One suspects that much civility and moral behavior is as much a function of vanity, self-image, and such, as fellow feeling, rationality, and common sense. As all human beings are, to one or another extent, vain, it seems likely that the feature has been selected for. Hominids, for example, frequently strut about, make a great deal of noise, leap up and down, and, in general, engage in display behaviors. And one often cares for one's appearance, selects clothes and accouterments with enhancement in mind, chooses homes, automobiles, furnishings, decorations, pets, and such with an eye as to how one will be viewed, and, occasionally, if sociologists are to be believed, one may occasionally select a spouse with similar notions in mind. In any event, few human beings are in much of a position to castigate other human beings on the score of vanity. In the human neighborhood there are many glass houses. This being the case it would seem morally acceptable for Helen to clone herself because she is so cute, or such a luscious dream, and for Bill to clone himself, as he is such a splendid fellow, such a dashing, handsome, roaring, virile bull of a guy. Why not? Perhaps they feel they owe the world more of them. There might be, of course, a number of reasons for having oneself cloned which are not connected with vanity, or, if so, at least less obviously so. For example, an individual might have a spinal injury, and, in order that something like himself might be able to walk and run, and enjoy life as he cannot, might have himself cloned. Deprived himself, he would give the gift of a prized life, such as he would like to have lived, to another. In the clone, he might see himself as he might have been. Too, some individuals, lonely, unmated, professionally absorbed, women past menopause, or such, might resort to cloning as a way to keep their genes in the gene pool.

One fear apparently associated with cloning, at least in the minds of some individuals, is that wealthy individuals, capable of affording the procedures involved, having the resources to hire batteries of host mothers, and so on, might have themselves repeatedly cloned. Would one want, say, four or five thousand Croesuses, Genghis Khans, or J. P. Morgans around? Whereas I, personally, would not look with delight on such additions to the gene pool, at least in such numbers, I do not feel that I am entitled to interfere in such matters. Society asks little of people, mainly that they behave themselves and pay their own way, and there are many individuals who fail to do either. If an individual should choose to have himself multiply cloned, say, five to fifty times, or more, can raise the clones, harms no one, does not burden the community, and such, there would seem no rational objection to his doing so. There would be, of course, objections. On the whole, these would be likely to result from hatred of the wealthy, envy of their resources, or such. Certainly envy is amongst the most common and powerful of human motivations. Envy, camouflaged under convenient moral rhetorics, as it always is, has fueled social movements, religions, revolutions, assassinations, judicial rulings, political campaigns, and legislative programs. On the other hand, a plethora of privately financed multiple clonings are not likely to come about. There are a number of reasons militating against such a development, even if it was not promptly prohibited by law, or social, psychological, and economic reasons. Wealthy individuals, as the rest of us, live with others, and are dependent on others. There is a pervasive desire, natural in a gregarious species, to be approved by others, and to conform to the values and views of one's reference group. Accordingly, supposing that social pressures would be negative in the case of multiple clonings, as seems likely, one would expect even the wealthy to be sensitive to such pressures, and govern their behavior accordingly. Even dictators and kings depend on others, without whose support they are helpless and vulnerable. Too, as suggested earlier, envy avoidance is a common motivation in human life, and this is the case, interestingly, even amongst those who have little likelihood of being envied. Psychological reasons, too, are likely to preclude excessive clonings, such as psychological saturation, perhaps in the form of either an incrementally increasing disinterest or the fear of reducing the value of a entity, or model, by, so to speak, over production. Economically, too, such things as the finite nature of resources, the problematical distribution of goods, possible legal problems connected with inheritance, and such, might act as disincentives.

Given the current rarity, or nonexistence, of human clones, it is difficult to speculate on the effects of cloning on the clones themselves. For example, if they should be subject to premature ageing, that would be a serious negative connected with the process. It is supposed that individual clones would be as indistinguishable in a general population as DI-conceived individuals. Accordingly, unless identified publicly, they should be subject to no more or less discrimination, positive or negative, than the donors from which they have been derived. Should the clone be apprised of his origin, one would expect the response to this information to depend primarily on the individual concerned, but, statistically, such a response might well be influenced by views imparted to him in his personal environment, presumably positive, or by social attitudes expressed in the community at large, possibly negative. Whereas clones derived from one donor would be likely to relate to clones derived from a different donor in ways not unlike those characterizing the various relationships of normal-origined individuals, there is a possibility that same-donor-derived clonal groups might have strong social bonds, that they would be more loyal to one another, and more supportive of one another, than would be likely to be the case with nonclonal groups. These notions are based on considerations of family, brotherhood, twin studies, and such. It is possible, as well, though this is exceedingly speculative, that it might be easier to impart information to identicals, that identicals might exchange information amongst themselves more easily than might nonidenticals, and that identicals might constitute good teams or task groups. To be sure, acculturation is important. To adapt for our purposes a bitter story, one might easily conceive of three clones, differently acculturated, who grow up and kill one another.

Whereas the discussion, to this point, has primarily concerned itself with the possibility of human cloning, it should be recognized that a variety of possibilities exist pertaining to animal cloning, beyond the obvious applications of multiplying desired types, for, say, commercial or experimental purposes. For example, a dying pet might be cloned, as a tribute to the animal or as a way of lessening an owner's grief. Cloning, too, might be useful in saving endangered species. The Chinese, for example, are interested in this sort of thing in connection with the panda. In the United States alone over 335 different kinds of animals are listed as endangered, for example, mammals, birds, reptiles, amphibians, fish, clams, crustaceans, snails, insects, and arachnids. And one day, as suggested earlier, the human species might have to be added to the list. In the profuse wastage of evolution by far the greatest number of species have become extinct. The human may have the unique distinction of being the first species which can manage its own extinction. More prosaically, it may be possible to clone animals with human-immune-system compatibility, which animals might be used as organ donors, or other animals which might produce useful substances in, say, their milk. Presumably, in time, other possibilities, not yet envisaged, will emerge, and be pursued. Animal cloning, one supposes, would tend to be less controversial than human cloning, if only because they have not yet been awarded the franchise, perhaps to be exercised by human proxies.

It is now time to examine a variety of arguments critical of, or suggesting reservations concerning, cloning.

Most of the arguments opposed to cloning, not all, seem to me to be based either on a misunderstanding of what would be involved or on unanalyzed *a priori* considerations, deriving from authoritarian sources uncritically accepted. For example, when one hears that cloning would "violate sacred family bonds" or turn children into "mere objects," one seems not only to be dealing more with assertions than arguments, but with assertions which seem more reflexive than reflective. What are these hypothesized sacred family bonds? Where did they come from? Surely there were families around long before it occurred to anyone that they were sacred, or ought to be sacred. Similarly, how would one prove that families were sacred, rather than just being very nice, wonderful, or such, and what is this "sacred" business, anyway? How would one test for sacredness? Does it mean more than families are to be respected, taken seriously, preserved within limits, or what? Who knows the mind of a god or gods? Perhaps they regard a cloned child with the same love, and favor, or more, than the accident in the next crib. Does divorce, annulment, separation, adoption, birth control, and such, violate "sacred family bonds"? If not, why not? What about single motherhood? Should we take their children away? Do we have here something other than an attempt on the part of self-serving, self-perpetuating, imminently practical, worldly-wise institutions, and their sycophants, to deepen and consolidate their power over others, by means of which they earn their living, a further attempt to make themselves ever more indispensable, by intruding themselves ever more deeply into human life, which did well enough without them, for hundreds of generations, and which might do as well, or better, without them? Similarly, who is entitled to say that an individual who desires a child by means of cloning thinks of the child as an object? Perhaps politicians or potentates, secular or ecclesiastical, regard the cloned child as an object, but the donor presumably does not, and, one hopes, the child sees itself as a human child, as meaningful and as precious as any other, not as an object. How disgustingly sanctimonious to so insult a child. How grievously must such a lack of charity reflect on those who would propound such dogmatisms. What ignorance and bigotry does it betray?

On the other hand, putting aside considerations of politics, of cruelty, and hypocrisy, it would be indeed a rare process, or policy, which did not involve trade-offs. Trade-offs are a part of life. The dollar spent here cannot be spent there; the hour spent one way is an hour not spent differently, and so on.

Cloning, or its conception, is apparently disturbing to many individuals. This distress is apparently real, and it would be arrogant and inconsiderate not to recognize it. It obviously counts as a negative, and not one to be lightly ignored. Even if one has the power to ride roughshod over such scruples, those of others, not one's own, one should refrain from doing so. Disagreement in such cases is not the sort of thing which should be contemptuously ridiculed or dismissed. Sincerity is not always a virtue, as one has learned from misguided dupes, manipulated pawns, homicidal psychotics, and such, but when one finds sincerity in men of virtue, one should listen, and be prepared to enter into dialogue.

We will deal, briefly, with six familiar objections to cloning.

2. An Analysis of Some Common Objections

2.1. The Gene-Pool Objection

Cloning would reduce the amount of variety in the gene pool, and thus reduce diversity, viewed as a value, and, concomitantly, imperil the gene pool's capacity to adapt to new conditions, survive new challenges, and such. For example, had such genetic diversity not existed in the gene pool, the human race might have long since become extinct, certain geneplexes being more or less susceptible to certain diseases than others.

Whereas there is much to be said for the harmless pleasures of aesthetic diversity, flowers in a garden and so on, and much to be said, ecologically, for genetic diversity, few human beings really approve of diversity in important matters. For example, amongst putative celebrants of diversity, it is common to find an almost rabid intolerance of real diversity, namely, intellectual or ideological diversity. Everyone is free to be just like them. Few people are genuinely open or pluralistic, and, even amongst those who are, their openness or pluralism is likely to have limits. For example, it would seem permissible to disapprove of an individual who wishes to mutilate, torture, and kill other human beings, perhaps for religious reasons, however sincere he may be in the matter. Too, diversity, ethnic, religious, cultural, racial, and so on, historically, has been divisive, and a perennial source of unrest and conflict, a source of hatred, riots, wars, massacres, and such. On the other hand, genetic diversity certainly contributes to the viability of a gene pool. Indeed, intersexual relations, with its genetic mixings, as opposed to parthenogenesis, appears to have been selected for in the course of evolution. A viable gene pool apparently requires something like two hundred persons, and, if the gene pool contain several billions, so much the better.

The value of this "genetic-diversity objection" to cloning would depend primarily on the extent of cloning in a given gene pool. Whereas it is empirically possible, should cloning, surprisingly, become the preferred method of gene replication, and this preference remain in place over several generations, that the gene pool might be damaged, the possibility seems remote. A population consisting of celibate, cloned Bills and virginal, cloned Susans seems unlikely, if only because the Bills might start looking good to the Susans, and the Susans to the Bills. Long before the advent of such an era it would doubtless be anticipated, and steps would be taken make sure it never arrived. Also, as noted earlier, statistically, few individuals seem interested in cloning themselves. Should cloning occur it is likely to be rare, and to have no appreciable effect on a gene pool, even one relatively small.

2.2. The "Overly Cohesive Group" Objection

Cloning would tend to increase the formation of overly cohesive groups, and thus would be likely to lead to exclusiveness, ethnocentrism, ethnic conflict, and such.

This objection, like the preceding, and as will several of the following objections, requires for its weight, or plausibility, the notion that cloning might become common, and widely spread, to the extent that it has a recognizable demographic impact on a population.

One supposes that "overly cohesive" is intended to carry negative connotations, so we let that stand. Cohesiveness, in general, might be taken to be a value. For example, one hopes a family would be cohesive, a military command, a sports team, a company's staff and employees, and so on. There seems no particular reason for supposing a clonal group would be "overly cohesive," unless, perhaps, it was discriminated against, or persecuted in some way. If one wants to produce "overly cohesive groups, groups, say, irrationally smug in their sense of superiority, isolating themselves from outside influences, controlling and restrictive in the behaviors permitted to their members, intolerant of other groups, and such, there are easier ways to produce such groups than cloning. Indeed, cloning, by itself, would scarcely suffice. Again, clonal groups would not be likely to be populous enough to affect demographics, certainly not to the extent that political, racial, or sectarian groups might do so. Again, clone groups of identicals, if only for economic reasons, would be likely to be extremely rare, though clonal clubs, social organizations, support groups, or such, might exist. Amongst identicals, clone relationships are less likely to be horizontal than vertical, less a matter of contemporaries than one of generations. Some clonal groups might be "overly cohesive," but so, too, are many other sorts of groups, to their own and their society's loss.

2.3. "Abuse" Objections

If cloning is regarded, in general, as some sort of abuse, perhaps as constituting a departure from familiar processes or something, then it would be anomalous to distinguish a subset of objections as "abuse" objections. On the other hand, it seems clear that some individuals might find cloning acceptable under some conditions and unacceptable under others. In accord with such a view, one might consider some applications of cloning which, to some individuals, might be regarded as unacceptable, or, at least, morally dubious. Needless to say, views as to legitimacy or illegitimacy, what is in good taste and what is not, and so on, might differ amongst individuals. A consensus might not exist. Certainly not all terrain is morally charted. Occasionally one ventures into new meadows. Under "abuses," we distinguish between those we might speak of as "particular," or "individual," and those we might speak of as "massive," or "general."

2.31. Particular Abuses

Earlier we noted one might clone a child, perhaps repeatedly, and then sell the embryos. Some might view this as a criminal act. On the other hand, if it is true that embryos have no legal protection, are without legal standing, are meaningless tissue, may be disposed of at will, and so on, then, it seems, the only problems involved would have to do with the child, who, however willing he may be, may not have attained the age of consent. Perhaps the parents or guardians might be regarded as having a power of attorney, or something comparable, in such matters, rather as they might use their best judgment on behalf, say, of an infant "model," used in advertising, or in arranging work for a child actor. The profits, of course, might be put in trust for the child, minus fees for management, and such. Legal rulings would presumably be in order.

Another possibility might be that enterprising celebrities, notable personages, scientists, artists, champion athletes, gold-medal winners, and so on, would market some of their cells, which may be easily spared as they have trillions, for denucleation. One could conceive of there being some interest in this sort of thing. For example, a woman might wish to bear the child of a movie star, loved from a distance. To be sure, the genes would be his, not hers. But reprogenetic techniques might manage a chimera, in which the genes might be shared. If the consent of all parties was obtained, or was implicit, nothing illegal would have taken place in such cases, at least in a state in which cloning, or perhaps certain applications of cloning, were not prohibited by law.

Alert attorneys might apply themselves to questions of custody, visiting rights, child support, inheritance, and such. New fields of law would be in the offing.

Cases more likely to be viewed askance would be cases in which cells were obtained surreptitiously, without consent, by theft, so to speak, as in a barber shop, a dental office, a massage parlor, a gym, a medical facility, or such.

It would seem likely that a black market would soon emerge, in which such goods would be available.

Doubtless fraud would frequently occur, in which cells certified as those of "x" would actually be those of "y," perhaps those of the salesman or a confederate, or an unwitting stranger.

A king might be cloned without his knowledge, by a particular faction within the kingdom, interested in disruption or power, hoping to profit from competitive princes. Too, would a clone of the king take precedence over a naturally conceived and delivered child, as it is more closely and uniquely related to the royal line? What if it appeared before the naturally conceived and delivered heir? Here we have work for royal counselors, or parliaments, or such.

A psychotic might clone an enemy, in order to have something to mistreat and abuse. A masochistic lunatic might have himself cloned for similar reasons. Perhaps a primitive assembly, with unusual theological views, might execute the clone of a missing heretic, as some sort of proxy for the original. Perhaps a king has a prince cloned, if not for having an organ warehouse on hand, for purposes of duplicity, to utilize the clone in dangerous situations, for example, doubling for the prince when assassins threaten, when the palace is being overrun by revolutionaries, or such.

One could think of a number of applications of cloning, plausible and implausible, which one would be likely to view with moral concern.

2.32. Massive Abuses

By means of cloning, undesirable types might be propagated, hundreds of offspring for psychopathic dictators, legions of compliant goons, suicidal simpletons, manipulable voters, mindless consumers, and such.

This is certainly possible but scarcely likely. For one thing, suicidal simpletons and mindless consumers, and such, are not so much bred as made. Certainly some genetic materials would be more likely candidates for one conditioning regime or another, but the conditioning regime would presumably be important. A cocker spaniel is an unlikely candidate for an attack dog but, properly conditioned, it will do its best.

One man's "undesirable type," of course, might be another man's "desirable type." A prophet, for example, will do better with battalions of unquestioning adherents than a squad of intellectual "stray cats," so to speak, going in different directions for different reasons.

It does seem possible that Narcissism might become somewhat more common in society. After all, to clone oneself one would presumably think rather well of oneself. This, on the other hand, luckily, is not a crime. Too, the Narcissist might decline cloning himself, as this might to some extent compromise or diminish his own inimitable singularity.

It is hard to know about these things.

In passing, one might note that one does not need cloning to produce undesirable types. For such purposes, ordinary methods of reproduction are available. Too, given the nature of cloning, the selectivity involved, and such, it seems more likely it would be employed for the production of "desirable types," at least as commonly understood, rather than undesirable types, at least as commonly understood.

2.4. The "Mental and Emotional Health" Objection

Cloning might prevent the formation of certain psychosexual relations which may be important from the standpoint of mental and emotional health.

In the absence of studies along these lines, not possible at present, it is difficult to assess this suggestion. It is particularly difficult to assess in a politicized environment, where ideology is likely to demand its say. In any event, presumably, in most cases, the clone would have a normal mother and father, at least from the point of view of social parenthood. Presumably the mother would supply unconditional love, assuring the clone of its fundamental value, and the father would add in authority, and discipline, set standards in terms of which the child is expected to measure up, and prove itself, and so on. The general point here is that the clone is a normal human child and how it is raised, traditionally or not, will presumably affect its mental and emotional health. If it is raised in a normal, or familiar, manner, then it will, statistically, be likely to develop in a normal, or familiar, manner, for better or for worse. It is a child, as any other. If problems with psychosexual relations exist, they would be more likely to exist with the donor. For example, if a long-term, loving relationship with a member of the opposite sex commonly contributes, statistically, to the mental and emotional health of a human being, then a donor parent, if it should lack this relationship, probably as any other person lacking such a relationship, might be expected, at least statistically, to be less mentally and emotionally healthy than his counterpart. Similarly, it is sometimes alleged that a woman who has failed to complete her biological cycle is more susceptible, statistically, to mental and emotional problems than one who has conceived, delivered, and mothered a child. As considerations such as these are empirical matters we surrender them gratefully to appropriate professionals, hopefully for a politically neutral resolution. All we need at this point is that the child is a normal child and, well raised, should have no more or less difficulties in life than other children. If difficulties attend donorship, they should become evident at a later time. Much might depend on how the donor views himself, or is viewed in the community. Is he respected as a parent, or despised as an eccentric, a freak, or criminal? There are some interesting possibilities, the effect of which on self-image, mental and emotional health, remain to be seen. For example, a woman might have herself cloned, and then have the clone cloned, and so on. In this way, she could find that she is, at the same time, a mother, grandmother, great-grandmother, and so on. In theory she could have, in her own lifetime, several generations of grandchildren and great-grandchildren. One could well imagine the conversations in nursery school, kindergarten, the first grade, the second grade, and so on.

2.5. The "Fewer Adoptions" Objection

If cloning becomes popular, fewer orphans will be adopted, will be raised in private, loving homes, and such.

This objection appears to be based on a mistaken view, namely, that there are large numbers of orphans which, for whatever reason, are unlikely to be adopted. This is incorrect. There are, currently, far fewer orphans available for adoption than there are couples hoping to obtain a child by adoption. There seem to be two major reasons for this, first, the legalization of abortion, and, second, the rise of single motherhood, to which less of a social stigma is now attached than hitherto. If there was a demographic impact on adoption currently, it would presumably be due less to cloning than to other reprogenetic techniques, in virtue of which many hitherto childless couples are now enabled to have families.

2.6. The "Incompatibility with Religion" Objection

As there are a great many religions, perhaps something over five thousand, one supposes that some would have no objection to cloning, and many others might be unaware it exists, or, if informed, would regard it of little, if any, religious interest. Further, presumably, if only to distinguish themselves as being *au courant* or such, one supposes some religions, if they thought about it, might actually approve of cloning, perhaps as a technique now provided by the god or gods for the use of human beings, to be gratefully accepted, rather like blood transfusions, organ transplants, and such.

On the other hand, there are doubtless some religions which would, and do, object to cloning, on one grounds or another. Insofar as their objections might be moral, they would not be essentially religious, and might be considered independently of religious teachings, traditions, dogmas, or such. Insofar as their objections might be religious, a number of considerations suggest themselves. First, is cloning really against the religion in question? Presumably, in most cases, cloning did not come up in their religious documents, written or oral, as no one had heard of it until recently. Thus, their religion could not have had a position on the matter. To be sure, there might be things it did have a position on which might be invoked, if necessary.

Second, who knows what the settled position might turn out to be? There were once religious objections to various things, for example, inoculation, anesthetics, birth control, lightning rods, the motion of the earth, the heliocentric theory of the solar system, evolution, and who knows what. Religious leader A might not agree with religious leader B, and either might change his mind, or be reversed at a later date by Religious leader C. Changeless churches, for example, change. Third, if one is not a member of a particular church, then its views on various matters are not determinative or coercive, even in principle. At most, they might be taken seriously as a precaution against being imprisoned, fined, mutilated, tortured, or killed, that sort of thing. Lastly, before the religious views on particular matters, if they exist on such matters, are to taken seriously, the church, say, should prove, first, that it is the one, true church, and, second, that it happens to be right on this particular issue, as it is not clear that the one true church has to be right every time, even if it says so. Analogously, the one true king might make a mistake now and then. Whether the one true king is infallible or not is not well left up to the one, true king. It is better left up to the way the world is.

It is certainly possible that Religion R might say something true. On the other hand, if what is said is true, it is not true because it is said by Religion R.

Accordingly, being compatible or incompatible with Religion R is irrelevant to the truth or falsity, the value or disvalue, of a belief.

2.7. The "Not Natural" Objection

I think that this is a reasonably common objection to cloning. The notion seems to be that what is not natural, or, worse, "unnatural," is somehow wrong. We dealt with this sort of objection earlier, in passing. Problems appertain, for example, to what counts as natural and nonnatural. Is the transition from organic evolution to superorganic evolution natural for a rational species or not? Is shaping stone and kindling fires natural or not? Certainly they seem good ideas whether or not they are natural. One might also wonder if what is "natural" is all that good. Presumably droughts, floods, fires, hurricanes, earthquakes, and such are natural, but they do not always seem all that good. Nature, from the backyard to the plains of the Serengeti, from the pond in the park to the depths of the ocean, feeds on nature. What could be more natural than the cobra and the barracuda, the crocodile and the boa constrictor, the virus and the bacillus?

Civilization, presumably a value, consists largely in a flight from nature. Nature, so to speak, is diverse. It is neutral, really, neither good nor bad in itself, no more so than tides and storms, trees and rocks, nor the universe, but it is surely not neutral to us, and it affects us, exalting us to the ecstasy of living and condemning us to inevitable decay and dissolution. In short, that something is natural, or unnatural, however we define these things, does not resolve these issues.

One sometimes hears that one should "not play God."

This is hard to understand. Who knows what a god or gods might be? Who knows what they might do or not do, favor or disfavor? What a colossal, unconscionable arrogance it is to claim to understand a god or gods, and to claim to speak in their name. If there were gods, would that not constitute the ultimate blasphemy? That the mongers of superstition thrive, and wax fat on the offerings of the faithful, the ignorant, and gullible, would seem to suggest that the gods do not exist, or, if existing, are disinterested in us. Or perhaps they find the spectacle amusing.

One wonders if the dentist plays god, or the surgeon.

So, is cloning a human good or a human evil?

Is there a general, sweeping answer to such a question? One supposes not. Life is complex. Situations are diverse. Context is determinative.

The most common feature characterizing simple answers is that they are wrong.

If one encountered a clone, one might put the question to him. Would it be better if he had never been born? Would he prefer not to have existed?

What would he say?

If you were a clone, what would you say?

CHAPTER THREE:
GENETIC ENGINEERING

1. Prologue

'GENETIC ENGINEERING' IS NOW A RELATIVELY familiar expression, so familiar that one is unlikely, now, to understand, or, at least, to emotionally react to, the radicality of the concepts involved. One more commonly thinks of engineering in terms of the analysis of, arrangement of, and manipulation of, physical objects, of liquids, gases, and solids, the design and testing of engines and artifacts, the construction of road systems and bridges, energy and communication networks, such things. Social engineering is a metaphorical extension of the concept, the selection of, propagation of, and enforcements of meme complexes, or memeplexes, to bring about desired results, social, religious, and political. The human, as any other animal, is trainable. There is little defense against such techniques, unless it be intelligence and the refractoriness of the materials worked upon, an evolved, stubborn, wary human nature. That nature, human nature, long the despair of theologians, imperils the work, as well, of his contemporary counterpart, the social engineer. Try as one will, from a silk purse, so to speak, it is difficult to produce a sow's ear. From such a nature, with its splendors and corruptions, its good and evil, it is difficult to produce the social artifact, designed with a specific good or evil in mind. Civilization is a garment, often ill-fitting the beast it is designed to redeem, transform, and cloak. Suppose, however, one could work upon such a nature from the inside. What then might one not do?

2. Some Revolutions

To date, though others loom awesomely beyond, the human race has been distinguished by five major revolutions. The first is the linguistic revolution. The greatest inventor in the history of our species is not the inventor of the stone knife, the lever or wheel, but the word, the sound, not a grunt, shriek, or squeal, but a word, a sound which could "mean." What an awesome, mystical, indescribable moment that must have been, that moment of insight, when some small, hirsute, shambling creature conceived the possibility of the word. That may be the point at which we might divide ourselves from our fellow hominids, when the first ape spoke. Let the word be the symbol of the beginning of a species, now different on one world from all others. The second major revolution was the transition from a hunting, gathering economy to one of animal husbandry and agriculture. In a hunting, gathering economy it takes something like twelve square miles to support a single human being. In an agriculturally based economy something less than an acre will suffice. With Malthusian dynamics a species proliferates. A division of labor becomes practical, complex social organizations may be formed, urbanization arises; for the cleverest or most ruthless leisure allows time for thought, for ruses, for devices of exploitation hitherto impossible, for festivals and holidays, for the construction of temples, fortresses, and palaces, for the formation of armies, for the founding of kingdoms and empires. If one wishes a symbol for such a revolution it might be the digging stick, scratching a hole for seed, or, later, the plow. Or, more bitterly, a blade of bronze, or one of iron, which would better hold its edge. Later, a musket or mine. This phase, like the first, would be measured in millennia, millennia in which the basic realities of life, for most of the human race, in its rhythms and labors, indexed to the seasons, would have been familiar to Hesiod or Cincinnatus. Later there takes place a major transformation in social and economic life, in which masses enter history and transform a world. Historians speak of this as the "industrial revolution." We take the steam engine as a suitable symbol for this development, as startling in its way as a climatic change, resulting in the readaptation of a species, adjusting itself to new conditions, to new ways of life. One might suggest a fourth revolution, one tumbling historically on the heels of the third, something one might call the technological revolution. I would prefer to choose for a symbol here, the computer, rather than, say, nuclear power, the hydrogen bomb, such things. The fifth revolution, what one might call the "biological revolution," might take for its symbol something likely to be surprising, the restriction enzyme. This is the tiny organic tool by means of which, under laboratory conditions, genetic engineering is accomplished. By means of such enzymes, genes, which are molecules of DNA (deoxyribonucleic acid), may be excised, or, if one wishes, relocated, attached to, or spliced amongst, other genes, whence they might, going about their business, produce interesting, or surprising, sometimes troubling, results.

There is a sense in which genetic engineering, of a haphazard sort, is familiar. For thousands of years plants and animals have been bred selectively. DNA tracing, for example, proves that all the varieties of dogs in existence, from the Pekinese and Chihuahua through the Great Dane and Saint Bernard are derived from a single variety of canine, the gray wolf. On the other hand, in such cases, one dealt with nature's packages, her givens, so to speak, mixing and matching them, breeding ever faster, keener-eyed falcons, faster, stronger horses, swift, responsive, agile beasts capable of carrying an armored rider distances with ease. In contemporary genetic engineering, on the other hand, one opens the packages themselves, and may split the givens, so to speak, a biological breakthrough comparable to the physicist's splitting of the atom. Within these opened packages, one may then, as one wishes, shift about or exchange their contents, producing not better falcons or horses, but perhaps things altogether different, even new forms of life.

3. Some Actualities and Possibilities[1]

Genetic engineering is an element within a broader field, which one might call DNA technology. It may be useful, to give some sense of the sort of things which are, or might be, involved in DNA technology, amongst which, of course, would be our element of major concern, soon to be engaged, genetic engineering.

---

1. In this study, earlier, now, and later, we reference a considerable number of actualities and possibilities, primarily connected with science and its projected future. The background data for this book has been acquired over many years, and was not gathered with the idea of publication. Accordingly, for better or for

One field of DNA technology is forensic DNA. DNA technology, for example, may be used for resolving questions of paternity. Similarly, DNA evidence has functioned in cases of rape, either to identify an assailant or clear an innocent man, wrongly accused. DNA can furnish evidence, as might fingerprints, of a perpetrator. Years following a crime, saliva on an envelope, inadvertently supplied by a suspect, might lead to arrest and prosecution. More benignly, if annoyingly, it might link a prominent political figure to a forsworn lover. Too, DNA can be of genealogical interest. It can suggest that widely separated individuals are related or have an ancestor in common. For example, let us suppose, to take a purely hypothetical example, rather than an actual, controversial example, we have several widely separated families all claiming descent from a particular historical figure, say, William, Duke of Normandy, William the Conqueror. DNA analysis could supposedly determine whether or not these families were related, and thus their likelihood of deriving from a common ancestor, whether William the Conqueror, or not. The case is easier with, say, the relation of contemporary varieties of a species to a common ancestor, particularly if the ancestral species is still about, as both ends of the DNA chain would then be available for direct comparison. This sort of thing is possible in many cases, with plants, insects, mammals, birds, and so on. We mentioned the case of the gray wolf.

One of the most significant applications of DNA technology is that of supplying an individual with genetic information. For example, genetic tests are available for over nine hundred diseases. Awareness of such information is likely to influence behavior, for example, in mate selection. A dramatic example of the value of

worse, I am unable, in the overwhelming majority of cases, to document a specific source for this fact or that alleged possibility. Presumably a reader, if inclined, might pursue the matter. Certainly there are a number of books, articles, internet entries, and such, which deal with such things. As an analogy, I believe that a molecule of water consists of two atoms of hydrogen and one of oxygen. On the other hand, I have no idea where I first heard this, and I would feel it somewhat embarrassing, both to me and a reader, for me to supply a source for the claim, perhaps from the report of a laboratory experiment. To be sure, the analogy is imperfect. Happily, the primary interest of this book is not in the listing of facts, or alleged facts, which seem to change , even alarmingly, from week to week, from month to month, so much as it is to deal with the sorts of things involved. Naturally, I am grateful to the many authors, scientists, lecturers, and others from whose work I have profited, and whose work, in the laboratory and otherwise, will do much, sooner or later, to awaken philosophy "from her dogmatic slumbers."

such genetic testing, acted on, has occurred in connection with Tay-Sachs disease. This is a genetic disease often associated with Ashkenazim, individuals of predominantly European Jewish ethnicity. In this community the disease is now, for most practical purposes, eliminated.

Genetic information pertains to all aspects of the genome, not simply to particular genes which might be regarded as potential threats to life and health. Given the nature of human beings, the shifting fashions in what is prized or not prized, one could conceive of potential mates, or, more likely, their parents, reviewing genetic profiles. What lurks behind that phenotype? One could suppose that some rather plain, unprepossessing individuals might count as genetic "10s," so to speak, whereas some rather imposing individuals might fare less well, counting perhaps as genetic bench warmers, awaiting a chance to play, if not exiles, banished from the game. One supposes this sort of information might qualify cultural practices, such as courtship and romance. On the other hand, traditionally, in most parts of the world, in most times, such matters have been subject to rational calculation. One supposes, too, such information would be likely to affect the self-image of the individuals involved, positively or negatively. Doubtless legal problems, concerns of privacy, and such, would become involved in such matters.

Such information could also be used for screening purposes, for example, in estimating insurance risks, in reviewing job applicants, in assigning military occupational specialties, and such. It seems to be currently accepted that alcoholism has a genetic component, not that there is a gene for it, *per se*, as human beings have been around long before alcoholic beverages, say, wine and barley beer, but that certain genetic predispositions might make one more susceptible to alcoholism. More controversially, there may be genetic predispositions to sudden and violent responses, something that may once have had selection value but is likely to be less prized in most contemporary cultures. These sorts of things, of course, might be said for almost any disposition. Presumably there is no gene for flashy ties but there might be some genetic predispositions which might make some individuals more likely to adapt billowing togas, "like sails," satin doublets, flashy ties, and such. Obviously one becomes soon involved in sophisticated causalities, what causes what, software vs. hardware, internal programming vs. external programming, individual quirks, rationality, intelligence, calculation, nature vs. nurture, and such. And, in sensitive cases, as with alleged genetic linkages to social problems, one will hear from the politics of the day, whatever it might be at a given time.

Within DNA technology, one finds genetic engineering, and that is our primary concern in this portion of the study.

One possibility associated with genetic engineering is gene therapy, which involves the treatment of genetic problems, rather than simply their detection. The first use of gene-cultured cells for the treatment of disease, in this case, Severe Combined Immunodeficiency, was approved by the FDA, the Food and Drug Administration, in 1990. As many abnormal conditions are genetic, or have genetic components, gene therapy may open new horizons of medical treatment. Gene flaws, for example, have been allegedly found for abnormalities as diverse as leukemia, breast and ovarian cancer, kidney cancer, sickle-cell anemia, cystic fibrosis, phenylketonuria (PKU), Huntington's Disease, obesity, schizophrenia, and manic depression. As a great many diseases, some claim most, some all, diseases, are either genetic or influenced genetically, gene therapies may revolutionize not only medical treatment, but the concept of medicine itself.

Genetic engineering is particularly common in food stuffs. Numerous crops, for example, are affected. Percentages of particular crops vary, but progressively. The trend seems likely to continue. Such engineerings improve the quantity and quality of yield, and, as far as can be determined, are without adverse effects. Such crops may be more disease resistant, and be less susceptible to cold, this extending their range and growing season. Further, some are more resistant to herbicides and some, interestingly, produce their own pesticides, supposedly harmless to humans. Genetically engineered bacteria are utilized in producing a growth hormone common in dairy cattle; specially genetically engineered bacteria are also involved in foodstuffs ranging from soy beans to cheese. A dramatic example of genetic mixings, between life forms, occurs in certain tomatoes, which contain enzymes from the Arctic flounder, this enabling them to withstand lower temperatures, which, as suggested earlier, increases their range and growing season. A familiar example of this general sort of engineering, though not dealing with food, is engineered insulin, producible in any quantity, from recombinant E. coli. Size may also be affected. For example, transgenic food fish have been produced which, quickly, attain sizes and weights several times that normal for their species. It has been claimed, as well, that genetic engineering might effect similar increases in the size and weight of cattle, for example, a single steer the size of a small truck.

It is well known that longevity tends to be associated with given family lines, which suggests that ageing is either a genetic matter or something on which the genome exerts a clear influence. One theory suggests that ageing is dependent on the action of regulatory genes which are dependent, in turn, on certain materials in the cytoplasm of the cell. Accordingly, it seems that if this theory, or something like it, is correct, then, if the cytoplasmic materials involved could be altered or controlled, ageing, presumably at a given point, might be slowed or stopped, altogether. There are various theories of ageing, but most involve genetic aspects, which, in theory, might be subject to manipulation. Along these lines it is interesting to note certain experiments with flatworms conducted in California, in which, by gene manipulation, mutations were produced, some of which doubled the worm's life span, and others of which increased it some six times, both with undiminished youthfulness. The fountain of youth, so to speak, if it exists, is more likely to be found in some laboratory than in the Everglades.

It is also apparently possible, at least in theory, to produce artificial genes. Indeed, in the case of a simple life form, a virus, this was done in 1976. An artificial gene was synthesized, and used to replace a defective predecessor. It is one thing, of course, to do something like this in the case of a virus, and another in the case of, say, a flatworm, or a mammal. On the other hand, the difficulties are presumably more those of degree than kind. Assuming the naturalistic hypothesis, that life is a mechanical process, that there are no mysterious vital forces, or such, involved, the question is mainly one of knowledge, means, and skill. One recalls, for example, that the tobacco mosaic virus, as long ago as the 1950s, was broken into its chemical constituents and then reconstituted, reassembled, so to speak, into a functioning life form. Presumably the right pieces in the right order produce a functioning watch. In theory, it would be possible to synthesize a human being.

The notion of artificial genes, of course, carries one well beyond even the "splitting of the givens," spoken of earlier. It is one thing to, say, mix the genes of a fish and a plant, perhaps those of the tomato and the Arctic flounder, and another to create a hitherto unconceived form of life. What would it be like? Would it be a small thing? Would it reproduce? Would it be benign, or dangerous, containable or uncontainable, controllable or uncontrollable? How fast would it grow? How fast would it move? On what might it feed? It is one thing to use old letters to spell new words, and another to invent new letters, a new alphabet, a new language, in which to spell altogether unfamiliar words.

4. Genetic Engineering: An Exercise in Application

There are few problems to the solution of which genetic engineering could not contribute, either directly or indirectly. Similarly, there are few problems which it might not, directly or indirectly, in one way or another, worsen. It is a key which makes possible the opening of the doors to a thousand futures, doors behind which it is not always clear what lies. I think a useful introduction to the nature of, and to the theoretical relevance and importance of, genetic engineering would be to consider some of its possible applications to a number of problems of global import, for example, population; resources, of food and energy; the environment; health; social problems; and even the human condition, itself.

4.1 Population

A variety of genetically engineered adaptations might permit humans to considerably extend their habitat. For example, human space could be considerably enlarged, both horizontally and vertically, if a portion of the species could return to the sea, that environment from which its ancestors once emerged. This might envisage a gilled human, capable of drawing its oxygen from water, with perhaps suitable adjustments for withstanding pressures at various depths, and temperatures in various depths and at various latitudes. Something like two thirds of the Earth's surface is covered with water, and, in many places, water of considerable depth. Thus, as the bird exploits the sky the human might exploit the sea. Utilizing this additional living space the planet could support a population several times its present capacity. This adaptation would presumably be more efficient than building vast artificial islands, or constructing pressurized underwater cities. This adaptation would also make possible a vast extension of aquaculture and aquahusbandry, which might produce food for billions of humans, beneath or above the surface. Interesting cultural divergences, over time, might be anticipated. The aquatically adapted humans, too, would, as other forms of sea life, be protected from wind and storm damage, and would be relatively safe from seismic disturbances. Presumably, too, unless engaging in wars, raids, and such, amongst themselves, they would be substantially protected, in their hundreds of underwater "Switzerlands," from the conflicts and social perturbations historically common on the surface of the planet.

Other adaptations, say, to cold and thinness of air, might further enlarge the human habitat. Extraterrestrial adaptations, too, are not beyond the possibility of an eventual empirical realization. At the time of this writing, something between six and seven hundred exoplanets, or extrasolar planets, have been discovered. The great majority of these planets, as would be expected, given the difficulties of detection, are quite large, and presumably unsuitable for life, certainly as we think of it now, but it seems highly probable that where there are large planets, there would be, as in our own system, smaller planets, as well. Such would be expected, given the usual forces supposed to be productive of planetary systems. These smaller planets, small enough to lose hydrogen and large enough to retain oxygen, if located in a star's habitable zone, neither too far from, nor too close to, their primary, if not immediately hospitable, might be robotically terraformed, seeded, and such, and, thus, over time, might support human life. They might be reached by multigeneration ships, perhaps carrying thousands of frozen embryos, which might be brought to term, serially, in large numbers of artificial wombs. Adjustments useful in adapting to such environments, if anticipated, might have been programmed into the embryos.

One might also, of course, decline to indefinitely expand populations, given a presumed finitude of resources. In such a case, genetic engineering might be used to reduce, or stabilize, a population. There are many ways in which this could be done, without engineering ethnic-specific viruses, or engineered airborne plague viruses, against which the home population might have been suitably inoculated. Certain adaptations common amongst social insects might suggest possibilities. One might, for example, engineer massive numbers of neuters, unable to reproduce and disinterested in doing so. Similarly one might reduce or postpone the period of fertility. There are many such possibilities. Such techniques might be applied selectively, in virtue of political or social ends.

4.2 Resources

With respect to food, current or new plants might be engineered, in a variety of ways, for example, to increase yield, to require shorter growing times, to be capable of thriving in wider ranges, to be more resistant to frost, to insect damage, and such. In animal husbandry, one might anticipate dairy cows that produce more and better milk, perhaps nutritionally fortified, hens that produce more and better eggs, cattle that are larger, heavier, healthier, and so on. Another approach to such matters would be to engineer a more omnivorous species. For example, it has been suggested that a recombinant E. coli bacterium might be produced which might synthesize cellulase. This protein can break down cellulose fibers in grass, in such a way that they may be digested. A human being with such bacteria in his digestive system could then feed on grass. Indeed, the natural possession of such bacteria in their systems is apparently what makes grass-eating animals possible.

Many insects thrive on diets which would might shock a garbage disposal unit. *Mastotermes darwiniensis* is a primitive termite, for example, whose appetite can reputedly challenge that of the cockroach. It can reputedly manage not only buildings and living trees, but, amongst a number of other things, ivory, leather, rubber, wool, salt, excrement, cable linings and billiard balls.

Since this is all a matter of chemistry, and chemistry can be manipulated, it follows that, in theory, what the digestive system of another organism can do, so, too, might the digestive system of a suitably engineered human organism.

This is not an anticipation that fast-food restaurants of the future might put out cable linings and billiard balls, but it is to note that a more omnivorous human species is not impossible.

Presumably, in working such things out, one might be forced to take account of, and engineer one's way around, unanticipated, unwelcome side effects or byproducts. It is a rare world in which one can do A and not have some effect on B and C, perhaps hitherto out of sight, desirable or undesirable.

With respect to energy, and its conservation, much might be accomplished. Here it might be judicious to reengineer the human species, or substantial portions of it, taking note of two of evolution's most common strategies, insulation and size. Two obvious possibilities suggest themselves, the furred human and the small human. A number of mammals do quite well in polar environments, for example, seals, polar bears, and Arctic foxes. More relevantly, furred animals turn on no heaters, build no fires, and so on. Similarly, they do well without expensive and elaborate dwellings, making out nicely with burrows, caves, and such. This adaptation would certainly allow one to do with a considerably smaller expenditure of fossil fuels. And when the last automobile ran out of gas and the last generator whirled to a stop one might think once more of horses and candles. Doubtless dandies would groom themselves carefully and belles would be concerned with the glossiness of their pelts. Smaller people, too, furred or not, would be likely to consume fewer resources. For example, they would take up less space, could accordingly get on well with smaller houses, smaller bathtubs, smaller furniture, and such, would require less cloth for their clothing, would eat less food, and so on. Such adaptations would postpone the exhaustion of fossil fuels for centuries. Too, it might be recalled that for thousands of years human beings got on quite well without fossil fuels. Croesus and Maecenas never missed them.

Another possible adaptation might be hibernation, which would allow human beings to be quiescent during periods of cold or times of limited food supply.

Some people might see advantages in the exhaustion of fossil fuels, for example, the emergence of more decentralized, closer-knit communities, a more rural milieu, or such. One would be likely to see a general return to a more agricultural economic base, a countryside of farms and villages, such things. Perhaps manorialism would be restored, with feudal hierarchies, and the necessity of policing might produce new forms of knighthood, a meaningful knighthood. It has been speculated, though it is difficult to believe it was reflectively speculated, that the exhaustion of fossil fuels would render war impossible. "Sorry, generals, you can't have your wars anymore." One supposes that Thutmose III, Alexander, Caesar, Genghis Khan, and such, would have been stopped in their tracks.

4.3 Environment

It would doubtless be possible to genetically engineer a human being who would be more tolerant of pollution, noise, dirt, fumes, foul air, and such. For example, one might reduce aural acuity, develop a transparent membrane to protect the eye, diminish the sense of taste and smell, narrow the apertures of the face and head, and so on. One supposes other solutions would be preferable, of course, reducing noxious gases, lowering noise levels, and such, but if such obvious solutions proved impractical, say, politically or economically, genetic engineering could lurk in the wings, armed with dreary devices.

More benignly, it should be possible to develop filtering plants, by means of which air might be purified. Similarly, bacterial scavengers or biodegraders might be developed, which might break down noxious compounds into innocuous, perhaps valuable, substances. It is not necessary for a world to wade through its own garbage. Similarly, one might bioengineer diseases which would selectively attack harmful organisms, for example, disease-bearing mosquitoes.

4.4 Health

Genetic engineering could make marvelous contributions to human and animal health. It offers the promise of eliminating genetic diseases altogether, and, in general, producing longer-lived, stronger, healthier organisms which are more disease resistant. Through an original, one-time germ-line intervention such endowments would be transmissible to offspring, through the generations. Another possibility is the engineering of hunter/killer bacteria designed to destroy specific disease-producing bacterial or viral organisms. Useful substances might also be synthesized, for example, human hormones, which might be used in the treatment of dwarfism. The recombinant production of insulin, earlier mentioned, has long been in place. More radically, it seems possible that genetic engineering might be used to affect physiological processes, such as growth and ageing. Such interventions, if actualized, would presumably precipitate a large number of economic and social problems. One might anticipate riots, crime, and revolutions. If such interventions were generally available, for example, administered universally, *gratis*, by a global state, perhaps in the form of inoculations, controls on population would obviously be in order. One might anticipate the sanctioning of abortion as a sacrament, the ritual cannibalization of infants, wide-spread sterilization, arranged "sport wars," legally imposed limits to length of life, or such.

4.5 Social Problems

Perhaps surprisingly, genetic engineering might eliminate or ameliorate a number of social problems. To be sure, the application of genetic engineering to such problems, as to many others we have previously considered, would seem to presuppose an obedient, docile population uncritically subservient to a totalitarian state. Such a population might possibly be brought about by means of judicious social engineering. The human being, statistically, conditions nicely. Exceptions might be dealt with, as defective in one way or another.

A unified race might be gradually brought about, similar in color and features. In order to minimize the ethnocentric and xenophobic predispositions of human beings, doubtless once of value in group survival, one would suppose a new color, new features, and such, would be desirable. Certainly one would not wish to select any current race as the paradigm, which would be likely to generate unease, if not actual ill will. Perhaps blue would be a nice color, with a small nose and large ears. The controllers could decide. It is possible, of course, that human beings have a need to discriminate or hate. A "uniform race," thusly, might eventually find itself divided up into new races, for example, with slightly different shades of blue, slightly smaller or larger noses, ears, and such. If human beings were so uniform that numbers might be required, placed, say, on the forehead, one would have to be concerned with differences between odd and even numbers, or such. Perhaps one could go after those with a different accent, or those who favored a different hand. Ambidextrous individuals might be able to pass, fitting in nicely with either the left-handed or right-handed race.

Perhaps a different approach to the race problem might be popular with those who favor diversity, at least of a nonideological sort. This would be to multiply races to the point where racism might become pointless, there being too many races about to make it practical to discriminate against any particular race.

Let us suppose that in the "multiplying races" approach, we have, contrary to all probabilities, something like fifteen hundred races. Is it really likely that this would render racism impractical? Might it not, rather, multiply racisms? One might well suppose that rivalries, feuds, vendettas, or wars might develop between, say, race 608 and race 1159, and between, perhaps, race 201 and race 1422. Such things can occur amongst families and tribes of the same race, so it seems likely they might occur amongst smaller races, as well. Groups of races might become involved, for example, lighter-colored races versus darker colored races, blond-haired races versus dark-haired races, heavier races versus lighter races, taller races versus shorter races, and so on. There would likely prove to be coalitions of races. There might even be unpopular "target races" singled out for discrimination, and then, when that race was eliminated or subdued, a new "target race" might be selected, and so on. Perhaps the best bet would be to have a small number of large races. But then one already has that arrangement.

We noted earlier that there might be genetic predispositions to sudden and violent responses. It is only a step from this possibility to speculating about genetic predispositions which might lead to violent crime. There are apparently studies suggesting the plausibility of this hypothesis. A conference organized to consider this possibility had to be removed to another venue, however, from one country to another, because of political opposition. Hypotheses may be evaluated on various grounds, such as precision, clarity, simplicity, generality, fruitfulness, and political acceptability.

Let us suppose, per hypothesis, there might be a genetic predisposition to violence. In such a case , genetic engineering need not be limited to eliminating or reducing such tendencies; it might, as well, be used to instill them, for example, in order to facilitate the implementation of aggressive national policies.

Much human conflict can be traced back to differences, which need not be racial. Many human beings are not very tolerant of differences, for example, in ethnicity, ideology, religion, culture, language, class, and intelligence. Combining social and genetic engineering one might, in theory, produce a substantially homogeneous human species, of whatever sort desired. Presumably this would dangerously reduce the adaptivity of gene pool and result in a mediocre, stagnant civilization, but it might, as a species approached its twilight, reduce social envy. There are always trade-offs.

Genetic engineering might also contribute to the solution of gender difficulties. Perhaps it might manage a more politically acceptable male, one, say, less virile, more sensitive, less ambitious, less possessive, more passive, smaller, weaker, more manipulable, and so on. The difficulties, after all, one gathers, lie with the male, who, in his present inchoate, barbarous state, seems to be the problem. To be sure, one might engineer a different female, as well, perhaps one more virile, less sensitive, more ambitious, more aggressive, more violent, larger, stronger, and so on. To be sure, much would depend on who was doing the engineering. In the deplorable grasp of a male chauvinist pig who knows how many Merle Oberons, Gene Tierneys, Sophia Lorens, and such, might turn up.

One might also, one supposes, manage a human being somehow capable of asexual reproduction.

It is difficult to anticipate how much demand there might be for this solution.

The many individuals who understand the nature of society and what must be done to improve it tend, on the whole, to think in terms of social engineering, conditioning regimens, and such, as being the most likely instrumentations in virtue of which society might be molded into something they anticipate they would prefer. As their thinking is the correct thinking, and their values are the correct values, it is understandable that everyone should think as they think, and that their values should be imposed, in one way or another, on others. The major obstacle to the success of their program, as to so many such programs, is the recalcitrance of human nature, particularly the inclination of some human beings to think for themselves, despite their inability to do so properly, despite their lack of enlightenment. Freedom of thought is thus the lurking serpent in their Eden, ever poised to strike. Accordingly, the social engineers should really look more closely at genetic engineering for it might prove to be an inestimable ally, by means of which their dearest projects might be brought to fruition. This is because it can work directly and effectively on that very refractory nature which has hitherto given them so much frustration and grief. This will relieve them, among other things, of the necessity to denounce disagreement, when it might incomprehensibly emerge, as the result of stupidity, evil, or insanity. To facilitate the conformity essential to their success, one need only engineer an improved human being, one of a reduced level of intelligence and a more than usual susceptibility to psychological suggestion. Such raw material, presoftened and malleable, should ask little other than to be told what to think and what to do. Lacking chunks and quirks it is ready to be poured smoothly into the waiting molds. Thus the combination of genetic engineering and social engineering might produce the outcome unattainable by social engineering or genetic engineering alone, as social engineering alone might founder on the rocks of human nature, and genetic engineering alone would lack a specific ideological content, which contents differ from one set of infallible social engineers to another. In this way, the social problems attendant on freedom of thought, judgment, criticality, and such, with their woeful consequences of divisiveness, disagreement, confusion, and disruption might never arise.

4.6 The Human Condition

A particularly weird possibility which might appeal to apocalyptic ideologies is that of the "ambush gene." This is a gene which, after several generations of transmission, by means of which it would be widely disseminated in the global population, rather like a genetic timer, can switch off life. This could obviously produce a great deal of damage in any population, but would be unlikely to result in the extinction of the human race without a global organization, perhaps the United Nations, insisting on and enforcing universal inoculation, doubtless under the pretext of promoting world health. In this way, omnicide could be accomplished, and the world might be abandoned to fish and insects. Whereas this monstrous scenario is apparently an empirical possibility, its enactment is surely improbable. End-of-the-world enthusiasts, committed misanthropes, religious lunatics, unsung and unimportant individuals who wish to be important, and make a real difference in the world, have a much easier way to bring about the extinction of the human species, thereby solving all problems, social and otherwise. They need only genetically engineer one or more hardy, conveniently transmissible lethal diseases, any one of which might have reduced the Bubonic Plague to the status of chicken pox. Many individuals, laboratories, governments, and such, already possess the technology, the means, and opportunity, to bring this about. The most dangerous weapons of mass destruction are likely to be inexpensive, easily produced, easily delivered, easily distributed, and invisible.

5. Some Considerations

In the preceding discussion we have alluded to many actual and potential aspects of genetic engineering, benign and noxious, attractive and alarming. It is reasonably clear that genetic engineering might prove a universal boon or bane, that it might make possible almost universal salubrities or catastrophic horrors. Doubtless the on-going reports, decade by decade, century by century, will be mixed. Neither a golden age nor a descent into universal misery and barbarism is likely. Both, however, and almost anything in between, are possible. The doors to the future are many and mysterious. Presumably one should move slowly and monitor carefully, always aware that the consequences of any action are likely to produce unanticipated effects, of both a contemporaneous and future nature. Sometimes what happened is not clear until long after it has happened. One thing is very clear, and that is that genetic engineering places within the grasp of a species its own evolution. Classically one thinks of evolution in terms of genetics "casting the dice" and the environment "selecting the winning numbers." Today, much of the environment is subject to human control, and the dice can be loaded. This dual intervention places the future of a species, for the first time in zoological history, barring cosmological accidents, at the disposal of the species itself. This is an awesome possibility, that one may decide one's own being, one's own zoological destiny. Who is so godlike, so all-knowing and wise, as to accept this task? Perhaps this is an option best left unnoticed, or best ignored, or, if not ignored, not exercised.

5.1 Containment Problems

Containment problems are particularly relevant when dealing with altered or designed microorganisms. The current recommendations involve a gradation of several biological levels, ranging from simplicities such as washing one's hands and mopping up spills, to safety precautions involving sealed environments, protective gear, and remotely operated manipulation devices.

As society cannot place a policeman on every corner, and a policeman to watch policeman on every second corner, society cannot place an inspector in every laboratory. Respecting precautions thus, for most practical purposes, devolves on the on-site personnel, the scientists, technicians, and others, who work in the laboratories on a daily basis. Naturally one supposes that such personnel are for the most part conscientious, and do their duty, if not as a matter of social responsibility, at least as a matter of prudential self-interest. Some facilities involved in hazardous work are located in remote areas. Also, some organisms are raised on special diets, and could not live outside the laboratory. There are always the unsettling possibilities, of course, as one is dealing with human beings, of mistakes, neglect, carelessness, sabotage, theft, and such.

5.2 Controls

Whereas it is possible that genetic engineering, as a whole, would be outlawed, that eventuality seems unlikely. Too much would be lost. Too, if it were outlawed in one state it could move to another state, and if, given a totalitarian global state, it were outlawed on the planet, it would presumably "go underground." Too, it would presumably be utilized by the global state itself, albeit not publicly. It is hard to conceive of a general ban on genetic engineering, anywhere, without supposing it to be motivated by misunderstanding, ignorance, and fear, presumably deriving from one or another religious establishment, solicitous for its vested interests.

Assuming then that genetic engineering is not likely to be prohibited as a whole, the question of controls emerges. Controls or not, and if controls, of what sort and to what degree, would probably depend on the particular aspect of genetic engineering in question. In many cases, presumably the most rational approach to these matters would be to have no controls, but to allow rather for common sense to have its say, and the inevitable self-adjusting mechanisms and constraints of an open market to guide research into the channels of greatest interest and demand. On the other hand, just as one would not wish to leave the manufacture of nuclear weaponry up to the entrepreneurial zeal of a number of private organizations, there would presumably be certain areas of genetic research which might be carefully regulated, if not forbidden altogether, for example, those interested in the production of lethal viruses.

Controls might be left, in most cases, to the individuals involved in the work, as is currently muchly the case. Private organizations, too, might impose their own standards and controls, those obtaining in their own facilities. Beyond such things, particularly in sensitive areas, despite the inefficiencies, and impediments usually associated with bureaucracies, the state might well involve itself. What level of governmental supervision to be appropriately emplaced would presumably depend on the matter in question, and, ideally, on which level of government would be best situated to exercise meaningful oversight, for example, the locality, the member state or the macrostate, whether it be that of a nation or a world government.

Common sense and context should rule in such matters.

One hopes they would.

Ultimately, responsibility rests with individuals, not an abstract entity remote from the vials and cages.

5.3 Acceptance Problems

We have noted, in the foregoing, that some remarkable divergences from the varieties of human with which we are familiar would become possible by means of genetic engineering, for example, marine adaptations, extraterrestrial adaptations, human neuters, more omnivorous humans, furred humans, small humans, indefinitely youthful humans, new races, new forms of men and women, and so on. One might also remind ourselves of the possibility of the engineering of enhanced life forms, hybrid life forms, and even new forms of life. For example, one might, in theory, considerably raise the level of simian intelligence, perhaps to form a new inexpensive, servant class, produce eleven-foot basketball players, or enhance wrestlers or acrobats with additional limbs; too, one might create chimeras, mixes of species, perhaps dolphins with hominid appendages for tool and weapon manipulation, dolphins which might be used for underwater salvage and repair, or the placement of nautical mines; or who might, with crossbowlike weaponry, patrol beaches, protecting swimmers from predatory marine life. The centaur suggests the crossing of human and equine genes, with the result of four-legged humans who might run with cheetahs and dogs; consider a night watchman or soldier who was part tiger; too, there might be new forms of life altogether, perhaps difficult to even conceive of now, perhaps bacteria engineered to break down metal and plastic, perhaps new forms of food plants, which might be carried with one, say, in trays or globes, which could produce edible pods or fruits as a chicken might lay eggs; perhaps a swift, serpentine creature designed to pursue and devour rats, perhaps exotic, air-rooted, crawling, heat-seeking, poisonous plants intended to enter enemy camps at night and strike like cobras, perhaps massive androidal creatures synthesized from artificial genes, faster, stronger, more prolific, more adaptable, more intelligent than human beings.

It seems likely that a number of these enhanced, combinatory, or original life forms might face problems of acceptance in civil society. Two of the most common human responses, so common it seems likely they once possessed a distinct survival advantage, are ethnocentricity, in which one prizes and favors one's own group, and xenophobia, in which one disprizes and disfavors, perhaps even resents, suspects, fears, and hates alien groups, different groups, strangers, and such. The predilections of most human beings to belong to one "gang" or another, ethnic, religious, political, or whatever, are easily understood; the individual is unimportant and weak, insecure and vulnerable, by himself; membership in a "gang" provides him with protection, a sense of importance, of meaningfulness, of identity, and power. He is likely to understand himself and define himself in terms of the gang, large or small. In a sense, he has no self-image except in such terms; without his membership, often in overlapping or concentric gangs, he is scarcely comprehensible to himself; he is scarcely even matter; certainly he is without social form, except perhaps as others categorize him, the inferior, the different one, the outsider, the pariah, the nonentity, and so on. Considering the pervasive nature of these dispositions, and their natural concomitant, intolerance, one anticipates that many genetically engineered life forms would not be easily integrated into human society. What differs from oneself is often experienced as immediately questionable, if not repellant, revolting, and potentially threatening. Genocide, ethnic cleansing, and such, are not modern phenomena. Gods themselves have been conceived of as not only tolerating such activities, but commanding them. And thus is sanctified the wholesale extermination of populations. The defense of "obeying orders" is one of ancient lineage. Such things being the case, it seems likely that acceptance problems might plague many conceivable life forms. Some such life forms might be incorporated into one gang or another, perhaps as allies. More likely they would be subjected to discrimination, socially and legally. Such groups would be likely then to band together, forming their own groups, hostile islands amongst hostile islands. Where practical, they might be segregated, or killed. If they might prove congenial to a given political group, they might be enfranchised, the anticipation being that they would then support the appropriate group. Sensitive to political power, they could eventually form their own block. One could conceive of blue minorities, multilimbed minorities, and so on. In all likelihood, however, such possibilities would not be actualized. Presumably such life forms would not come into existence, having been prohibited by law. Perhaps some might be brought into existence as a matter of experimental interest, to function as examples of scientific progress and prowess, or perhaps some might be created for commercial purposes, to be exhibited as freaks, housed in zoological gardens, or such.

5.4 Disposition Problems

Presumably a good deal of wastage would accompany efforts to produce enhanced, combinatory, or original life forms. Dr. Wilmut's triumph, Dolly, was the result of something like 277 attempts, resulting in thirteen embryos, and that involved cloning, not genetic engineering. In cloning one would have, it seems, a much simpler procedure, as one is dealing there with same-species, preformed cells and nuclei, not, say, with interspecific gene mixes or the synthesis of artificial genes. If one is interested in producing winged frogs or ten-legged snakes, presumably the likely wastage involved would provoke little public outrage, except on the part of activists who have discovered or invented bird rights, frog rights, snake rights, or such. On the other hand, if human embryos are being utilized, one would anticipate objections, and possibly active opposition. Whereas the disposal of mistakes, of unused tissue, and such, would presumably take place privately, it is not difficult to suppose that the matter would eventually be brought to public attention, with orchestrated consequences arranged by interested parties. And if one can plausibly anticipate public resistance to procedures likely to be confined to table-top laboratory glassware, it is almost certain that the disposition of results more advanced in size and maturity would be of even greater concern. One supposes, for example, that few people at present would approve discarding the results of every experiment which turned out badly, say, a botched five-year-old human-canine mix, perhaps designed originally to locate contraband substances or plastic explosives, or a handicapped, eleven-year old, four-legged human, whose legs were of unequal length, perhaps originally developed for competition in some new international sport. In any event, one would have to concern oneself with failures, rejects, "obsolete models," and such.

Many such problems, of course, presuppose that something like contemporary moral views would continue to obtain, which, given a candid view of such matters, human history, and such, could not be guaranteed. For example, it might be maintained that the entities in question were mere biological artifacts, lacked "souls," were not members of the moral community, or such. A species which has enjoyed beast fights, gladiatorial bloodshed, public executions, drawings and quarterings, public torture, the merciless extermination of entire groups, men, women, and children, and even the asses, goats, sheep, and cattle of such groups, is certainly capable of looking upon the destruction of unwanted life forms with equanimity.

5.5 Deciders

Many decisions in these matters will be made privately, here and there, from time to time, by one individual or another, or one group or another. For example, the "Dolly phenomenon" was apparently not anticipated by the general public. It was not, for example, heralded by the media, subjected to general scrutiny, debated in parliament, or such. The world was confronted with a remarkable *fait accompli*. Dr. Wilmet's funding was promptly canceled, absurdly enough, but this was not done until after the fact. One supposes that something of this sort, such "flights under the radar," for example, might occur in the future, as well. What could be done here could be done, for the most part, unobtrusively, without great expense, and without a considerable outlay of visible resources. No "Manhattan Projects" are necessary. Medical research could proceed quietly, and apace. More frighteningly, the bioengineering of lethal viruses could take place in someone's basement, unsupervised, without adequate safeguards.

Given the nature of this type of research, the first line of defense against possible catastrophe is individual responsibility. This may be a frightening consideration, but, in the long run, responsibility always devolves on individuals, some individuals. To be sure, psychological and political considerations must be noted. An individual intrigued by, and excited by, a possible line of research is likely to pursue it, even obsessively. It has never been done. Can it be done? We will see if it can be done. It is easy to rationalize almost any endeavor, by means of principles at hand, or devisable, by means of which almost anything can be justified. If worse comes to worse, one assures oneself, the results can always be contained or destroyed. Political considerations are even more alarming, particularly in totalitarian states, where an established power elite, above effective internal criticism, can do much as it pleases, openly or "in the shadows," so to speak. Resistance to authority may be an affordable luxury in a democracy. It is less easily purchased in a tyranny. When the options are obedience or death, it is easy to see obedience as prudentially, and possibly morally, appropriate. "Must," so to speak, implies "ought." Heroes may be admirable, but they are often dead, as well. In any event, the first line of defense against possible catastrophe, is individual responsibility, however fragile or rare it might prove to be.

Who is likely to decide questions here, and who should decide them, and how should they be decided?

One fears it may be the state.

Is it not the state which possesses the *fasces*, the rods and ax of power?

Given the mindless, hysterical response of certain religious and political leaders to the mere cloning of a sheep, the condemnation and banning of cloning in various states and countries, and such, the prospects of genetic engineering, at least as an open project intelligently pursued, seems dim, indeed.

One is reminded of the ignorance, and the fear for vested interests, that led others, in other times, to denounce theories, to proscribe books, to condemn views perceived as at odds with their own, and then, in sanctimonious consistency, utilizing the resources of compliant authority, to torture, imprison, and kill. One might naively have thought that such self-serving bigotries were a thing of the past, but, it seems, they may have merely been quiescent. Perhaps they will be with us as long as the drug of power and the fear of exposure continue to exert their baneful influences over the pretentious, the fraudulent, and parasitic. Let them not be rearmed. Space is large; the Earth orbits the sun.

In our time, given the *ethos* of the day, in particular, the unwillingness to assume personal responsibility, and its habitual, now almost reflexive, abdication on the part of most human beings, one supposes that important decisions in such matters, cloning, genetic engineering, and such, are likely to devolve by default on the instrumentalities of the state. After all, the state, however ignorant and uninformed may be its ministers and representatives, and however subject these may be to particular temptations, influences, and pressures, from others as ignorant and uninformed as themselves, is all wise, and may be depended upon to protect and save us all. Is that not the received wisdom? How naturally the steel cornucopia of the state is looked to, to solve all problems, to shape the future. Surely it is the *deus ex machina* by means of which the culture will achieve its salvation. Further, happily, the state is armed, and possesses a legal monopoly on coercion and violence. We know this, for it tells us so. The power of the state, thus, accepted humbly by those who have relentlessly sought to attain it, may decide the issues involved, discriminating amongst possibilities for us, deciding futures on our behalf, what may be and what may not be, what may come about and what may not come about. We are at its mercy. Law by law, order by order, ruling by ruling, ever more intrusive and powerful becomes mighty Leviathan.

One fears the state.

The state darkens the future.

Liberty, it seems, will become illegal, but then it may merely conceal itself.

What is important, interesting, attractive, and valuable is likely to be brought about, here or there, sooner or later, in one way or another.

Biotechnologies may eventually become a determining factor in the wealth and power of nations. One should think carefully before one closes a door to the future. That door may be opened sooner or later, by someone else.

CHAPTER FOUR:
ARTIFICIAL INTELLIGENCE

1. Prologue

WHETHER ARTIFICIAL INTELLIGENCE (AI) EXISTS
or not depends on how one will conceive its nature. It may be
defined very broadly, so broadly as to accept electric eyes opening
doors and thermostats responding to temperature changes as
instances of artificial intelligence, for example, electric eyes
"noticing" objects, thermostats "recognizing" it is time to
switch the furnace back on, and so on. If one is this generous
then it is obvious that artificial intelligence exists. Even if one
sets the requirements for artificial intelligence considerably
higher, for example, the capacity to engage and possibly defeat
an organic competitor in an intricate board game, artificial
intelligence must be said to exist. To be sure, much depends on
the criteria held in mind. For example, if chess is in question,
it is not clear that the machine is playing chess. For example,
it does know what it is doing; it is merely functioning; it does
not, for example, know if it has won, lost, or drawn. It does not
anticipate the game; it does not know what is at stake; it does
not worry; it does not concentrate; it is not sensitive to time
pressure; it does not mind losing, nor does it take any pleasure
in winning. It does not even know when the game is over; it
does not know anything. A considerable amount of intelligence,
of course, went into the design of the program, the construction
of the machine, and such, but then a considerable amount of
intelligence goes into the construction of a refrigerator or an
airplane engine, and they do not think, either. The general
point here is that whether artificial intelligence, "machine
thinking," or such, exists, could be reduced to a matter of its
truth conditions, those conditions under which we would take

the claim to be true or false. Under one set of truth conditions, say, adjustment to conditions, response to stimuli, or such, the thermostat thinks, and under another set, it would not, as the truth conditions might require more than, say, responsiveness to stimuli. For example, a rock kicked down a hill, or a plant turning toward the sun, responds to stimuli. Whereas the question of artificial intelligence could be semantically trivialized, the interest in such a question suggests that a great deal more is involved than an inconsequential ruling. Self image, self-respect, species chauvinism, fairness, objectivity, rational assessment, the notion of a person, the perimeters of the moral community, the range of legal obligations, and so on, could all be involved. Presumably one does not want to define artificial intelligence into existence, by setting the truth conditions for such a claim so broadly that, say, automatic house lights, responding to a timer, would count as brainy, nor in such a way that nothing whatsoever might count as artificial intelligence, whatever its properties, or seeming properties. One does not want to get into a situation where one puts artificial intelligence into the category of square circles or the last natural number, always out of reach. To be sure, where to put the goal posts is an interesting question, and it is not clear that one should always have them in the same place. What seems a judicious distance at one point might, as one learns more, and encounters new devices, say, chess machines, seem less judicious later. As in many conceptual questions, one begins with an inchoate mass of habits, practices, prepossessions, and intuitions, and tries, to one extent or another, to refine and improve them. At a certain point, the bell rings. At that point one has either at last uncovered the gold for which one has long sought, which was there in the ore all the time, waiting patiently to be extracted, or, more likely, one has invented a conception with which one is now satisfied. In either case, this discovery or invention, whichever we take it to be, may or may not close the case. It may well prove to be only "a temporary resting place for thought," to quote a phrase from the great R. G. Collingwood, a pit stop, so to speak, in the philosophical journey, a journey commonly negotiating the terrain of "conceptual geography," a terrain partly explored, a terrain partly created.

2. Mental and Physical

The Mind/Matter distinction is, *prima facie*, a sensible, even obvious, distinction. Certain properties are often associated with mind, typically such things as consciousness, intentionality, subjectivity, and understanding.[2] It would be very unusual to associate such properties with tire irons, electric toasters, elevators, chess machines, and such.[3] Some typical mental events would be thoughts, memories, dreams, images, of one sort or another, feelings, emotions, and such. Some typical physical events, on the other hand, would be such things as sound waves, electromagnetic waves, impingements of stimuli on sensors, excitations of nerves, electromagnetic impulses in the brain, the firings of synapses, and such. Two familiar differentia with respect to physical and mental events are, first, that physical events are locatable in space, and mental events are not locatable in space, and, second, that physical events are publicly observable, at least in theory, and mental events are not publicly observable. Without accepting these criteria uncritically, one can certainly see their

---

2. "Consciousness" is difficult to explicate, as it seems to be a primitive conception, at least as usually understood, not defining it in terms of behaviors, whereby, say, the electric eye or thermostat might count as conscious. For example, if we are looking for the light switch, find it, and turn it on, that is most likely a conscious act, on at least some level, an act of which we are aware, at least on some level. On the other hand, if the light is turned on automatically by an electrical timing device, that is not a conscious act; indeed, it is not even, strictly, an act, at all. It is an occurrence. How far down the phylogenetic scale consciousness extends is an interesting question. One supposes there are degrees of consciousness. Dogs and cats are pretty clearly conscious; the matter is less clear with paramecia and amoebas. "Intentionality" is the property of "aboutness." For example, a mental state has a content or focus. It is about something, say, a tree, whereas the tree is not about being a tree; it is a tree, and not about anything. "Subjectivity" is the property of being self-aware, of having a sense of oneself, having an inner life, so to speak. Once again there are levels or degrees of this property. "Understanding" is probably a primitive conception, like consciousness. For example, the moment of insight, the seeing that something must be the case, as in seeing that vertical angles in Euclidean geometry must be equal, is an instance of understanding. A familiar example of understanding would be that of making sense out of marks or noises, as in the comprehension of meanings. A machine might scan the ink marks 'cat', process them, and produce a definition, printed or spoken, but, if it is a machine, as we now know machines, it does not know what it is doing. It does not know a cat from a mouse or dog, no more than a dictionary or encyclopedia. Indeed, it does not know anything.
3. An interesting question arise in connection with ascribing such properties, consciousness, and such, to unlikely objects. Consider 'The olive tree is thinking of

point, and their plausibility. A distinction of some sort, however we might eventually wish to conceive of it, certainly exists between, say, the image of an elephant in consciousness, either in seeing, imagining, or remembering an elephant, and whatever is going on in the brain which might account for, or accompany, the seeing, imagining, or remembering. For example, the image seems relatively simple and static but the producing, or accompanying, brain activity is likely to be extremely complex and a veritable fireworks of activity. And clearly, as a simple matter of logic, A and B can be identical if and only if they have the same properties, and this A and this B would fail that test. They clearly differ in several properties. Returning to the proffered differentia, the brain activity is clearly located in space, there in the brain, but it is not clear that the image is in space, for example, it does not seem to be there, look for it though we may. Secondly, the brain activity is public in the sense that it can be observed, in a broad sense, for example, monitored, recorded, measured, and such, by sophisticated devices, but the image of the elephant does not show up on the graphs; that is private.

Such considerations simply confirm what seems to be obvious, a difference between the mental and the physical, obvious enough between an image and brain activity, and obvious to the point of absurdity between, say, an elephant and the image of an elephant. Few are likely to confuse elephants with elephant thoughts. Elephants eat peanuts and elephant thoughts do not. Elephant thoughts trample no one.

Athens' or 'The olive tree is thinking of Minneapolis'. Interestingly, the first locution seems more plausible than the second locution. Also, one might imagine the first locution occurring in a poetic context, and actually touching one's feelings, perhaps with some wistful, gentle poignancy. Poets can do things like that. This sort of effect would seem much less likely with the second locution. Some individuals might consider both locutions to be nonsense. Generally, however, that analysis seems injudicious. Given contemporary neurological theory, suggesting the unlikelihood of thought without a central nervous system of a given sort, one supposes the locutions are best interpreted as expressing a false statement. Accordingly, as truth-value possession is a sufficient condition for meaningfulness, and both locutions express a false statement, one must regard such locutions as meaningful. Another sense of 'nonsense', say, that which is highly unlikely to be true, is not relevant in this context, for example, one might dismiss the claim that Saint Francis robbed the poor box as nonsense.

3. Some Views on the Mind/Body Problem

It would be a mistake not to recognize the sorts of differences we alluded to in the preceding remarks. They are a matter of fact. They exist, undeniably. On the other hand, should they be taken as somehow endorsing or implying a radical divergence amongst substances, that reality consists of at least two sorts of fundamentally different substances or radically different kinds of being, one is shortly embroiled in a number of paradoxes and improbabilities, violating some of the most commonly shared presuppositions of causality, say, that like can affect only like, and the principle of the conservation of matter/energy. As normally conceived, and seldom reflected upon, mind is supposed to be intangible and nonspatial, whereas matter is supposed to tangible and spatial. Viewed in this familiar way, one stands upon the brink of a no man's land, from which few have returned unscathed, that of the mind/body problem, the mind/matter problem, the spirit/ matter problem, or such. For example, how can two such different substances interact with, or affect, one another? Analogously, how could a geometrical theorem paint the garage yellow or give Fido a bath, and, similarly, how could one give the theorem one or two coats of paint, yellow or not, and what would it have to fear from a vigilant Fido, disgruntled after an unwanted bath, should it inadvertently find itself in the backyard? Different orders of reality would seem to be involved. How then could the intangible, not-in-space mind bring about changes in in-space objects, such as human bodies with their measurable cubic content, and how could human bodies, or the physical brains of such bodies, both weighing such and such, bring about changes in objects that are not even around, not being anywhere, so to speak. The not-in-space, for example, cannot be in the room, or in the house somewhere, even in the universe, as these are all in space. There are several other problems involved, as well, for example, having to do with energy, which is presumably required to do work, with the conservation of matter/energy, with the modality of decision, with the location of a decision maker, and so on. Obviously this is not the time or the place to enter into these matters, but it is the place to point out that serious problems, conceptual and scientific, attend hypotheses of either dualism or pluralism. This is the brink of the no-man's land we referred to earlier. It is no wonder that many turn back at such a troubling, formidable border. One may, of course, and there is much to say for this notion, as it saves much philosophical wear and tear, note the apparent interaction of mind and body, accept it as a fact, and not worry about it, giving it up as a mystery which is beyond human comprehension, which, if such interaction is the case, it certainly seems to be. For example, Jones decides to stagger to the bar and his body staggers to the bar; Jones then places an additional quantity of alcohol into the same body, that which staggered to the bar, and his views of the world, ideas, and such, undergo a change, likely to last until he awakens the next morning. Certainly the mind seems to affect the body, and the body the mind.

There are, of course, a number of proposed "solutions" to the mind/body problem, or problems, other than an unanalyzed, default dualism, but most of the "more-than-one-substance" solutions would invoke a god or gods, which, though nonspatial themselves, are supposed to have no problems, being gods, with interacting with material substances. In pre-established harmony, for example, a god or gods, presumably utilizing omniscience, set up and synchronize two independent tracks of events, one mental, one physical. Thus there is no interaction whatsoever but things go along together, the god or gods having arranged it from all eternity, for example, that when Jones decides to stagger to the bar his body will stagger to the bar, and when the physical alcohol enters his physical body, his views of the world, ideas, or such, as arranged, again from all eternity, will undergo an alteration. In "occasionalism," on the other hand, we have an arrangement in which the god or gods need not have prevision, thus avoiding the problem of reconciling prevision and free will, but perform on-site miracles constantly, on this occasion or that, for example, moving Jones' body to the bar when he decides to approach it, and, presumably with another miracle, altering his mental outlook after the additional alcohol has entered his physical system. These occasions must occur frequently, and the god or gods must be very busy, except that this is no problem for an omnipotent entity, or entities. There seems to be a moral problem involved here, in both pre-established harmony and occasionalism. For example, if Jones decides to shoot someone, he is incapable of doing so, unless the god or gods, enable him to pull the trigger. This would seem to make the god or gods an accomplice. More plausible solutions to the mind/body problem are the "only-one-substance" solutions. Examples of these are idealism, materialism, and psycho-physical parallelism. In idealism, as commonly conceived, material reality, matter, or such, does not exist, but a god or gods are required. In one familiar version, reality consists of spirits and the ideas, or experiences, of spirits, supplied for the most part by another spirit, the divine entity. In some other versions, one might have a divine pantheism, in which, it seems, ultimately, there is only one spirit, other seeming spirits being aspects of the one spirit, or such. This suggests a sort of multiplex divine schizophrenia. As idealism saw everything as essentially spiritual, in one way or another, so materialism, as usually conceived, would see everything as material, in one way or another. Actually, in materialism, one might have a material god or gods, or a more normal sort of god or gods, spiritual or such, as long as mind, that is, human mind, and such, was seen as material. Generally, however, this view, historically, has proceeded without the notion of a god or gods.

It is difficult to see how idealism could explain anything. For example, how could a divine entity produce the idea of, say, a bar, in Jones' mind? To be sure, divine causality is difficult to conceive of in any case. For example, how would a divine entity create a world? How would it create the stuff to begin with, and how would it doctor it into approved configurations, acting on, say, atoms or molecules. Would it act on, say, 800 thousand species of insects serially, or just pop out atoms to get to work on the project, in one fantastic shot? And how would it pop out atoms, to begin with, etc. In any event, given a material world at some point, material causality seems easier to understand than ideal causality. One rock or billiard ball bouncing into another rock or billiard ball, and moving the second rock or billiard ball, seems easier to understand, at least if one has not read Hume. On the other hand, how a nonmaterial spirit, without recourse to matter, could produce, say, the idea of a tree in another spirit, an idea such that the other spirit might bump into that idea, so to speak, perhaps bloodying an idea of a nose, seems harder to understand. It is at least less familiar. To be sure, it also seems difficult to understand how a material reality, atoms, and such, could produce consciousness, resentments, affections, theorems, numbers, and such. Certainly there seem considerable differences between hydrogen and hope, mercury and meaning, plutonium and poetry, and so on. This type of consideration may have suggested the one-substance theory known as psycho-physical parallelism, not to be confused with pre-established harmony. This theory is usually associated with the remarkable rationalist, Benedictus Spinoza. Whereas this theory can be interpreted in more than one way, a common way of understanding it is that matter and mind are not substances, but attributes of, or expressions of, an underlying substance. In short, mind does not produce the idea of matter, nor does matter produce ideas, but, rather, both owe their being to a *tertium quid*, a third thing, which is something other than either matter or mind. Ultimately there is only one kind of stuff, but it expresses itself variously, indeed, in an infinite number of attributes, only two of which are accessible to us, namely, mind and matter. This approach does not reduce our problems, but multiplies them. For example, there is no empirical evidence for the existence of the hypothesized underlying substance, there seems no reason to hypothesize its existence, and, if it did exist, it is not clear how it could produce attributes, nor what its relationship might be to the attributes it produced.

Another view, one more ambivalent, is epiphenomenalism. This is the notion that consciousness, mind, and such, are epiphenomena, effects, or produced phenomena, caused, but not causing. Some common metaphors are that mind is no more than a "shadow" cast by matter, or that mind is analogous to the smoke emitted by a steam locomotive, a mere byproduct of fundamental processes. Thus, supposedly, matter acts on mind, produces mind, but mind, being inactive, an epiphenomenon, does not act on matter, has no effect on matter. Causality is unilateral, nonsymmetrical. What seems ambivalent on this view is whether mind is material or not. If it is not material, one is returned to the interaction problem, e.g., how could the material produce the nonmaterial; if it is material, there seems no reason why, being material, it might not affect, and participate in, material processes. Some worries pertaining to epiphenomenalism, aside from the dubious nature of the nonspatial-product view, are, first, the seeming implausibility of the claim that consciousness is without causal efficacy, as it seems to be causally efficacious, e.g., Jones' decision to move to the bar seems to have something to do with Jones' moving toward the bar; second, relatedly, the view is incongruent with cybernetic, or feed-back, theory, for example, noting that the car is heading toward a tree is likely to result in an effort to change the car's direction; and, thirdly, consciousness is apparently widely spread throughout the animal kingdom, and it seems unlikely that it would have been so pervasively selected for if it were not without value with respect to survival, gene replication, or such. Consciousness is either effective or not. Suppose it is not effective. Then, epiphenomenalism, if true, would seem to be an inexplicable anomaly, involving an incomprehensible waste of biological resources, using them to produce a meaningless consciousness, with all its meaningless feelings, thoughts, and such. And, on the other hand, if consciousness exists so pervasively, presumably it would serve some purpose, have some distinct value, in short, would be effective. The more seriously one takes evolution the less seriously one is likely to take epiphenomenalism.

The preceding remarks have a very serious relevance to questions of artificial intelligence.

Given the several difficulties seemingly affecting "more-than-one-substance" theories, or pluralistic theories, it would seem that the best solution to the mind/body problem is likely to be some "one-substance theory," some monism, say, idealistic, materialistic, or Spinozistic. If this is the case, the way is cleared, in theory, for forms of artificial intelligence which would have to be understood, on any fair assessment, as "actual intelligence" or "true intelligence," at least if human intelligence is to count as actual intelligence, or true intelligence.

In the following, for the sake of simplicity, we limit our alternatives to the two most likely candidates for the monism in question, namely, idealism and materialism. Similar remarks, however, if one wished, could be made for, say, a Spinozistic monism.

1. At one time there was no human consciousness and, at a later time, there was human consciousness. (Presumed scientific fact.)
2. Let us suppose we wish to be monists. (In this way, one would avoid the Interaction or Mind/Body Problem.)
3. On this approach, there was at one time unconscious x and, at a later time, there was conscious x.
4. If we wish an interpretation of x, two candidates suggest themselves:
   (1) Unconscious matter, then conscious matter.
   (2) Unconscious mind, then conscious mind.
On this approach the distinction between calling x matter and

calling x mind is a matter of indifference. Calling it mind costs matter nothing. Calling it matter costs mind nothing.

Everything stays in place. Nothing is lost.

This approach gives us a same-genus phenomenological dualism. One thus accounts for the experiential or phenomenological dualism, which obviously contrasts rocks and thoughts, trees and memories, without rejecting monism. In this sense, we have two species, but only one genus, whatever it might be. This preserves the distinction between unconscious x and conscious x. The same x is involved, whether conscious or unconscious. If we adopt the materialistic construal, or terminology, then all is matter, there being unconscious matter and conscious matter; and, if we adopt the idealistic construal, or terminology, then we have unconscious mind and conscious mind. Alternatively, one could retain "unconscious x" and "conscious x." In any case, on any of these approaches, the way is cleared for a serious view of the possibilities of an authentic artificial intelligence. For example, on the materialistic construal, if we can manage consciousness with atoms and molecules, then there seems no reason why a machine might not do so, as well. For example, if inorganic materials, such as silicon, cannot manage the matter, organic molecules might be utilized, or synthesized. The resulting product, or machine, or such, might or might not resemble a human being.

4. The "Other Minds" Problem

Considering that it is logically possible, for all one could ever tell, that one is the single center of consciousness in the universe, or even the sole entity in existence, it is not surprising that a peculiar, interesting, hopefully relatively innocuous problem should also exist, in this case, the "other minds" problem.

It is altogether possible that certain metaphysical views might be genetically programmed into an organism. For example, a particular epistemological view is obviously programmed into organisms, and that is the indispensable, invaluable, life-preserving, life-enhancing, false view of naive realism. It seems we open our eyes and see an outside world, in which we move, pursue goals, and such, and whereas one could certainly call this "seeing," if one wishes, what is involved is enormously complex and has little resemblance to, say, putting up a shade and looking out a window. Stimuli impinge upon the organism, transductions of energy take place, and, in virtue of brain activity, one has certain experiences. These experiences, strictly, are all internal to the organism. What is going on is in the brain. The same sort of thing takes place in dreams and hallucinations. Do you not see things in your dreams? Does the rapt devotee not see Apollo, or Dionysus? Happily, there seems to be a close, topological relationship between the brain-dependent, internal world of experience, and the presumed external, brain-independent world of, say, rivers, rocks, trees, cliffs, and saber-toothed tigers. On this bet evolution paid off big.

Similarly, it seems that certain metaphysical views are genetically programmed into the organism. Three such views would appear to be that an external world exists, that the future will resemble the past, and that other people, and doubtless many other forms of life, are conscious, have minds, and such. These views are so much taken for granted that many people do not even realize they have them. Certainly they would seldom be articulated, and perhaps for good reason. Consider informing a stranger, perhaps on the train or bus, that you believe in an external world, that fire will be hot tomorrow, and that he has a mind, despite the lack of evidence. Who but philosophers would address themselves to such questions, and everyone knows about those fellows.

On the other hand, the "other minds" problem does have some relevance to our topic, as we shall see. Briefly, the problem is, how does one know that the other person has a mind, not that anyone, in most situations, would doubt it. For example, in our own case, we have a direct awareness of our own mind. As psychologists have taught us, this awareness of our own mind does not have to be infallible, at all, but even they, if they are not behaviorists, of a particular stripe, admit that we have something going on in there of which we are aware. For example, if we are not Wittgensteinians, we know from our internal states whether or not we are pleased, angry, in pain, or such. For example, if we are in pain we can usually determine this from our internal states, without listening carefully to see if we are screaming and groaning, or watching ourselves in the mirror, to see if we are grimacing, twitching, writhing about, or such. We do have this sort of direct awareness of our own mind, or of our own thoughts and feelings, even if we might occasionally misinterpret this awareness.

On the other hand, we have no such direct awareness in the

case of other minds. They are private, as we would expect by now. The classical view is that they are enclosed in their own boxes, so to speak, and we conjecture their existence, and states, from the behavior of the boxes. What is wrong with the classical view, if anything, is that it is overly intellectualistic, suggesting that our belief in the existence of other minds is a matter of logic, of inference, from A, behavior, to B, mind. It is more likely that our view of other minds is the result, at least primitively, of a genetically conditioned, preconscious, unquestioned assumption, a default presupposition, or "given," so to speak, which goes along with our form of life. It seems unlikely that the infant, for example, calculates the odds on Mommy's having a mind, or infers that she is conscious, on the basis of her behavior. It probably does not occur to him to think about it, let alone wonder about it. In a sense he takes it for granted; in another sense, he does not even take it for granted; the question does not even come up. There are two questions which must be clearly separated here, and they are the origin of an idea and the justification of the idea. An excellent example is the common reliance on induction. Given the pervasiveness of this reliance, it seems to have been selected for in the course of evolution. Those who bet on induction, so to speak, tended to survive. Induction is wired in. Evolution has seen to it. Humean antelopes, unwilling to stereotype lions, recognizing that logic permits the next one who shows up to eat grass and turn out to be the best of friends, are rare. The origin, then, of induction, taking it for granted that the future will resemble the past, or acting as though the future will resemble the past, and such, is presumably zoological. The justification of induction, on the other hand, is an entirely different matter, and one of considerable philosophical interest. Analogously, the origin of the idea of other minds is one question, presumably with a zoological answer, if it even comes up, and the justification of the idea of other minds is another question, presumably one calling for reasons, not causes. This distinction is particularly important when one deals with questions of artificial intelligence. In dealing with artificial intelligence, at least at the present time, the default position, namely, taking it for granted that the "other" has a mind, is not available. In the case of something which seems human, we would want evidence that it does not possess a mind, perhaps that it is an electronically controlled manikin, or something; in the case of a machine, on the other hand, we would want reasons for supposing that it does have a mind. In the case of the human we would want a justification for not attributing mind; in the case of the machine, we would want a justification for the attribution.

It will be instructive, however, to examine the "justification question" in the human case, as it has obvious applications to the question of intelligence, or mind, in the case of the artifact, or machine. As noted earlier, it is logically possible that one is the sole center of consciousness in the universe, and, indeed, that one may be the sole entity in existence. These are requirements for weak and strong metaphysical solipsism. Schopenhauer, perhaps in a typically surly mood, observed that solipsism required not a refutation, but a cure. That is fortunate, for it is irrefutable. Suppose now that we wish, merely as a philosophical exercise, we trust, to justify our belief that Jones has a mind. How would we go about it? Supposing that we do not wish to define mind in terms of behavior, which is unlikely to be taken seriously by anyone who lacks an unusual theory, we would presumably concern ourselves with behavior and analogical reasoning.[4]

Before we proceed to do this, however, one should note a general caveat, to the effect that it is often difficult, even on the

---

4. There is another approach to these matters which we might note. This would be the notion that an expression such as 'Jones', or, better, 'human being', logically entails the possession of a mind. Thus, no justification is necessary, other than, say, a quick inspection of one's relevant linguistic intuitions. For example, as 'x is a circle' logically entails 'x is round', so 'x is a human being' logically entails 'x has a mind'. On this general approach to philosophy, a great deal of time and effort is saved, because the answers to philosophical questions lie conveniently at hand, embedded in one or another natural language, in the meanings, rules of language, uses of language, or such, depending on the school or subschool involved. Where there is much to be said for this sort of approach to philosophy in many cases, I do not think there is much to be said for it in this case. Putting aside its welcome simplicity and charming arrogance, its usual notion that philosophy is a mistake, and that philosophy's job is to save itself from itself, and go out of business as soon as possible, it does not seem of much help in the present instance. For example, linguistic intuitions differ. If 'x is a human being' really entailed 'x has a mind', the question would not have come up in the first place. For example, we do not have many people in doubt as to whether or not circles are round. Second, given the fact that one might be the sole center of consciousness in the universe or the sole entity in existence, which are logical, if bizarre, possibilities, it does not follow that what we normally take to be human beings would need to have minds. Given the first possibility, that in which one is the only center of consciousness in the universe, all human beings other than oneself, namely, what one would call human beings, would lack minds. For example, if, following a Cartesian analysis of animal life, animals were automata, dogs would still be dogs, and cats cats, and so on, even if they had no feelings. On the second possibility, that in which one was the sole entity in existence, the word 'human being' would refer to aspects of one's own

presumption of mindedness, to know what might be in another mind. If this can be obscure to the individual himself, as one gathers from certain suppositions of depth psychology, it should be difficult, as well, from the point of view of an "outsider."

A dramatic example of this sort of thing is the logical possibility of the "rotated spectrum." This goes well beyond likely differences in consciousness, which might be associated with moods, health, acculturation, and such. Certainly the "mind world" of the deaf and blind differs considerably from that of the more normal individual, and, to a lesser degree, so, too, would the world of the tone deaf and color blind. In the case of the "rotated spectrum," one considers the possibility that color experiences might vary in a uniform, and thus undetectable, manner, from perceiver to perceiver.

Let us suppose we have six individuals and six colors, and that the following relationships obtain, as suggested in the table.

I_1	I_2	I_3	I_4	I_5	I_6
Red	Purple	Blue	Green	Yellow	Orange
Purple	Blue	Green	Yellow	Orange	Red
Blue	Green	Yellow	Orange	Red	Purple
Green	Yellow	Orange	Red	Purple	Blue
Yellow	Orange	Red	Purple	Blue	Green
Orange	Red	Purple	Blue	Green	Yellow

consciousness, to sense data, or such, and, just as figments of one's mind, so to speak, would lack minds, so, too, would those aspects of one's own consciousness. Similarly, in that world dogs would be dogs, just as, in a world of dreams, dogs would be dogs. Experientially, everything would be the same; in reality, much would be different. To be sure, we would never find out. It would seem anomalous to rule human beings out of the solipsistic worlds, particularly in view of the fact that the world may be solipsistic, and, particularly, as well, from the ordinary-language orientation, in which things are what they are called, for example, that what makes a chair a chair is that it is called a chair, and so on. If one should disagree, what right would one have to impose the views of one world on those of another world? Further, languages can differ, and languages can change. Language $_1$ might, in theory, have it a matter of meaning that human beings had minds, and language $_2$ might not have it so, and either language might change from time to time. Indeed, we might have a language in which it was a matter of meaning that human beings did not have minds. Such might be a language of a solipsistic world. The original problem, of course, would remain, namely, whether or not a certain life form had what we now think of as mind. That is not a verbal issue; it is a question of fact. There is also a danger that the issue of machine intelligence might, similarly, be misunderstood as a question of language, for example, that no machine can be intelligent, and thus, if such and such is intelligent, it cannot be a machine, despite, say, the fact that it was put together in Jones' basement.

Thus, on the first line, what the first individual sees as "red" will be seen by the other individuals, in order, as purple, blue, green, yellow, and orange. Similar remarks would obtain for the other possibilities, for example, what Individual$_2$ sees as yellow would be seen by Individual$_4$ as red, and so on. It should be noted that none of these individuals are "color blind"; all would pass color tests with flying colors, so to speak, and none of them would ever discover, given the uniform variation supposed, that they were not seeing the same colors as everyone else. If we asked any one of them to bring us, say, the red ball, the yellow toy, the blue napkin, or such, each of them would bring us the same ball, toy, napkin, or such. Their diverse color worlds would be a secret, but one they did not know they were keeping.

Analogously, a conscious machine might have an inner world other than ours, but one which might be correlated with ours, in such a way that neither we nor the machine were aware of the fact, and the fact would not much matter, just as the same sort of thing would not much matter amongst diverse human beings.

The common justification for the belief that other human beings have minds proceeds, as mentioned, by means of behavior and analogy. To be sure, 'behavior' in such contexts is used broadly, for example, "appearing" counts as a way of behaving, rather as, in some languages, literally translated, a tree greens, a woman beauties, and so on. Similarly, sitting down, keeping still, sleeping, and such, would count as ways of behaving. Something which merely behaved like a human being, in an narrow sense, say, a moving advertisement, a cartoon figure, a talking doll, or such, would not be likely to be credited with a mind. Similarly, something which appeared much like a human being but did not behave like a human being, for example, running on gas, emitting fumes, honking a horn, and such, might not be credited with a mind. Accordingly, in the argument from "behavior" and "analogy," "behavior" involves both looking like a human being, seeming to be a human being, and acting like a human being. And, implicitly, it means being pretty much like ourselves, at least as far as we can tell. That being the case, the argument from analogy proceeds, schematically, rather as follows:

Self	Other
A	A
B	B
C	C
D	
	So, D.

As we see above, we note that we have several properties in which we agree with the "other." Since we have in common properties A, B, and C, and we have property D, we hypothesize that the "other" also possesses property D. This is not a deductively valid argument, but, if the similarities are sufficient, it is a good argument, one which, if not conclusive, is at least likely to be rationally persuasive. One often infers from externals, such as appearance, likely internal states, such as dispositions. One is likely to infer, plausibly, that the lion seen only once, and never again, is conscious, predatory, dangerous, carnivorous, and so on. Similarly, in the case of another human being, we will never see his mind, not once, but we are familiar with something that looks and acts much like him, namely, ourselves, and may thus legitimately hypothesize that the "inside" of that one is probably not much different from the "inside" of this one, namely, ourselves. In a sense, this is merely another instance of the "outside-to-inside" inference, with which we are familiar, in a number of ways, on a daily basis. It is certainly logically possible that we are mistaken, namely, that the other lacks a mind, but that he possesses a mind, consciousness, feelings, and such, seems a very good explanation for what he does, says, and so on. It accounts for the data in a relatively clear, plausible, economical manner. It is a good bet, and one, it seems, we would be unlikely to lose.

What obviously emerges from this is that if we have no way of knowing, in a very strong sense of 'knowing', that another human being has a mind, we would certainly have a similar problem with knowing that an artifact, a machine of some sort, has a mind.

Could we know in a "weak sense" that a machine has a mind?

In a strong sense of knowing, we know little, if anything. To know in a strong sense of knowing, the belief would be rationally immune from counterevidence. Namely, the belief is such that no possible evidence could convince us, rationally, to repudiate the belief as false. On this approach, the best candidates for knowing would analytic propositions, namely, logical truths, such as circles being round, two and two equaling four, and such. Very few synthetic, or empirical, propositions, would be known. For example, we might be sure that Fido is a dog, but one could conceive of counterevidence to this belief, such as observing Fido in the back yard removing his dog suit, conversing with small green creatures, recharging his ray guns, and such. The best candidates for incorrigible empirical beliefs would presumably be things like the belief that one seems to be looking at a red patch now, that one seems to remember now that one was looking at a red patch earlier, that one seems to have a headache now, that one seems to remember now than one had a headache earlier, and so on. And even then, what if one were confused about the meaning of 'red', 'patch', 'now', and so on? And, similarly, worries might appertain even to allegedly analytic propositions, for example, was it circles or squares which are round, and what if one made a mistake in addition, coming up with four, and such. So, in order to bring knowledge into some practical relation to life and thought, one weakens the requirements for knowing. Without entering into a number of interesting epistemological issues, a familiar explication of knowing is that knowing requires only a true belief which is adequately supported. This analysis is beset with many difficulties, ranging from obscurity to just being wrong, difficulties which, for our current purposes, may be ignored.[5] Despite its technical problems, of harrowing importance to epistemologists, the traditional, or classical, analysis will do for most practical purposes.

---

5. Some sociologists of knowledge seem willing to use 'knowledge' of widely spread beliefs, such as knowing that the Earth was the center of the universe in the sixth century, and such, but normally one requires truth for knowledge. The work of Edmund Gettier, and others, has made out an excellent case that there are well-warranted true beliefs which we would not count as knowledge. Others have noted that lotteries provide counterexamples to the usual analysis of knowledge. For example, Jones will lose the lottery, we believe that he will lose the lottery, and we have overwhelming evidence that he will lose the lottery, e.g., his chances of winning are only one in several million, but, presumably, we do "know" that he will lose. One can conceive, too, of an individual seeming to know, in spite of a lack of belief and evidence, as in, say, always predicting the right card in thousands of drawings. Too, there is animal belief, which, to some extent, we must share. For example, birds know where their nest is but it is not obvious that they have beliefs, in any usual sense, certainly not articulated beliefs, and, further, it is not obvious

Could we know, then, at least in a "weak sense," that a machine has a mind?

It would seem so.

For example, in a "weak sense," it seems we could know that another human being has a mind, namely, we could have a true belief to that effect, which was well-warranted.. It seems that we have good evidence for the belief, based primarily on analogical inference. In theory, one might even go beyond obvious similarities in behavior and monitor brain activity in ourselves and others. If consciousness in our case is associated with brain activity of such and such a sort and the other has similar brain activity, that would suggest consciousness exists in that case, as well. To be sure, it is not obvious that our mental image of, say, a blue elephant, would be accompanied by the same or even similar brain activity in the other, when he claims to be entertaining the image of a blue elephant, but, presumably, something is going on. His blue-elephant image, if he has one, would presumably be constructed from different memories, laid down at different times and with different backgrounds, and this would doubtless make a difference in brain location and activity. Similarly, one supposes, for one reason or another, that the two images would differ in a number of ways, as well. For example, what sort of elephant is it, African or Indian, large or small, what shade of blue is it, what direction is it facing, and so on.

The distinction between "weak" and "strong" knowledge is reasonably clear, as the notion of strong knowledge is reasonably clear, and weak knowledge would be knowledge which is not strong knowledge. On the other hand, the distinction between knowing, in a weak sense, and not knowing is extremely unclear, and likely to be quite controversial. It is at this point that the major problematicity affecting knowledge proves itself to be unavoidable and acute. What would count as evidence, and what would count as good evidence, and when would the evidence be good enough, to warrant a rationally justified knowledge claim to the effect, say, that an artifact or machine was intelligent, conscious, had a mind, or such?

---

that their beliefs, if they have them, are a result of the marshaling of, or evaluation of, evidence, at least in any usual sense. The usual analysis, too, would seem to require that one "know" that the evidence is good evidence, which would render the analysis circular, a logical fault. Perhaps the most obvious fault with the classical analysis is the obscurity of the evidence requirement, e.g., what counts as evidence, what counts as good evidence, when is the evidence good enough, and so on.

Whereas one would not wish to require a knowledge standard for machine mind which could not even be met by human beings, what sort of standard might be rationally required or justified is obscure. Even in ideal circumstances, it seems a great deal might depend on subtleties of context, which would be difficult to spell out in general. Much would depend on the machine in question, and the behaviors of the machine in question. One could conceive of cases in which one might wish to attribute mind to one machine and not attribute it to another machine, which might differ from it only slightly, perhaps failing to exhibit fear, perhaps lacking an ability to add even numbers, or such.

The major problem we encounter here seems to be the failure of that analogical inference which was so important in connection with human beings and the "other minds" problem. That line of argumentation breaks down in the machine case. Consider the following schematism:

Self	Other
A	E
B	F
C	C
D	
	So, D.

Let us suppose that the self has several A and B properties, but the "other" lacks most of those properties, but possesses several E and F properties, most of which are quite different from those of the self. The only verifiable properties in common are the C properties, or, more likely, on the machine's part, some subset of the C properties, perhaps the capacity to respond to certain stimuli in ways identical to those of the self. Let us suppose the same question is put to the self and the "other," something like multiplying six times five. The self is conscious, intends to do the multiplication, has it in mind, is aware that it is multiplying, and understands what it is doing. The machine arrives at the same answer, perhaps more quickly. If the machine is one of a sort with which we are currently familiar, say, some sort of hand-held calculator, no one would attribute mind to it. On the other hand, if the machine is far more sophisticated, and its C properties are very similar to those of the self, say, as similar as would be likely to be the C properties obtaining between two human beings, the question of consciousness, ideas, mind, and such becomes much more interesting. On the other hand, the argument from analogy, so powerful in the interhuman case, is much weaker in the human/machine case.

Indeed, the argument from analogy would be likely to work against the mind-attribution in the machine case. Since the majority of properties are dissimilar in the two cases, this would seem to increase the probability that the machine lacks the D property, the mind property. As it lacks most of the other properties, why should it not lack the D property, as well?

The self and the "other," in this case, the machine, resemble one another very little. Even if the machine were constructed to resemble a human being, it would differ from a human being in a number of respects, for example, it was built and not grown, it is formed of different materials, and it is not the sort of thing with which we are likely to associate mind. Indeed, it is exactly the sort of thing with which we would not be likely to associate mind. Furthermore, as far as we know, ruling out gods and such, consciousness is associated with familiar sorts of nervous systems, the results of long-term evolutionary processes, extending backward over several billion years.[6]

One should not, however, suppose that a given outcome, say, mind, must necessarily be the result of any particular single process. It is logically possible, and it may be empirically possible, that the seemingly inexplicable miracle of consciousness, seemingly so zoologically pervasive, and so unaccountably emergent from certain forms of quantifiable physicalistic interactions, might be the outcome of other forms of quantifiable physicalistic interactions, as well. If animal consciousness can arise from so

---

6. The question of artificial intelligence should not be confused with a different question, namely, the possibility of synthesizing an organism. For example, given a naturalistic world, it is empirically possible to synthesize, say, a dog, a cat, a human being, which, then, being canine, feline, or human, would presumably be a conscious organism. Naturally one supposes that such a project could never come to fruition, given the orders of difficulty involved. Such creatures emerged from evolutionary processes lasting billions of years, involving unbelievable tragedy, wastage, conflict, and horror. If a human being could, in theory, be synthesized, so, too, it seems, might be an android, or humanlike being. It would be more likely to begin with something simpler, perhaps an amoeba, a coelenterate or annelid worm. We noted earlier that a virus, the tobacco mosaic virus, was broken apart, and reconstituted, as long ago as the 1950s. In any event, the synthesized human, the android, and such, would not normally be thought of as machines, except in the general sense that any naturalistic life form might be regarded as a machine, and that sense of 'machine' is not relevant in our current context.

unlikely a base as certain arrangements of atoms and molecules, it seems possible machine consciousness might, in theory, arise from other arrangements of atoms and molecules. In passing, it might be noted that given the size of the universe, its countless galaxies, and their countless stars, and, doubtless, the countless planets in the habitable zones of millions of these stars, and, given the sameness of the materials constituting the universe, and the uniformity of its laws, it is quite likely that intelligence has evolved at least thousands of times in the universe. And, if that is the case, it seems unlikely that it would have been achieved in each of its instances in precisely the same way.

Who is to say it is impossible that, in a sufficiently complex machine, of enormous subtlety, approaching that of a human brain, a sudden awareness might arise, an awareness not even understood as an awareness. Might there not be a birth cry, alone and silent, somewhere, sometime, not even understood as a birth cry?

In any event, the best argument, the behavior/analogy argument, as normally conceived, fails in the case of machines.

That brings us, naturally enough, to the question of what, if anything, might convince us, rationally convince us, of machine intelligence, or mind.

And that brings us, naturally enough, to the Turing Test for machine intelligence, or, better, to Turingtype tests for machine intelligence.

5. Turingtype Tests for Machine Intelligence[7]

The sort of test we have in mind here, suggested by A.M. Turing, is based on a British party game, the "imitation game." One gathers this game can be played variously, but a well-known

---

7. A. M. Turing was one of the astonishing minds of the 20th Century. He was perhaps the most influential and productive of the early pioneers in computer science. He was also a brilliant cryptologist, who contributed significantly to the Allied cause in World War II, by breaking a major German code, a secret kept by the allies until after the war. It is a bit hard to know how to read the "imitation game" approach to machine intelligence, as I suspect Turing wrote it in a somewhat light-hearted mood, but, on the other hand, the sort of thing involved, namely, indistinguishability, may be, at least until further analyzed, a *prima-facie* plausible test for machine mind. The classical reference is "Computing Machinery and Intelligence." *Mind*, Vol. LIX, No. 236 (1950). As is generally the case in this study, I am primarily

version apparently involves a man and a woman, and two other players, who might be of either sex. The schematism below suggests the nature of the game.

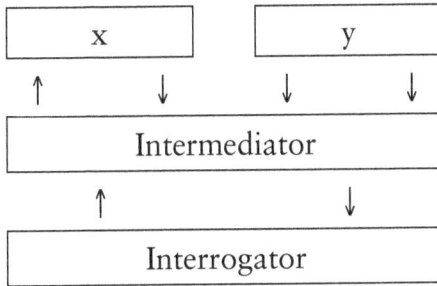

```
  ┌──────────────┐      ┌──────────────┐
  │      x       │      │      y       │
  └──────────────┘      └──────────────┘
   ↑        ↓          ↓           ↓
  ┌─────────────────────────────────────┐
  │            Intermediator            │
  └─────────────────────────────────────┘
        ↑                    ↓
  ┌─────────────────────────────────────┐
  │            Interrogator             │
  └─────────────────────────────────────┘
```

In the game, it seems, as often played, x and y would be individuals of different sexes, and the object of the game is for the interrogator to discover which individual is male and which is female. He supplies questions to the intermediator, who refers these questions to x and y, and then relays their responses to the interrogator. If the interrogator, after a time, identifies which mystery player is the man or the woman, he wins, and, if not, he loses. One of the mystery players, as the game may be played, may be required to answer truthfully, and, in this way, attempt to assist the interrogator, while the other, in his responses, attempts to deceive the interrogator. There would be various ways in which such a game might be played. Now, let's suppose that instead of x and y being a male and a female, one of them is a human being, of either sex, and the other is a machine. The point of the Turingtype test is that if one cannot distinguish between the responses of the human, to whom thought, intelligence, mind, and such, is rationally attributed, and the responses of the machine, then one would be well warranted in ascribing thought, intelligence, mind, and such, to the machine. I think the best way of understanding Turing here is not so much as claiming that the

---

concerned with the problems, and not with the explication of particular thinkers. Accordingly, I am more concerned here with "Turingtype" argumentation, the sort of testing he was suggesting, rather than any particular test. This will also be the case, substantially, when we deal with John R. Searle's "Chinese Room Argument," namely, that we are primarily concerned with the type of arguments involved. Both of these arguments, or sorts of arguments, the Turingtype arguments and the Searle argument, have generated literatures, which, naturally, are beyond the scope of this study. Interestingly, these literatures, complex and narrow, barely glimpse related philosophical vistas of the future.

machine is thinking, whatever thought might be, as that it would be justifiable under such circumstances to attribute thinking, and such, to the machine. Indeed, is this not pretty much what we do with one another? We do not, strictly, in a strong sense of "know," know that Jones has a mind, but we would seem to be rationally entitled to suppose that he has a mind, to attribute mind to him.

If something like this would not entitle us to ascribe mind to a machine, what would? One of the attractive aspects of Turingtype tests is that they suggest that the issue of machine intelligence may be understood as an empirical question, to which empirical data are relevant.

The essence of such tests is, initially, response indistinguishability, and, later, more broadly, activity indistinguishability, as might be the case with, say, a robot's behaviors, or actions.

One supposes that success in an "imitation game" context might not be the last word on machine intelligence, but it seems a useful way to begin a conversation. Several years ago, for example, K. M. Kolby and his associates in the Stanford University Intelligence Project, designed a complex program intended to simulate the responses of a paranoid mental patient.[8] By means of a keyboard, an individual could conduct dialogues with "Parry," the "paranoid-imitating system." A number of such dialogues, together with dialogues conducted with actual paranoid patients, were collected and sent to a randomly selected sample of American psychiatrists, who, for the most part, were unable to discriminate between the human and machine responses. Some people, one supposes, might see this as evidence that Parry can think. Others, one supposes, might see it as telling us something, possibly something alarming, about American psychiatrists. In any event, this would seem, in its way, a successful response to an "imitation game" test.

The "imitation game" type of test is, in its way, a symbol for, or a metaphor for, indistinguishability, which is the crux of the matter.

We shall later concern ourselves with the relationship of indistinguishability to machine intelligence.

First, however, let us consider, briefly, four general objections to "machine thinking."

---

8. Cf. Yorick Wilks, "Your Friends and your Machines." (Mind, Vol. LXXXIII, No. 332, October, 1974.)

6. Some General Objections to "Machine Thinking"

The following objections, ranging from terrible to interesting, are worth noting, as each, in its way, sheds some light on the question.

6.1 The Definitional Objection

The notion here seems to be that someone's notion of ordinary language is all that is required to resolve the issue. The idea is that machines cannot think, and, thus, if something thinks, it is not a machine. In other words, machine thinking is ruled out by definition, rather as one might rule out square circles. The first response here would seem to be that this alleged analyticity, that machines cannot think, by definition, might be peculiar to particular speakers. Many individuals, for example, seem capable of imagining conscious, thinking machines, and imaginability seems to be a sufficient condition for logical possibility. It is not like imagining, or drawing a picture of, a square circle. Too, the meanings of words can change, and often do. Even if it were analytic in a given language at a given time, which it is not, that machines could not think, that would not preclude a change in the future, in the light of, say, advances in robotics. More importantly, whereas it is obvious that semantics is of considerable philosophical importance, it seems just as obvious that it seldom, if ever, holds the answer to philosophical questions in its statistical, transitory grasp. If it did, the question, if it came up in the first place, which seems doubtful, might be promptly resolved. One would simply consult one's linguistic intuitions, inquire of native speakers, check the dictionary, or such, and be done with things. Lastly, let us suppose that we have before us an object which thinks. We do not claim to know that it thinks, in some strong sense of knowing, but suppose that it does, in fact, think, whether we know it or not. Also, let us suppose that this object has been constructed in a laboratory or workshop, contains a great deal of silicon, is powered by electricity, is square, weighs forty pounds, is cased in metal or plastic, and sits on a table. It certainly looks like an artifact or a machine. It would seem anomalous then to claim that since it thinks it is not a machine. As an analogy, let us suppose we have a claim that no chimpanzee can think, and then we encounter a chimpanzee who is moral, erudite, toilet trained, and chats brilliantly and amiably in Amslan about a number of matters, including American literature, even if somewhat controversially, particularly in the case of Herman Melville. Would one then say that it is no longer a chimpanzee? What would he think of that? Perhaps he is proud of his simian ethnicity, heritage, and such. And can you manage branches with the same agility? I think we would say, and he would probably insist on it, that he is indeed a chimpanzee, and one who probably knows more about American literature than we do.

6.2 The "Richness" Objection

The essence of this objection is that thinking is a rich, complex concept, and that something like playing the "imitation game," or such, is too narrow an achievement to count as thinking . What about worrying, hoping, planning, liking or disliking, loving or hating, joking, suspecting, speculating, noticing, remembering, misremembering, calculating, anticipating, playing about with ideas, imagining, fantasizing, dreaming, lying, deceiving, teasing, taunting, insulting, talking, celebrating, feeling depressed, feeling happy, and so on?

This objection, however, seems to ask the machine to be, in effect, a human being, perhaps in an unusual form, and that is a great deal to ask of a machine. For example, the machine was not raised in a family, did not experience sibling rivalry, never went to kindergarten or the first grade, was never punched by a bully in the schoolyard, and so on. The basic question here should be whether or not the machine thinks, not the richness, complexity, subtlety, or such, of its thought. For example, anyone who is familiar with dogs or cats, with the possible exception of René Descartes, will not doubt that they are conscious, and that they have an inner life, and that they think. By parity of reasoning, if a machine could manage as much, it should count as thinking. Similarly, if we set the bar for thinking too high, we might rule out a number of human beings, and a great many young children, both of whom we suppose to have minds, to think, and such.

In its way, this objection, rather like the first, seems to seek victory, though more subtly, by means of definition.

Let us suppose we had levels of thinking, or, perhaps better, what might be regarded as thinking, ranging from, say, $Level_1$ through $Level_{100}$. We shall suppose that there is no doubt that thinking takes place between, say, $Level_{90}$ and $Level_{100}$. Let us then suppose that a machine manages, on this scale, to do something which, if a human did it, would count as $Level_{60}$. One might then rule this out as thinking, requiring that, say, at least $Level_{70}$ is necessary for thought. If a machine seems to manage $Level_{70}$, then the requirement might be raised to $Level_{80}$, and so on. A variety of problems seem to appertain to this approach. For example, how much is needed? Where, for example, on this continuum, does one set the point which divides thought from nonthought? When, in the spectrum, does blue shade into nonblue, is blue so different from nonblue, and aren't they all colors?

Lastly, what if the machine seemed to function at $Level_{100}$? Would the requirement then be moved to, say, some $Level_{100+n}$?

I think the "richness" objection fails, as it seems to confuse thinking, *per se*, with a particularly rich, complex sort of thinking. This seems a bit like saying painting and composing do not count as painting and composing unless it is very good painting or composing, indeed, perhaps up there with Rembrandt and Beethoven.

6.3 The "How" Objection

The notion here is that the crucial difference between thinking and nonthinking, or at least one important difference, is how a particular end result is brought about, or achieved. Process is seen as important here. Thinking delivers in one way, nonthinking in another way. For example, let us suppose we decide to torment Jones, and ask him to multiply 9999 times 9999. He sets to work, and, if not careless, he is likely to eventually come up with 99,980,001. On the other hand, let us pop those same numbers into a calculator and, in short order, in about the same time it would take to flick a light switch, we come up with the same number. If thinking is involved, we would have to say that the calculator is doing this better than Jones, or, at least, is doing it much more expeditiously. On the other hand, there is no doubt that Jones, sitting over there, sweating with his pencil and paper, is thinking. Few people, on the other hand, except an occasional cognitive scientist, or a fellow with an unusual theory, would claim that the calculator was thinking. The calculator would more likely be thought of as operating, functioning, or such. Analogously, when we flick the light switch, making the connection, we do not suppose that the wiring is thinking about turning on the light, or such. In a case like this it seems clear that process is important. It seems clear that the process in the human, which we call thinking, is quite different from the process in the calculator, which most of us would not call thinking.

As suggested earlier in this study, artificial intelligence can be defined in a broad manner, indeed, so broadly that electric eyes and thermostats might be credited with thought. Presumably very few people would wish to so extend the concept of thinking, or intelligence. On the other hand, some enthusiasts for artificial intelligence, to whom it is important, for some reason, that machines can think, as it seems to be important for some others that machines cannot think, have a very low threshold for thought, a threshold likely to admit some perhaps dubious candidates. Let us consider, say, Jones and the calculator. Certainly there are some common denominators, or genera, involved. Both come up with the correct answer and both do so in virtue of a process. A small argumentative presupposition seems to be involved, at least on the part of a low-threshold AI enthusiast. The argument might go something like this, put in terms of embarrassing simplicity.

Thinking is a process.
Functioning is a process.
Thinking is functioning.

This argument is obviously invalid, and even commits a familiar fallacy, of a type noted even in syllogistic reasoning, the fallacy of undistributed middle. This is the sort of fallacy in which one might infer from the fact that both giraffes and raccoons are animals, that giraffes are raccoons. Nothing, of course, is this simple, but it does seem likely that the low-threshold AI enthusiast does have a very broad sense of process in mind, under the canopy of which might be subsumed, in some cases, both programs and minds. In any event, this objection to artificial intelligence, namely, that not simply the results are important, but how the results are achieved is important, wishes to emphasize a perceived, or alleged, difference, between, say, minded thinking and mindless functioning.

If there is a difficulty involved in the "how" objection,

other than a huffy, chauvinistic preference for biological over electronic processes, it seems to be the supposition that there is only one legitimate "how" where ratiocination, thought, mind, intelligence, and such, is involved. One supposes it is possible that, empirically, rather than logically, there may be nomological reasons militating against machine intelligence, namely, that, given the laws of nature, anything we might be likely to call thinking, even in a most open-minded mood, cannot be achieved except in the biological manner, worked out over eons by the processes of evolution, from unicellular irritability to sophisticated cosmological speculation. On the other hand, that seems to be an open question. Certainly logically there is no impediment to machine intelligence. And, perhaps in some laboratory or workshop, someday, somewhere, a machine might wonder, like Adam in paradise, "how came I thus, how here?"

This consideration leads naturally to the fourth objection.

6.4 The "Consciousness" Objection

The notion here is that thinking presupposes consciousness, that consciousness is a necessary condition, either empirically or logically, for thinking. Informally, an argument might go like this:

> Thinking presupposes consciousness.
> <u>Machines are not conscious.</u>
> Machines do not think.

More formally, utilizing the material conditional, and the following predicate constants:
Tx: x thinks
Cx: x is conscious
Mx: x is a machine
              an argument might go like this.

1. (x) (Tx É Cx)
2. (x) (Mx É Cx)
3. Ty É Cy 1, universal instantiation
4. My É Cy 2, universal instantiation
5. ~Cy É ~Ty 3, transposition
6. My É ~Ty 4, 5, hypothetical syllogism

7. (x) (Mx É ~Tx) 6, universal generalization

1.   (x) (Tx ⊃ Cx)
2.   (x) (Mx ⊃ Cx)
3.   Ty ⊃ Cy   1, universal instantiation
4.   My ⊃ Cy   2, universal instantiation
5.   ~Cy ⊃ ~Ty   3, transposition
6.   My ⊃ ~Ty   4, 5, hypothetical syllogism
7.   (x) (Mx ⊃ ~Tx)   6, universal generalization

The premise-set and conclusion of the argument could be read as:

> For every x, if x thinks, then x is conscious.
> For every x, is x is a machine, then x is not conscious.
> For every x, if x is a machine, then x does not think.

As the argument is constructed to be valid, our only concern is whether or not it is sound, namely, whether or not it has both a valid form and a true premise-set. Attention here is thus focused on the premise-set. Utilization of the material conditional does not affect the validity of argument.[9]

One must distinguish between three claims, or assertions. First, machines are not, in fact, at present, conscious, which claim seems likely to be accepted. Second, it is logically impossible for machines to be conscious, which claim is clearly false, as there is no contradiction in the supposition that a machine might be conscious, as one can imagine a machine to be conscious, and so on. No conscious machine would be in the "square-circle category," so to speak. Third, it is empirically impossible for a machine to be conscious, as that would contravene the laws of nature. Clearly it is the third claim which might be in dispute. If one wishes to regard nonconscious machines as thinking, that is

---

9. A material conditional is true if and only if either its antecedent, its first term, is false, or its consequent, its second term, is true. It is thus false if and only if its antecedent is true and its consequent is false. Such conditionals are to be distinguished from logistic or entailment conditionals, whose truth or falsity depends on meaning, and nomological conditionals, whose truth or falsity depends on both meaning and the laws of nature. 'If something is a circle, then it is round' would be a logistic or entailment conditional; 'If something is a material body free of impressed forces, then it tends to travel in a rectilinear path' would be a nomological conditional.

a different problem. The "consciousness" objection presupposes, first, that no machine is conscious, and, second, that thinking, intelligence, and such, require consciousness. I do not see the "consciousness" objection as ruling out the possibility that an intelligent machine is precluded by nature, but as merely claiming that thinking, and such, would require consciousness.

With these thoughts in mind, let us consider the premise-set in our argument. The first premise would be regarded as false by some AI theorists. Presumably, no one supposes that a calculator or a chess-playing machine, or chess program, is conscious, but, depending on how one understands "thinking," it might be claimed that the calculator, or the chess-playing machine, or chess program, are thinking. This is the sort of claim whose truth conditions depend on how one chooses to understanding "thinking." If this counts as thinking, then machines, or programs, think. Many, however, are understandably reluctant to lay down their arms, or surrender their reservations, or abandon their uneasy scruples in this matter, particularly in the light of so seemingly convenient and prompt a victory. One tends to be suspicious of "one-sentence refutations" purportedly demolishing seemingly informed, complex, plausible positions. It seems likely that the "thinking is easy" position tends to be unsatisfactory to many individuals because they tend to think of thinking along the lines of what they do when they do what they are likely to think of as thinking. For example, recalling the multiplication tables, or figuring out from scratch what six times five amounts to, seems very different from tapping some buttons on a calculator. If they thought the calculator was really doing what they do, when they worry about six times five, they would probably grant that the calculator is thinking, but they do not think the calculator is doing what they do, and they have very good reasons for thinking it is not doing what they do. Indeed, the calculator does not know it is calculating, and, in a sense, it is not even calculating; it is just functioning. Rightly, or wrongly, the average human being is going to associate thinking with consciousness, with intentionality, with subjectivity, with understanding. That is the sort of thing he associates with thinking. It might be claimed that those are mere accidental accompaniments of thinking, idiosyncratic to biological beings, but if it were not for such things, the process would not be likely to begin, take place, or make sense. Does one not have to think, even to turn on the calculator? Presumably "thinking" is not well understood as always being out of sight, or "off camera," so to speak.

Whereas it seems clear that the average human thinks of thinking in terms of conscious thinking, justifiably or not, there is another possibility, other than chess-machine, or chess-program, thinking, or such, under which one might take the first premise, 'For every x, if x thinks, then x is conscious', as false. Depth psychology has suggested the possibility of unconscious thought. It should be noted, in passing, that the data, as one would expect, on which the hypothesis of unconscious thought is based, slips of the tongue and pen, motivated forgetting, dream distortions, insights suddenly occurring, and such, admit of alternative explanations. Too, one might speculate that thought which is not accessible to one level of consciousness might be accessible to a different level of consciousness, in short, that thought which is not conscious at Level B might be conscious at a lower level, say, Level A. This would preserve consciousness as a necessary condition for thought, and, presumably, the calculator or chess-program is not conscious at any level. On the other hand, let us suppose that there is unconscious thought which is not conscious at any level. What then? Then, it seems, the "consciousness" objection would fail. It does not follow from this, however, that the calculator or chess program, or such, must then be regarded as minded, intelligent, or thinking (albeit, unconsciously). The fact that they are artifacts does not rule them out as thinkers, but it does not rule them in either. It seems quite plausible that a human being, a dog, a cat, and so on, think, and might, at least part of the time, do so unconsciously, but they have in common billions of years of evolutionary heritage on which to build, which renders them, at least *prima facie*, disanalogous to artifacts. Thus, granting unconscious thought to a human, dog, or cat, if one should wish to do so, does not necessitate granting unconscious thought to a program, or machine. In short, the question remains open.

If the first premise, 'For every x, if x thinks, then x is conscious' might be false, what of the second premise, 'For every x, if x is a machine, then x is not conscious'? That premise seems true, at present. If, on the other hand, the premise was 'For every x, if x is a machine, then x cannot be conscious' the truth value would be problematic. The force of the 'cannot' is crucial. If that was a logical "cannot," as in 'A circle cannot be square', then the revised premise, as we saw earlier, would be false. On the other hand, if the 'cannot' is nomological, e.g., 'A cheetah cannot run faster than twenty miles per hour', or 'A cheetah cannot run faster than 200 miles per hour', the answer, true or false, would depend on the laws of nature, what would be compatible with them or incompatible with them. Thus, the truth value of 'A machine cannot think' would depend on the laws of nature, what would be compatible with them or incompatible with them.

7. Some Remarks on the Imitation Game

Whereas the Imitation Game is best viewed as a metaphor for indistinguishability, which, robustly understood, is doubtless the best test possible for machine intelligence, it may be helpful to note that, narrowly understood, successful "playing" of the Imitation Game would be neither a necessary nor a sufficient condition for thinking, or the attribution of thinking, for any entity, machine or otherwise.

Let us begin with some entity, which might be a person, an animal, a machine, an alien, an unusual being or object, and so on. In examining the matter, a dichotomizing approach may be helpful.

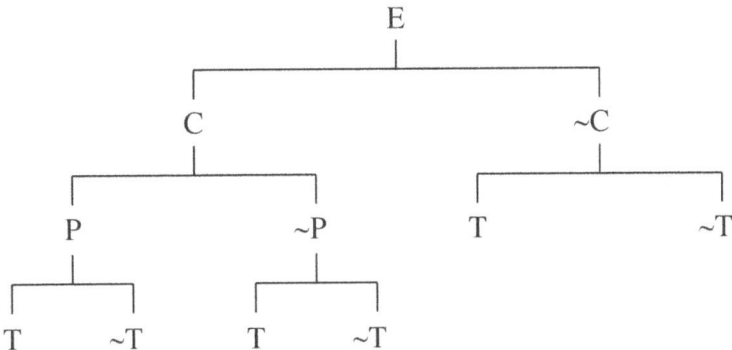

I. The entity is capable of playing the game.
    A. The entity plays the game.

1. The entity thinks.
2. The entity does not think.
   (Perhaps a good program is operating; perhaps coincidence is involved.)
  B. The entity does not play the game.
    1. The entity thinks.
      (Perhaps the entity is too proud to play, thinks the game is silly, is too inhibited, etc.)
    2. The entity does not think.
      (Perhaps it is turned off, is running another program, the right software is not loaded, etc.)
II. The entity is not capable of playing the game.
  A. The entity thinks.
    (Perhaps it is psychologically incapable, perhaps it is pathologically fearful of competition, perhaps it is too young, can't understand the rules, etc.)
  B. The entity does not think.
    (This would be the case for most entities, e.g., trees, rocks, can openers, etc.)

As should be clear from the schematism and the appended, interpreting outline, successful playing of the imitation game is neither a necessary nor a sufficient condition for thinking, or intelligence, or the attribution thereof. For example, it is not a necessary condition for intelligence, as some intelligent individuals might not understand the game, but, presumably, they would realize they did not understand it. Indeed, they might not know the language of the game, or, possibly, any language, as, say, a feral child. Too, an intelligent individual might refuse to play, because he disapproves of games, regards them as beneath him, or such. Perhaps, for moral reasons, he becomes ill whenever he thinks of something that might be fun. Perhaps he is paralyzed, and can think, but cannot communicate, and so on. Thus, playing the imitation game, or being capable of playing it, is not a necessary condition for intelligence, as many might not play, might refuse to play, might be unable to play, and so on, who are nonetheless intelligent. Similarly, and more important in this context, successful "playing" of the game would not be a sufficient condition for intelligence, as the entity's responses might be a simple matter of being well-programmed. Too, there is always the possibility that some unusual coincidence, or set of coincidences, might have taken place. Under such circumstances a human interrogator might easily misidentify the "mystery player" without the mystery player's thinking, being intelligent, or such. Something along these lines apparently took place in the case of Parry, the "paranoid-imitating system," in which case a number of psychiatrists were apparently unable to distinguish between the responses of Parry and those of genuinely paranoid individuals. Presumably, one would not wish to regard Parry, the program, as human, thinking, or paranoid.

On the other hand, if one takes the "imitation game," as suggested, as a metaphor, or symbol, so to speak, for indistinguishability, it points in the direction of a brilliantly serious, and plausible, approach to machine intelligence, one likely to be attractive to machine-intelligence enthusiasts, annoying to biological chauvinists, and provocative, at least, to neutral observers, should any exist.

Let us suppose that a machine exists whose responses, within appropriate ranges, are indistinguishable from those of a human, whom we grant is minded.

Would that count as sufficient for attributing intelligence to the machine?

Presumably the answer to this would depend on the machine, the complexities, the responses, involved. For example, one would presumably not wish to grant intelligence to the calculator, in spite of the fact that there would be likely to be considerable overlap, that is, indistinguishability, between its output and that of a human, within the appropriate range of activity, say, calculating sums, percentages, and such.

A machine which could play the imitation game would be a more serious candidate for machine intelligence but, as we have seen, successful playing of the "imitation game" need not turn the trick for machine intelligence. Indeed something analogous to, though simpler than, the imitation-game scenario already exists. Conceive of a human playing chess over the internet, or by means of an intermediator of some sort, who believes his opponent to be another human being. Or, to vary the scenario, he is playing chess with two opponents, one human, one machine, or, perhaps, two machines, or two humans. Given the sophistication of contemporary chess programs, one supposes that our human would have no way of distinguishing amongst these opponents, whether both were machines, both were human, or one was human and one a machine.

On the other hand, many, as we have suggested, would be reluctant to ascribe thinking to the machine, or program, for a number of reasons, though, doubtless, an enormous amount of intelligence went into the design of the machine, or program. Similarly much intelligence goes into the design of a building or bridge, but we do not suppose buildings or bridges think.

In these matters indistinguishability is obviously important, if not decisive, but the level of the indistinguishability and the nature of the indistinguishability are also important. For example, the calculator level of indistinguishability is likely to seem too low for the attribution of intelligence, and the chess-machine, or chess-program, form of indistinguishability is likely to seem too narrow for the attribution of intelligence. On the other hand, what seems important here is that thinking takes place, not its level or nature. This suggests that more than indistinguishability is involved. For example, one could conceive of a program the responses of which were indistinguishable from those of an ignorant, misinformed, simple, confused, even maniacal individual, rather along the lines the Parry situation, without thereby hurrying to ascribe thought to the program, say, ignorant, misinformed, simple, confused, even maniacal thought.

Presumably the sort of indistinguishability involved would be that which would go beyond calculation and problem solving, even beyond playing an "imitation game." The imitation of thought, after all, need not count as thought itself. To imitate a dancer may be to dance, and to imitate a thinker may be to think, but the insect which looks like a leaf is not a leaf, and the photograph of the Eiffel Tower is not the Eiffel Tower.

A major problem here is to have in mind something or other which would be acceptable as serious evidence of, perhaps rationally conclusive evidence of, if not logically conclusive evidence of, machine intelligence. If, say, a sophisticated chess program is not enough, what would be enough? One does not wish to rule out machine intelligence, *a priori*. That would return us to the failed program of "victory through definition," which is not so much a triumph as an acknowledgement of bankruptcy and defeat.

We noted that one of the arguments against accepting a chess machine, or program, as intelligent, was that it did not know what it was doing. It did not know if it was winning, losing, or drawing. It did not even know it was playing chess.

But what if a machine could understand things, really understand them? What if it could understand, say, stories?

8. The "Chinese Room"[10]

The following discussion is based very much on the sort of thing suggested by John R. Searle, in particular, his now-famous "Chinese Room" argument. In our small discussion we combine elements from the basic argument, together with some independent and related points. Our primary concern, as is doubtless apparent by now, is less scholarly and expository, than it is philosophical and analytical. In short, our primary concern here, for better or for worse, is less with Searle than with the matters to which he so shrewdly and interestingly addresses himself.

Searle's "Chinese Room" argument is directed against claims he identifies as those of "Strong AI," or "strong artificial intelligence," claims he regards not only as extreme, but demonstrably false. "Strong AI" may be contrasted with "Weak AI," or "weak artificial intelligence." In "Weak AI" all that is required is that a machine or program simulates intelligence, so to speak, though, technically, if Searle is correct, it is not clear that a program could even simulate. For example, does a calculator simulate adding and subtracting, or a chess program simulate playing chess, or do they just work, or function?[11] In any event, Searle's primary concern is with "Strong AI," which claims, following Searle, that programs are identical with minds.

10. As mentioned in an earlier footnote, literatures exist pertaining to the artificial-intelligence controversy, so one has replies, and replies to replies, and such. Also, as before, our primary concern here is to give the reader a sense of the issues, not to examine, in detail, or attempt to deal with, particular contributions of particular authors, pro and con, to the controversy. A nice place to begin, for those interested in entering into these issues, is John R. Searle's' seminal article, "Minds, Brains, and Programs." (Behavioral and Brain Sciences, 3, 417–424.) Portions of the material above are derived, as well, from a variety of sources, including lecture and conversation. Lecture and conversation are informal sources, and so, should any discrepancies exist between anything in the above and published material, priority should be assigned to the published material.

11. The sort of point one has in mind here was developed by Keith Gunderson, in the context of a delightful thought experiment. "Can rocks imitate?" Presumably the answer to that question was in the negative, at least until tackled by Gunderson.

According to Searle, the following equation is accepted by those who subscribe to strong AI.

$$\frac{\text{Mind}}{\text{Brain}} = \frac{\text{Program}}{\text{Hardware}}$$

The equation may be read as, as minds are to brains, so programs are to hardware. If Searle is correct, the "strong-AI" theorist would maintain then that minds are identical to programs, and, perhaps, if we accept the equation, that brains are identical to hardware.

An analogy might be:

$$\frac{1}{2} = \frac{2}{4}$$

The analogy, however, fails significantly. One half might be identical to two fourths in some mathematical sense, but one is not talking mathematics in discussing either minds or programs, or brains and hardware. The only form of identity which would be relevant in such cases would be some sort of functional identity, as in bicycles and motor cars are identical in being able to get one from point A to point B, or generic identity, in the sense that raccoons and giraffes are identical in both being animals, having a cellular structure, doing metabolism, possessing DNA, and so on. There is a sense in which anything that exists is identical with anything else that exists, for example, in existing, but this is so general as to be absurd. Certainly brains and hardware might be analogous, in some ways, but so, too, are wind and water, spades and spoons, and so on. Minds and programs, and brains and hardware, are not even identical in the sense that two automobiles, of the same make and model, might be said to be identical, and so on. Accordingly, one supposes that the AI theorists, who are

---

Suppose we have a game, a sort of "imitation game," in which an individual places his foot through a hole in a wall or screen. Someone, perhaps a man or a woman, he does not know which, you see, is on the other side of the wall or screen. The other individual then steps on his foot. Was it the man or the woman? Perhaps the man steps lightly, or the woman steps heavily. It is doubtless difficult to know. Was that Bill or Susan stepping on your foot? We now become philosophical. An apparatus is constructed, perhaps with cables and springs, in which we may place a rock. Now the contestant must determine if the "mystery player" is a human being, male or female, or a rock. If he cannot tell the difference, must we not now acknowledge that at least one rock can imitate, and might there not be a similarly competent second rock, and a third, etc.?

doubtless highly intelligent minds (or programs) do not really mean what Searle takes them to mean, if we are right about what he takes them to mean. If they do mean what Searle seems to suggest they mean, their claim is literally false, at least to the extent it is comprehensible. We shall suppose then that what they actually mean is that there is some sense in which minds might be thought of as programs, and vice versa, and, perhaps, some sense in which brains might be thought of as hardware, and vice versa. And, hopefully, this is in some more reasonable sense than the sense in which bicycles might be thought of as unusual motor cars, and vice versa, or that raccoons might be thought of as short-necked giraffes in some sense, or giraffes be thought of as long-necked raccoons, and such. In any event, putting aside alleged identity claims, let us address ourselves to what seems to be at issue, namely, machine intelligence, machine thought, machine understanding, that sort of thing.

Searle calls to our attention what he takes to be a crucial difference between minds and programs, a difference which would seem sufficient to discredit the strong-AI thesis, in so far as it is intelligible, even if it were not understood in terms of identity, but only in terms of significant similarity, or such. The difference is, following Searle, that minds possess a semantics (presumably an understanding of meanings, and such) and that programs lack a semantics, and thus that minds are not identical with programs, assuming one still wishes to think in terms of identity. The argument, stated very informally, might go as follows:

1. Programs are purely formal (are purely syntactical).
2. Minds have semantic content.
3. Syntax ^{1} semantics.
4. Programs ^{1} minds.

One is then, in virtue of the above argument, or a similar argument, given the nonidentity of semantics and syntax, supposed to infer that strong AI is false.[12]

This sort of argument can be put into a more precise form, for example, as follows, in this case, in the second-order predicate

---

12. The formulation in the text is from Searle, or based closely on a formulation by Searle. My notes suggest that it, or something much like it, came up in a popular lecture given by Searle several years ago at Queens College, of the City University of New York. Cf. Footnote 10.

calculus, in which, at least, it is provably valid. The predicate constants might be read as follows:

Mx: x is a mind
PRx: x is a program
Sx: x possesses a semantics

Then:

1. (x) (y) (P) [(Px & ~ Py) ⊃ ~ (x = y)] Tautology.
    For any x, y, and property P, if x has that property and y does not, then x is not identical with y.

2. (x) (Mx ⊃ Sx)
    For any x, if x is a mind, then x possesses a semantics.

3. (y) (PRy ⊃ ~ Sy)
    For any y, if y is a program, then y does not possess a semantics.

4. (y) (P) [(Px & ~ Py) ~ (x = y)] 1, universal instantiation
    For any y and for any property P, if x has that property and y does not, then it is not the case that x is identical with y.

5. Mx ⊃ Sx 2, universal instantiation
    If x is a mind, then x possesses a semantics.

6. PRy ⊃ ~ Sy 3, universal instantiation
    If y is a program, then y does not possess a semantics.

7. (P) [(Px & ~ Py) ⊃ (x = y)] 4, universal instantiation
    For any property P, if x has that property and y does not, then it is not the case that x is identical with y.

8. (Sx & ~ Sy) ⊃ ~ (x = y) 7, universal instantiation
    If x possesses a semantics and y does not, then it is not the case that x is identical with y.

9. Mx & PRy	Assumption
10. Mx	9, Simplification
11. Pry	9, Simplification
12. Sx	10, 5, Modus Ponens
13. ~ Sy	11, 6, Modus Ponens
14. Sx	12, 13, conjunction

15. (Mx & PRy) ⊃ (Sx & ~ Sy) 9-14, conditional proof
16. (Mx & PRy) ⊃ ~ (x = y) 15, 8, hypothetical syllogism
17. (y) [(Mx & PRy) ⊃ ~ (x = y)] 16, universal generalization

18. $(x) (y) [(Mx \& PRy) É \sim (x = y)]$ 17, universal generalization
   For any x and for any y, if x is a mind and y is a
   program, then it is not the case that x and y are
   identical.

As the argument is valid, what would remain to be considered
is the truth or falsity of the premise set. The first premise is
obviously true, as it is a tautology. The second premise and the
third premise seem well understood as true, with the result that,
if they are true, then one has a sound argument, namely, a valid
argument with a true premise-set. A sound argument is often
thought of as a proof. To be sure, pragmatic considerations may
be relevant, and some sound arguments would not be regarded
as proofs.[13]

It might be argued, of course that the second and third premise
need not be regarded as true. For example, if a program is
understood as a mind, without a semantics, then it would be false
that all minds have a semantics. If it is claimed, somehow, that
a program is a mind and has a semantics, then such a program
would be compatible with the second premise. Similarly, if it is
maintained that a program has a semantics, then the third premise
would be false. Much then depends on how one understands mind
and what might count as having a semantics. Presumably Searle,
or an individual of related views, would require a semantics as
a necessary condition for mind, and the lack of a semantics as a
necessary condition for a program. Illustratively, a mind would
require understanding the word 'cat' whereas a program might
spell out many things about cats but would not have any idea
what it was doing. Incidentally, Searle and those of related views
are pretty obviously concerned with minds of a certain level. For

---

13. Here are two arguments. Both are valid, being substitution instances of the valid
argument form, *modus ponens*. As the first premise in both arguments is true, and,
as a sufficient condition for the truth of a material conditional is the truth of the
second term, or the consequent, one of the following two arguments has both a valid
form and a true premise-set, and is thus a sound argument. Thus, if a proof required
only a sound argument, one of the following two arguments is a proof.
   Albany is the capital of New York State.
   If Albany is the capital of New York State,
   then Jones is guilty.

   Albany is the capital of New York State.
   If Albany is the capital of New York State,
   then Jones is not guilty.

example, a dog or cat, or a feral child, would presumably have minds, but would, one supposes, lack a semantics.

Let us suppose now that Searle, or anyone else, encounters an AI theorist who maintains not only that programs are minds, or at least some programs, but that some of those programs, or minds, can actually understand things, or some things, for example, understand stories, or some stories.

How do we know that Junior understands a story, say, "Jack and the Bean Stalk"? Unless we have some doubts about the matter, we don't even worry about it. He looks like he understands, he reminds us of a part we left out, he asks about how the giant's castle is suspended, perhaps with balloons, he might inquire into the botany of unusual bean plants, and ask for some clarification as to goose physiology, in particular, a reproductive system that produces metal eggs, and such. In many ways, it does not even occur to us that Junior does not understand the story. On the other hand, if Junior was a machine, our curiosity might be aroused. One could ask the machine questions about the story. Ideally, these are not questions as to what was in the story, for example, what did Jack get for his mother's cow, what did he do with his dubious gains, how he get to the giant's castle, and so on, for all that is in the story, and, if the right buttons were pushed, the tape could be replayed, and so on. Rather, we might ask Junior, People-Junior or Machine-Junior, questions about the story, the answers to which do not occur explicitly in the story. Some such questions might be, is Jack the sort of fellow to whom you would entrust the sale of your automobile, is Jack's mother good at parenting, does the giant's personality and attitudes help us to understand why he lives alone, and so on.

We now come to the machine, or program, which is supposed to understand stories, at least stories about hamburgers. In order to find out if the machine, or program, understands these stories about hamburgers, we ask it certain questions, and, ideally, questions the answers to which are not given in the stories themselves. For example, consider the following two stories, in both of which a fellow orders a hamburger in a restaurant. In the first story, the hamburger is burned to a crisp, the man does not pay for it, exits abruptly, and leaves no tip. In the second story, the man is pleased with the hamburger, and he leaves a big tip before paying his bill, and leaving. The same question is asked of both stories, namely, "Did the man eat the hamburger?" In neither story, note, is it spelled out whether or not he ate the hamburger. Supposedly the machine, or program, responds negatively in the first case, informing us that the man did not eat the hamburger, and affirmatively in the second case, informing us that the man did eat the hamburger.

So, does the machine, or program, understand the story?

The first thing to note is that neither story is likely to find a publisher. Perhaps the publishers are overstocked with hamburger stories. On the other hand, neither scenario is really a story, or much of a story. What seems to be asked for is a guess, based on the likely outcome of a simple situation. For example, let us suppose a poor man drops his last dollar on the sidewalk, or a rich man spots a cigar stub on the sidewalk. Are these stories? Presumably the poor man will retrieve his dollar, and the rich man will not pick up the cigar stub. Also, it is hard to suppose that the machine, or program, is to be credited with anything more than supplying a typical outcome. Certainly the machine's, or program's, answer is not infallibly correct, or the only possible rational answer. How does it know, for example, that, in the first case, the fellow did not eat the hamburger? Perhaps he is terribly hungry and eats the hamburger, though he does not like it. Further, having no money he rushes out, not leaving a tip, and so on. Perhaps he likes overdone meat, perhaps he is a masochist, perhaps he is unwilling to waste food, and so on. In the second case, perhaps the fellow does not eat the wonderful hamburger, but takes it home to give to his crippled grandmother, or perhaps to add it to his hamburger collection. To be sure, this may seem quibbling.

So, does the machine, or program, understand the story?

It is at this point, finally, that we find ourselves at the door of John R. Searle's famous "Chinese Room."

Searle tells us that he does not know Chinese, and we believe him. Indeed, he informs us, incredibly enough, that he cannot even tell the difference between Chinese and Japanese script. Both, to him, are so many squiggles. On the other hand, he does know English, and can certainly tell the difference between this squiggle and that.

He enters the "Chinese Room.

In the Chinese Room there are many squiggles, of diverse categories, doubtless files of squiggles. For example, a large number of squiggles constitute what one might call a "script," if one knew Chinese. It contains a great deal of "background information," in squiggles, of course, which might prove useful in the contextualization or interpretation of possible accounts, or stories. For example, if we were dealing with Jack and the Beanstalk it might supply us with information about human dispositions, needs, habits, and customs, Medieval marketing practices, exotic horticulture, views of the nature and character of giants, and such. We have here almost an encyclopedia of information, including many of the things an average person would be likely to know. This might prove a valuable resource. Secondly, also in squiggles, we have one or more "stories." Thirdly, in an actual case, we would have framed questions, also in "squiggles," or, if we knew Chinese, in Chinese. Then, we have the capacity to output answers, also in squiggles, to the questions asked about the story, which themselves are, of course, also in squiggles. Altogether then, we have (1) the script, (2) the story or stories, (3) the questions asked about the story or stories, and, lastly, (4) the answers to the questions. All of this is in Chinese and Searle, in the room, knows no Chinese whatsoever. On the other hand, he does know English, and can discriminate amongst squiggles, based on their appearance, size, shape, and so on. Now, he has instructions in English for correlating the script with the story, and he has instructions in English, further, for correlating the script and story, or stories, with the questions, in order to produce the answers. In short, given Sets 1 and 2 of squiggles, related to Set 3 of squiggles, he puts together and outputs a Set 4, of squiggles, which, to a Chinese speaker, is an answer in Chinese to a question asked in Chinese. Outside the Chinese room a Chinese speaker would have no reason to suspect that within the Chinese Room, instead of Searle, there was not an intelligent, informed Chinese speaker. Indeed, if the fellow in the room was Chinese, he would know what he was doing, and it would all make sense to him. To Searle, of course, though he follows his instructions, relating squiggles to squiggles, he does not have the foggiest idea as to what is going on. He does not understand either the script or the story, or stories, nor does he understand the questions asked, nor does he understand the answers which he is providing. In this case, a Chinese speaker in the room would have both a syntax and a semantics. Searle, however, has only, so to speak, the syntax. To him it is all meaningless. So, does Searle, limited to syntax, understand the story, or stories? No.

He doesn't even know, possibly, that he is dealing with stories, at all. From his point of view, he is just following instructions, arranging squiggles in what are, to him, meaningless patterns.

At the point at which the "Chinese Room" was invented, or constructed, Searle was ready to grant the machine or program a syntax, rules for relating signs to signs, in, say, well-formed formulas. Later, with great plausibility, he came to believe this was excessively generous, and a mistake, and that the machine or program did not even have a syntax, certainly not in the sense that a native speaker would recognize that, say, "See Spot run," was all right and "Spot see run," was not all right. The machine, or program, then, lacked a syntax in the usual sense of 'syntax'. The machine, or program, then, was simply working, simply operating or functioning, as, say, might a washing machine or a clothes drier, though, to be sure in a more complex manner than such devices.

There is no doubt, of course, as suggested earlier, that an incredible amount of intelligence, organization, planning, and such, would be invested in the creation of the sorts of machines, or programs, which are in question here. But it is one thing for thinking to produce such devices and another for such devices to think.

In any event, what Searle seems to have shown here, and which squares, one supposes, with the usual human intuitions in such matters, at least at the present time, is that what the relevant machines, or programs, can output does not require understanding, at least in the usual sense of understanding, because, in theory, the same output could be produced without understanding, in the usual sense of understanding. Searle has shown that in the Chinese-Room argument. Presumably the machine is not conscious, and it is taken for granted that understanding requires consciousness. To be sure, this brings us back to the "consciousness objection" to machine thinking, earlier discussed. Naturally, engineers or AI enthusiasts are at liberty to redefine 'understanding', if they wish, but, if that is done, the subject has been changed. One might, as well, one supposes, claim that the thermostat "understands" that it is time to turn the furnace back on, the electric eye that it is time to open the door, and such. To be sure, the possibility of "unconscious thought" is interesting, but there are two difficulties there, which might be pertinent, first, perhaps unconscious thought does not exist, as alleged evidence for unconscious thought might be otherwise interpreted, and, second, that unconscious thought might exist in an organic brain, with billions of years of evolution behind it, does not entitle us to infer that it exists in a very different environment, say, in a machine, or program. To be sure, one's intuitions in such matters might, over time, undergo change. Words, as pointed out by Nietzsche and Wittgenstein, are like pockets; that can contain more than one thing, and different things at different times.

As a last remark here, one might speculate on some likely differences between human understanding and "machine understanding." For example, would the machine respond identically to "Did the man eat the hamburger?" and "Was the purchase consumed by the male human?" Probably that would depend on the sophistication of the programming. One does suppose, however, that the machine would respond automatically and invariably, whereas human responses might be less reliable, and more diverse. A human, for example, might understand the story, and, for one reason or another, perhaps to be difficult, answer improperly. Presumably such a vagary would be unlikely in a machine. And, if it were programmed in, by occasional recourse to a random-selection procedure, that would be quite different from being motivated by an intention, say, to be difficult. A human, too, might misunderstand, because he does not catch on, but the machine would not misunderstand; it would just be unable "to compute."

To be sure, perhaps this is quibbling.

The question is really, does the machine think, not, does it think like a human.

Searle seems to presuppose that thinking requires understanding, and that understanding requires knowing what you are about, that it requires a consciousness which is an understanding consciousness.

A conscious machine now would be another matter altogether.

Suppose, a machine were conscious. How could it convince you it was conscious?

Similarly, let us suppose that one has a machine civilization, a civilization developed by, and populated by, conscious machines. Presumably this might come about in virtue of an extremely improbable quantum fluctuation, or such, the sort of unusual event which is not ruled out by quantum theory, or, more likely, the origins of such a civilization were lost in the civilization's antiquity. Long ago, one might suppose that self-replicating, conscious machines were devised by an organic race of creatures which subsequently became extinct, perhaps in virtue of disease, or perhaps because they were eventually exterminated as vermin by their superior, intolerant, "racist" machines. All signs, we shall suppose, of the machines' embarrassing origins, as a matter of machine pride, have been systematically eradicated. Furthermore, over time, say, millennia, the machines have developed creation stories, religions, theologies, and such. Presumably the machines regard themselves as having been produced by a cosmic machine, that they are constructed in its image, and so on. Further, let us suppose that the machines have a rather Cartesian view of organic life forms. For example, such life forms, lacking the proper wiring, circuitry, and such, cannot be conscious. Consciousness, then, is invariably associated with its familiar forms, machine forms. It is unthinkable that consciousness could be found as a feature of accidental "organic slime." That is inconceivable. Accordingly, other life forms are regarded as pure automata, no more conscious than clocks, automatic lubricators, electronic buffers, used for cosmetic purposes, and such. Suppose, then, you land on this planet, and find yourselves in the midst of the machine civilization. Fortunately you land in a community of electronic liberals, tolerant, or naive, devices who refuse to recognize your possible menace to their self-image, way of life, history, future, religion, and so on. If they thought you were conscious, they might spare you. On the other hand, they are planning on turning you into mulch for the local botanical garden. How could you convince them you were conscious, and not, say, an anomaly of sorts, or an ingeniously constructed automaton, designed to cleverly simulate consciousness?

You seem to understand stories, and such, but that might be expected in such an artifact.

Indeed, might not a skeptical Searle 73X99A construct an ingenious "Protoplasm Room" Argument to cast doubt on your consciousness?

How could you prove you were conscious, were the pervasive presumption being that that was impossible?

So, how then could one know that an entity was conscious, or, say, intelligent, or could one know that?

9. An Inquiry into Intelligence

A number of entities might be involved, depending on circumstances, in such an inquiry, for example, humans, animals, machines, plants, mineral formations, even atmospheres, or planets.

9.1 Regarding Access to the Intelligence of Others

In a sense one does have access to the intelligence of others, for example, by observing them, talking with them, and so on. What more could one want?

This, however, is indirect access, if it is access at all. This becomes clear when one realizes the solipsistic possibilities mentioned earlier, and the possibility that one might be hallucinating, experiencing an illusion, confronting a holographic image, interacting with the realistic simulacrum of a human being, and so on. In a strong sense of 'access', given the "privacy" property of consciousness, one would not, presumably, have access to the consciousness of another. Only the individual himself, or itself, might have access, that direct access, which is currently in question. This seems to be a matter of fact, and the question is not helpfully resolved by recourse to current speech habits, such as, what we mean by having access to one's thoughts is asking him what he is thinking, and so on. We might receive a stream of noises from a manikin, which we, not the manikin, might understand as, "I am currently thinking about raccoons and giraffes."

It is not analytic, of course, that one might not have access to the thoughts of another. Rather, it seems to be a matter of fact, for nomological reasons. For example, a god or gods might, one supposes, give one a direct awareness of the consciousness of another, by a miracle, or such. A more likely possibility, would seem to be something like telepathy or "mind reading." Putting aside the lack of convincing evidence for either telepathy or mind reading, one might consider the following:

In telepathy, one does not really have access to the consciousness of another; rather, one receives messages from the other. For example, it comes into one's mind that the other is thinking of the six of clubs or the ten of hearts. You are not experiencing his mind; you are experiencing your mind. It is more analogous to hearing something said, or getting a telegram or letter, than sharing consciousness.

In "mind reading" we have something much more germane to our interests, because here one might have a sense that one was not just learning about what was in another mind, but that one is within that mind itself, sharing its actual experiences. One might feel another's fear, the same fear; feel another's hatred, his actual hatred, and so on.

Let us suppose, *per hypothesis*, that mind reading is empirically possible, and that, moreover, instances of mind reading, or, perhaps better, consciousness sharing, actually take place.

A number of difficulties would still attend the hypothesis that any particular, seeming instance of mind reading (MR), or consciousness sharing (CS), was an actual case of MR or CS. How could one validate the claim of veridicality? How could one discriminate between a seeming case and an actual case? Perhaps what is going on is only in one's own mind? One has data, surely. But how should it be interpreted? How would one know it is A's mind one is reading and not B's mind? Perhaps C reads B's mind, and merely transmits similar signals to one's mind? Perhaps a coincidence is involved? Perhaps a god or gods are putting certain experiences into one's mind? Perhaps they are lying to one, or playing a joke on one, for their own amusement, and so on. Two experiences might well be very similar, and not identical. Indeed, they might be identical in one sense, and not in another, as two copies of the same page might be identical in one sense and not in another. They are copies of the same page but they are not the same copy. Sheet one is not sheet two. One would not wish to rule out either MR or CS, but it seems that, even if they were possible, one could never be sure they had taken place.

In any event, it seems probable, one, that one does not have access to the consciousness of another, and, two, even if one had it, one could not know that one had it.

9.2 Intelligent or Under Intelligent Control

One might conceive of an animal, human-appearing or not, with an anesthetized brain, being stimulated electronically, perhaps by radio signals, to behave in seemingly intelligent ways. The entity itself, while living, would not be conscious. Thus, its own intelligence, if it possessed such, would be in abeyance, and it would be acting in terms not of its own intelligence but in terms of the intelligence of another, its operator. Similarly, one might suppose a mobile machine, perhaps human appearing, with electronic sensors, being operated from a remote point. Perhaps it is a decoy, being used to lure humans, who might then be targeted by an alien life form. One might suppose similar scenarios, in which one might draw a distinction between intelligence and being under intelligent control, perhaps in terms of post-hypnotic suggestion, angelic or demonic possession, and so on.

9.3 Performance Would Be Our Criterion

One might draw a distinction between intelligence, as a dispositional property, and consciousness as an experiential property. As we usually think of these things, intelligence would presuppose consciousness, but consciousness need not presuppose intelligence. For example, it seems unlikely that an anemone or a sponge is intelligent, but, one supposes, being living forms, they have some type of sentience. Feeling presumably goes far down the phylogenetic ladder, but it is less clear how far down that ladder one might find what we usually think of as intelligence. Analogously, one could conceive of a sentient machine, one that might have some sort of feeling, some sort of consciousness, whether alien to ours or not, but might, as presumably some life forms, lack intelligence. On the other hand, intelligence is not all that rare. Certainly mice are capable of observational learning, as experiments have demonstrated. More surprisingly, other experiments have suggested forms of learned behavior in fruit flies, and even flatworms. When one thinks of intelligence, of course, one usually has more in mind than reinforceable behavior, and such. In any event, putting aside such considerations, ascriptions of intelligence are normally indexed to what an entity does; it is ascribed on the basis of its performances, on the basis of how it behaves. Interestingly, this is also our criterion for attributing intelligence in our own case. Only we, of course, can monitor our internal performances, as well as our external performances, an advantage we lack with others; so, in dealing with those others, one must make do with external performances only, what can be seen, heard, and such. Accordingly, as one has access to "internal" performances only in one's own case, we will be concerned, henceforth, with "external" performance only, hereafter denominated simply as 'performance', and will be concerned with typical "outside-to-inside" inferences, so to speak.

9.4 But Performance Would Be Neither a Necessary Nor a Sufficient Condition for Intelligence.

9.41 Performance is Not a Necessary Condition for Intelligence.

An entity might be intelligent and not perform. It might be the case either that it simply does not perform, that, say, it refrains from performing, or that it cannot perform.

In the first case, there might be many reasons why an entity might refrain from performance, at least in certain venues. One might suppose an intelligent alien, or an unusual life form, which, in a form of natural camouflage, rather like that found in the case of some insects or fish, which resemble leaves, twigs, rocks, or such, resembles human beings. It keeps to itself, of course, and avoids contacts with humans as much as possible. And humans, who never deal with it closely nor examine it, take it to be no more than, say, an eccentric, antisocial recluse. Close inspection, which the creature avoids, would betray its differences. Perhaps, it is found dead one day, in its tenement flat, and its true form of life is discovered, perhaps to human horror and consternation. Who knows what life forms might be amongst us, unnoticed?[14]

There might be many reasons why an intelligent entity might choose not to perform. Perhaps it might not perform from a sense of duty, or because of loyalty to a vow. "They must not know we are here." An entity might refuse to acknowledge its presence, perhaps from pride, perhaps from contempt for what it regards as an inferior life form. Perhaps, as many forms of life, it relies upon stillness as an approximation to invisibility. "Don't move...," and so on.

---

14. This example is suggested by "Mimesis," one of the great short stories in science fiction. Its author was the late Donald A. Wollheim, the founding publisher of DAW Books, and a much-missed friend.

It might also be the case that an entity is intelligent but it is incapable of performing. Perhaps it is an anemone-type intelligence, observant and recollective, but sessile; perhaps it could once perform, but now cannot do so, as it is paralyzed; perhaps it is a form of life which has quiescent periods, dependent on temperature changes, the intensity of light in the environment, or such. Perhaps it has restorative periods, similar to sleep, which, in its case, require immobility, perhaps for weeks or months at a time. Some individuals have attributed sentience or responsiveness to at least certain forms of plant life, and it seems not impossible that somewhere in the universe, on some worlds, evolution might have selected for complex forms of life which we might see as essentially botanical, either terrestrial or aquatic, which might be capable of some sort of feeling, some form of intelligence. Some such forms might be incapable of performance, as we think of it. On the other hand, other such forms might have some form of motion, have some sort of sensors, visual or otherwise, might have appendages capable of grasping objects, and so on. In the latter case, they would seem capable of performance, of a sort at least intelligible amongst themselves, if not to us. More exotically one might suppose "imprisoned gaseous intelligences" interlaced, say, in natural crystalline structures. Such intelligences might be incapable of making their presence known. Each might be essentially alone, for thousands of years. Certainly, as we shall see later, life, and perhaps intelligence, might not be restricted to life, or intelligence, as we know it. Indeed, perhaps our form of life, and intelligence, might prove to be only one species of life, and intelligence, within a larger genus.

9.42 Performance is Not a Sufficient Condition for Intelligence.

An entity might perform and not be intelligent. This is in no way surprising. Trivially, for example, machines are designed to perform. The thermostat turns the furnace on, the electric eye opens the door, and so on. When the chess machine, or program, produces a move, that is a performance. Few would maintain that such performances are indicative of intelligence.

On the other hand, in the case of certain machines, or programs, such claims have been made. As we have seen, an AI theorist might claim intelligence, even understanding, or such, for a given machine, or program. John R. Searle, and others, have questioned such claims.

All one needs at this point is the understanding that one might have performance without intelligence. For example, in many cases, it might be difficult or impossible for a human being to tell the difference between the responses of another human being and those of a sufficiently sophisticated, well-programmed machine. One supposes that, sooner or later, the machine might, so to speak, betray itself, but how much sooner or later would depend on the device, and, if the device were sufficiently sophisticated, perhaps this betrayal, or revelation, would never take place.

Within narrow, or circumscribed limits, of performance and time, even a simple machine need not be recognized as a machine, for example, a chess machine, played in another room, or over the internet. If a series of players, on the other hand, taking turns, were to keep the machine busy in game after game, and the machine provided moves unflaggingly, this might suggest that it was a machine, not stopping to eat, rest, go to the bathroom, and so on. On the other hand, perhaps it is not a machine but a team of humans on the other end, taking turns, too. And the machine need not always respond similarly. Perhaps it is programmed to take more or less time, once in a while. And it need not always make the same eleventh move in Evans Gambit, and so on. Once, however, one were permitted to transgress circumscribed limits, the programming would have to be extremely complicated. Let us suppose it has been programmed to handle questions such as, "Do you like oysters?" and "Do you like chocolate sauce?" Perhaps it has been programmed to respond "Yes" to both questions. It may not, however, have been programmed to handle something like "Do you like oysters with chocolate sauce?" If it responds "Very much," you are probably dealing with a machine. Similarly, it is difficult to suppose that a human and a machine would respond similarly to questions such as "What did you have for lunch," "What works best with Susan," "Where were you on the night of January 16th," and such. You might also inquire as to why the chicken crossed the road, or why firemen wear red suspenders. Also, machine responses to remarks such as "Hannibal was a avid stamp collector" and "Cleopatra invented the steam engine" might start a machine thinking, if it was not doing it before. To pass an "imitation game" of such challenging complexity, it seems a machine would not only have to be pretty much like a human being, but, for most practical purposes, would have to be a human being. Or better than a human being. For example, individuals who have serious reservations about human beings, perhaps regarding humanity as a "plague species," and looking forward eagerly to its extinction, often have a soft spot in their heart for artificial intelligence, and artificial life. Might not machines be better, stronger, smarter, gentler, kinder, more moral than human beings? Would the world not be well surrendered to a higher, nobler form of life, or being?

Another possibility is that performances might be misattributed, nonintelligent entities being taken as intelligent entities, as a voice might be transmitted through a robot, doll, manikin, or such. Let us suppose that a devotee in an Alexandrian temple in the first century believes himself to be conversing with a god, the god transitorily inhabiting its statue, the member of the faithful not realizing that a priest is producing this illusion by means of a speaking tube. Here there is a performance but it is due neither to a god, who is presumably elsewhere at the time, if existing, nor to a statue. Doubtless the devotee is enraptured. And the god actually speaks his language, and with an Alexandrian accent, as well.

Another possibility is that of coincidence, in which something appears to be intelligent performance, but is the result of chance. Several years ago, when answering machines were new, and unfamiliar, it was often the case that an individual addressed himself to the machine, not realizing it was a machine. Similar mistakes might occur with a phonograph record or a radio in another room. Imputations of significance have often been given for one thing or another, say, for the flights of birds, unnatural births, the look of livers, the appetites of chickens, the color of cats, the turns of cards, the dispositions of tea leaves, the configurations of stars, and such. Human beings look for meaning, and what one looks for, one is likely to find. Let us suppose a set of cards, on each of which is written either "Yes" or "No." The cards are shuffled, and then, questions being posed, are looked at, one by one. Conceivably the correct answer to various questions might turn up on the cards. Presumably this is a coincidence, but it might not be interpreted as a coincidence. It might be a good idea to prolong the run, to see if some false answers turn up. Consulting such cards would presumably not be a good way to resolve a question such as, "Should I invest my life's savings in an octopus hatchery," but some people do this sort of thing. Hopefully, they will become rich. More dramatically, let us suppose a couple, say, Mr. and Mrs. Smith, visit the zoo and, in a monkey cage, one notes a rhesus monkey playing with blocks. The blocks happen to spell out "Your apartment is on fire," and, later, Mr. and Mrs. Smith discover there was a fire in the apartment. Presumably this would be a coincidence, but, too, it might shake the world view of many a primatologist.

Many times, of course, it is difficult to know if intelligence is involved or not. Occasionally the matter seems reasonably clear, as in Ouija boards, small weights on chains, swinging over answer boards, and such, where muscular contractions control the indicators. In such cases, it seems the indicators, for most practical purposes, move themselves, but, given the laws of physics, they don't. This sort of thing can be seen as evidence for subconscious manipulation, but, presumably, alternative explanations might also account for the data, perhaps conscious but immediately forgotten goal seeking, acting belatedly on a previously conscious thought or desire, now forgotten, and so on. In any event, intelligence, in one way or another, does seem involved in cases of this sort.

In certain other cases, it seems reasonably clear that intelligence is not involved. The trap-door spider builds his insect trap in several steps. On the other hand, if the spider is interrupted at a given step, it does not continue on, picking up with the next step. It starts over, from the beginning. This suggests that the spider's work is a remarkable example of intricate, but primitive, genetic coding. A similar hypothesis seems likely in the case of the Italian sand fleas, which move to the water after hatching. The west-coast sand fleas move west to the sea, and the east-coast fleas move east to the sea. On the other hand, transplanting the west-coast sand fleas to the east coast, and the east-coast sand fleas to the west coast, results in both the west-coast and the east-coast fleas moving inland. The transplanted west-coast fleas, now on the east coast, still move west, and the transplanted east-coast fleas, now on the west coast, still move east. This makes it reasonably clear that the fleas are not seeking the sea consciously, or sensing its presence, or such, but are merely acting out their genetic coding. The fleas which went the wrong direction in the past did not reach the sea, and so did not replicate their genes. Presumably a number of remarkable animal behaviors may be similarly explained, such as the return of salmon to their original spawning grounds, and the Bamberg warblers to their original nesting sites. The salmon are apparently genetically coded to return to their spawning sites, and manage to do this in virtue of the "taste," so to speak, of the waters involved, in reverse order to their earlier descent to the sea, first, finding the right river, and then, in order, the appropriate tributaries, smaller streams, and such; and the warblers, as has been demonstrated in a planetarium experiment, with a sky reversal, home in on a particular star or configuration of stars. One of the most famous examples of this sort of thing, at least from the insect world, are the remarkable "bee dances," whereby a bee, having found a trove of nectar, returns to the hive and, by motions within the hive, correlated with the orientation of the sun outside the hive, indicates to his hive mates the direction of the trove, which they then seek.

Such things are marvelous and valuable performances but presumably do not involve intelligence in the sense that one normally thinks of intelligence.

On the other hand, there are innumerable animal performances which do suggest intelligence. A stellar example of such would be chimpanzees and gorillas which communicate in Amslan.

In any event, performance is neither a necessary nor a sufficient condition for intelligence. Some entities which are intelligent might not perform, and some entities might perform, which are not, as we usually think of such things, intelligent.

9.5. An Entity's Performance Might Be Undetectable to Us.

Unless the human race becomes a multiplanet species, it is doomed to stagnation, and inevitable extinction. Life, sooner or later, will become impossible on Earth. That is not a theory; it is a fact.

Let us then suppose that a minority of human beings heeds the summons of the stars, addressing itself to the seas of space, searching out distant, viable shores. It is not clear what it might encounter.

Science fiction, not surprisingly, provides an exotic ecology of logical and empirical possibilities, the contemplation of which exercises the muscles of the imagination and opens the eyes of the mind. That being the case, it is not surprising, either, that there occasionally exists a startling, electric connection between that *genre* of literature and the philosophical project, certainly as it tests the waters of the future. Both, in a sense, are concerned with the maps of tomorrow, with the hazards of speculation, with new ways that things might be.

Surely it is unlikely, even given the happy reassurances of convergent evolution, that the patterns of life are narrow, minimal, familiar, and prosaic.

Is a new planet uninhabited?

That might not be clear.

One might conceive of large, rocklike entities, which might take a hundred years to form a thought. Mining might be murdering. Perhaps there might be tiny things, so swift as to be almost invisible, whose life processes might be more akin to those of computers than familiar carbon-based life forms. To them, if they noticed us, it might seem that we took a hundred years to form a thought. Communication with such life forms would seem unlikely. One might not even understand the possibility of communication in such cases.

One might conceive of a life form which communicated by means of light in the ultraviolet range of the spectrum. This light would be visible to, say, bees, but not to us. It would be instrumentally detectable, and something might be managed, once we understood actual signals were involved. One supposes it would be logically possible that another life form's signals would not even be instrumentally detectable, perhaps taking place in another dimension, one of those alien dimensions concerning which physicists speculate. And perhaps communication in our dimension would be inaccessible to them. Two life forms might "pass in the night," so to speak, and neither be the wiser. Indeed, one hypothesis for the lack of confirmed signals from outer space, as in the SETI project, the Search for Extraterrestrial Intelligence, is that such signals, which seem likely, might be transmitted in a way or medium we fail to detect, let alone understand. For example, radio signals, if broadcast to hominids in the late Pleistocene, would have been unnoted. Drums or flares would have worked better. Similarly, if there were a life form which communicated by means of some form of mental telepathy, its signals might well be unnoted by minds not evolved to receive them.

Not all creatures might have vocal cords. Accordingly the "phonemes" in some forms of exolinguistics might be quite different from those with which we are familiar in human languages. Perhaps a pelted creature moves its fur in certain ways, or communicates by body language, perhaps twitching the skin of its left shoulder, or by facial expressions, or perhaps by grunts, hisses, and snarls; perhaps mistakenly taken as signs of hostility; perhaps it paws the ground in code, or communicates in ripples of color, as some languages do with ripples of sound. Perhaps another form of life arranges sticks and pebbles, which it carries about, or cracks rocks together in meaningful patterns. If there were colony creatures, perhaps the formations or groupings of the creatures might constitute a social discourse. Might the humming of clouds of insects similarly communicate? An aquatic creature might communicate with permutations of its body shape, these constituting its "phonemes." Something else might do something with a "talking stick" or with weapons, as in a manual of arms. Perhaps that might be seen as threatening. Another creature, vocal-corded, might sing. Another form of life might communicate by means of instruments, as with flutes or small drums. Many of these things might not even be taken for communicating, but they would be the communication itself. If a form of life could communicate telepathically one might, as noted, be unable to receive the signals. Suppose something humanoid is seen, carrying on its shoulder what one takes to be a bird. Perhaps the birdlike creature, on the other hand, is using the humanoid creature as a mount, and telepathically signals its symbiotic associate to utter vocalizations. We would presumably take the humanoid creature as the intelligent entity, and the bird as, say, a pet, thus mistakenly reversing the situation as it might be in reality.

9.6. What Sort of Performance Would Be Relevant?

Whereas it seems that performance would be neither a necessary nor a sufficient condition for intelligence, it does seem that it must be our criterion for intelligence.

That being the case, it seems important to consider the sorts of performance which we might associate with intelligence.

'Performance' here, as might be recalled, is used broadly, to cover what an entity might do. The word itself is not intended to presuppose intelligence. In this way, 'performance' and 'behavior' might be used rather interchangeably, rather than performance being regarded as a species of behavior, or, as some might prefer, as an alternative to behavior or, perhaps better, as an alternative to "mere behavior." It seems odd to think of a rock's behaving, as it rolls down hill, or a tree's performing, as it blossoms, or the weather behaving or performing, as the wind blows and the rain falls, but one does need some generic category, within which to attempt to distinguish between, say, minded performance and nonminded performance, or minded doings and nonminded doings.

9.61 Evidence of Tool Use (Technological Competence)

Wolfgang Köhler's classical experiments having to do with the mentality of apes demonstrated the capacity of apes to use tools, and to even form tools, in order to accomplish simple tasks. Even in the wild a chimpanzee might break off a leafed twig, soak it with saliva, and then use it as a swab, wiping it inside a rotted branch or trunk, in order to snag termites. Another species of monkey and at least one bird species uses a stick to drum on wood, to attract mates. It is interesting that we are likely to attribute intelligence to the monkey and less likely to attribute it to the bird, which is perhaps merely acting out a genetic coding, analogous to that of the salmon or sand flea. Perhaps one is a hominid chauvinist. At any rate, the apparently meaningful use of a tool to obtain a meaningful goal, say, using a wet, leafed twig to trap termites, using a stick to attract a mate, using a stone to sharpen a stick or shape flint, seems a familiar, and excellent, ground on which to ascribe intelligence.

It is surely possible that certain tools, say, Tool 1, on an assembly line, might make use of other tools, say, Tool 2 and Tool 3, but, in such a case, intelligence would be more likely to be ascribed to the designer of Tool 1 than to Tool 1 itself. A more familiar analogy would be a Tool 1, a timer, which turns on a Tool 2, say, a light, or coffee machine.

A more interesting case is that of self-replicating robots set to terraform a planet, to prepare it for eventual colonization. These entities would presumably use a number of devices and various sorts of equipment to pursue their goal, might mine and exploit the environment for required materials, and would see to their own replacements, that the work might be carried to fruition, perhaps five hundred years in the future. If there were inexactitudes in copying programming, one supposes, analogous to variations in DNA replication, natural selection might eventually produce a robot population quite different from that which was originally placed on the world in question. Different robot races might develop; there might be competition for resources, might be robot wars, and such. When humans came to the world they might find it more congenial to robotic existence than human existence, and might be exterminated as aliens, as invading interlopers. In theory, this might take place without thought, or intelligence, just as, on Earth, some marvelously effective adaptations have taken place, apparently without thought, or intelligence. On the other hand, just as natural selection apparently favored intelligence in organic life, perhaps some random mutation in programming, might have produced irritability or primitive responsiveness in sufficiently complicated machines, and from so inauspicious and inchoate a beginning might a new species of being begin its journey on the road to rationality.

One supposes that we would be more likely to assign intelligence to an alien life form which was quite different from us, say, long-necked, scaled, tentacled, and quadrupedal, utilizing primitive tools, say, sticks and stones, in simple ways, than a sophisticated robotic form of being, utilizing modernistic tools in a complex manner. Presumably this would be because we are more familiar with organic life than, say, robotic life. Indeed, we might suppose that robotic life, a form of "artificial life," was impossible. If we did encounter such machines, we would be likely to suppose that they were tools carrying out the plans of, and under the management of, creatures analogous to ourselves. Ethnocentricity runs deep.

9.62 Discourse

Discourse might be either verbal or nonverbal. In 9.5, above, we suggested some nonverbal forms of communication, in terms of motion or color, for example. One would not wish to make it analytic that nonverbal communication was impossible, as, say, that any unit of communication, audible or silent, must count as a word or sentence, and must thus be verbal, as that would sacrifice an obviously interesting distinction in modalities of communication. Similarly, communication is much broader than words, however conceived. Bee dances communicate but not in words. Body language communicates, but not in words. The monkey with his courtship drum communicates, the dog with its snarl, the cat with its hiss, all communicate, but not with words.

An entity capable of entering into discourse is likely to be intelligent. On the other hand, as we have seen earlier, a well-programmed machine, perhaps engaging in some sort of "imitation game," could be taken as discoursing. Parry, for example, the "paranoid imitating system," enters into conversational exchanges indistinguishable by many from interhuman conversational exchanges, and it is easy to suppose a more sophisticated program, which might enter into "dialogues" wholly indistinguishable, without further information, from those of normal human beings. It might take years to develop such a program, but it seems within the realm of possibility.

Thus, discourse, or its facsimile, would seem a weaker criterion for intelligence than, say, tool use.

9.63 Assent and Dissent Signified

Assent and dissent might be signified either verbally or nonverbally, along the lines of 9.62, above. This criterion would be weaker than the preceding. as it requires less than discourse. Intelligent assent or dissent, however, would require understanding, which would require intelligence. Gradations are involved in such things. One would not usually think of the squirrels and pigeons in the park as assenting to eat the food offered to them, but it seems likely that Fido, after some thought, puts aside his reservations, and accepts the dog biscuit. Whoever has medicated a cat, or attempted to do so, is likely to be aware of dissent, or resistance, and such. On the other hand, the "queen" in the alley accepts one or another of the circle of her admirers.

It is easy to conceive of a living planet, as in the science-fiction novel, *Solaris*, choosing to communicate pleasure or displeasure, and such, by means of natural phenomena. More realistically, a god or gods might choose to confirm or disconfirm religious hypotheses, or clarify their viewpoints, by signs or miracles. Similarly, a god or gods have been frequently associated with diverse sorts of phenomena, floods, storms, volcanic eruptions, earthquakes, lightning, and so on. ("Please thunder once for "Yes," twice for "No.")

We noted, however, in 9.42, above, how a deck of cards, or such, or, implicitly, a random-selection device, might be taken as being intelligent, or, more likely, being the avenue through which an intelligence might express itself.

Too, if a program could be devised which would simulate discourse, it would seem easier to devise one which could simulate a seemingly rational assent or dissent, in the absence of intelligence.

All in all, this criterion, would be weaker than one involving actual discourse. It might be useful, however, in dealing with alien life forms, individuals with whom one does not share a language, animals, small children, and such. Assent and dissent, acceptance and repudiation, "yes" and "no," are probably where communication begins.

9.64 Problem Solving

One of the best criteria for ascribing intelligence is that of apparent problem solving. Whereas much "problem solving" may not betoken intelligence, and may be genetically coded, as, presumably, the various problems solved by the trapdoor spider, building its trap, the Italian sand fleas, getting to the sea, the salmon, returning to the spawning grounds, the Bamberg warbler, getting to the nesting sites, the honey bee, informing his hive mates of a find of nectar, and such, other forms of problem solving seem to clearly transcend the behavioral dispositions associated with one's DNA. When one of Wolfgang Köhler's apes puts together sticks in such a way as to bring fruit into reach, that seems clearly an act of intelligence. Certainly we would ascribe such an act to intelligence, if it were performed by a child, to bring, say, a cookie or toy within reach.

The basic problems faced by a species are solved variously, food, water, shelter, security, reproduction, and so on. One of the distinguishing characteristics of an intelligent species is the flexibility, or variability, of its behavior. Learning of a sort has been demonstrated even in flatworms and fruit flies, and it seems likely that some sort of conditioned responses might be instilled in even simpler organisms, but anticipation, imaginative projection, preparation, planning, and such, tend to be found higher on the phylogenetic scale. Civilization, with its economic bases, divisions of labor, exchanges, cooperation, languages, laws, architectures, social arrangements, and such, is largely a record of problems solved. That one of the weakest, slowest, most fragile, least prolific of the Earth's species has come to the top of the food chain is clearly a testimony to the value of one of evolution's more unusual results, intelligence. Interestingly, the feature that makes possible building cities does not seem, as yet, capable of resolving how one should live in them. Intelligence, with sublime neutrality, has made considerable progress. By its means we may both preserve and destroy life. It makes possible, with open-minded equanimity, an assemblage of diverse utopias and the competitive garbage of a thousand hate-filled hells, each desirous to consume, claim, and remake the Earth in its own image. Aliens may somewhere debate the existence of intelligent life on earth.

Whereas it seems a simple fact that a species may be more or less intelligent, and intelligent in some ways and not in others, even tragically so, as the human species, we need only note here that a performance in which a problem is apparently solved, by thought, and not by programmed coding, organic or electronic, is a very good test for intelligence. It ranks with, and is often associated with, tool use, with technological competence. Human problems, as one learns from history, have often been solved by blood, by killing, by murder, by massacre. by war, and so on. There are many ways to solve problems. Those worked well in the Pleistocene. There is, after all, intelligence and intelligence.

We ourselves might find ourselves to be someone's, or some thing's, problem.

If an alien fleet of extraterrestrial starships lands in England, or New Jersey, and begins to methodically seek out and destroy human life, it would seem anomalous to doubt that intelligence is involved.

That inference may not be logically conclusive, but it is likely to be rationally coercive.

Problem solving is difficult to mistake, or fake.

9.65 Our Own Responses and Those of Other Organisms or Things

As I understand it, there is no seriously persuasive evidence for either mental telepathy or telekinesis. This is not surprising, as both would seem to be incompatible with the laws of nature, at least as currently understood. On the other hand, it is easy to conceive of data which, if honestly and uncontrovertibly obtained, would lend itself to explanation in terms of such concepts. Our concern here, of course, is not with the plausibility or implausibility of such concepts, but with the analysis of grounds for the ascription of, or the withholding of, the ascription of, intelligence.

One grants, happily, of course, that such things, mental telepathy and telekinesis, might exist, if not here, somewhere, elsewhere, perhaps in a species otherwise evolved. Man is not the measure of all things, not even Protagoras. And surely the world is a very mysterious place, though it may not be as mysterious as some folks suppose.

Let us suppose that we are in the presence of an unfamiliar entity. We have never seen anything quite like this before. Our Baedeker for this world has not prepared us for the entity. It is perhaps rare, or, for some reason, has not been included in the guidebook. In its presence, we "feel" answers. They come into our mind, rather forcibly, and do not seem much like our familiar thoughts. Perhaps we feel informed, somewhat irritably, that it is not bulbous, really, but dashingly trim for its sort, and that blue is a very nice color. We then seem to conduct a mental dialogue with the creature, rather as though we were conversing back at the lodge, but not aloud. We conduct a test or so on one another, giving one another problems to solve. We ask it to climb on a nearby rock, and it, from this coign of vantage, which we had not anticipated, notes some nearby berries, which it requests that we might pick and share with it, and so on. Pretty clearly intelligence is involved, as discourse of a sort seems to be taking place. More alarmingly, our responses might be physical, almost as though produced by some irresistible post-hypnotic suggestion. Perhaps telekinesis might be involved. Perhaps we find our body forced to pick the thing up and set it on a nearby rock, from which it discerns a berry patch, and we find ourselves moving to the patch, picking berries, and then bringing them to the creature on the rock. We note that it leaves some for us, but it does not force us to pick them up and eat them. It leaves that up to us, before it slides off the rock and disappears into the brush. We were used to solve one of its problems, one not uncommon to a short organism. We take it to be intelligent. And we begin to suspect why it was not included in the guidebook. The berries, taken to the lodge, and subjected to chemical analysis, proved to be edible.

Similarly the performances might not be our own, but might be produced in other organisms or things, perhaps people, machines, plants, animals, puppets, dolls, or such. In such a case, one would suppose the organism or thing was not itself communicating, but was being used as the means whereby something else might communicate, used as the voice by means of which it would speak, the pen by means of which it would write, and so on. We would be confident we were dealing with intelligence, in the face of such performances, but its location and nature would presumably be obscure.

9.7 Note: Performance and Intent

Performance with intent would normally be taken as intelligent performance. One of the reasons why intelligence might not be ascribed to, say, a chess machine, or program, is that it lacks intent. It does not intend to do what it does, no more than an automobile engine intends to move an automobile, or a can opener intends to open cans.

9.71 Performance Compatible with Intent.

The normal situation is self-initiated physical performance. We decide to pick up the pencil and pick up the pencil, and so on. On the other hand, one might have "other-initiated physical performance," which is perfectly compatible with intent. For example, one of the classical solutions to the mind/body problem is Occasionalism. It might be recalled that in Occasionalism a god or gods are invoked to solve the problem of, say, picking up the pencil. Jones wills to pick up the pencil, presumably an act of a nonspatial mind, unable to interact with matter, and this is the occasion of a god's or gods' performance of a miracle. Jones' body moves and the pencil gets lifted. Here then one would have a situation in which the act is "other-initiated," in this case, hypothetically, by a god or gods, but it is nonetheless an act compatible with intent, an act for which, one supposes, Jones may be held accountable, despite the fact that he, personally, was unable to perform the act. One might also suppose, interestingly, a complex set of random quantum fluctuations in the material continuum, which happen, as a matter of improbable coincidence, to accord with the intentions of Jones' nonspatial mind. He decides to pick up the pencil, and, behold, a lucky quantum fluctuation occurs, and the material pencil gets picked up. Here, one would not need a god or gods, or miracles, only an unusual world likely to appeal to a particular sort of physicist, prepared to take stochastic astonishments in stride. In this scenario, of course, Jones would not be held accountable for what the pencil does, only for the intention, which, in this case, is a material blank cartridge, having nothing to do with what actually takes place.

9.72 Performance Incompatible with Intent

Here one might distinguish between self-initiated physical performance incompatible with intent and "other-initiated" physical performance incompatible with intent.

Self-initiated physical performance incompatible with intent is rare. To be sure, much depends on how broadly or narrowly 'self-initiated' is understood. Presumably one would not wish to count reflexes, coughs, sneezes, and such, as "self-initiated" in a relevant sense. A subtler case would be that of an individual who, perhaps in virtue of psychological suggestion, displays allergy symptoms in the presence of artificial flowers, not realizing them to be artificial. A clearer case would be a brain-damaged individual who desires, say, to utter a given sentence, but only garbled noises are emitted, or desires to move one limb but another moves. A case which seem to count would be that of an individual under hypnosis, or under post-hypnotic suggestion, whose body moves in one way or another, without his assent. As the hypnotist, or "control," is external to the organism, it seems most judicious to regard this as a self-initiated act, while acknowledging the ambiguity involved, and noting the role of the hypnotist or "control" in the performance. A possible case, too, would be Freudian phenomena, such as slips of the tongue or pen, where one means to say one thing but says another, or intended to write one word but discovers he has written another, and so on. It could be maintained in such cases that intent is actually involved, subconscious intent, or such, but, certainly the slip is not consciously intended, and would thus seem to count as a case of self-initiated performance incompatible with intent.

"Other-initiated" performance incompatible with intent, in the primary sense in question, is conceivable, but would presumably be rare, and perhaps nonexistent. In the primary sense in question, one would not, for example, wish to regard it as a relevant performance of B, if he were shoved by A into the path of C, or such. In such a case, B's "performance" would be essentially indistinguishable from that of a wielded club or thrown rock.

Similarly, performing an act under duress would not be relevant in this context. Such an act, strictly, would count not as an "other-initiated" act, but a self-initiated act, for example, giving the bully one's lunch money, or such. Surely one could refuse to do so, and accept the consequences. A subtler case would be responses under chemical coercion. Yet here, too, the act would be internal, and thus self-initiated. Similar remarks would presumably hold, too, for the sibyl's ravings, writhing, drunken or tortured, on her tripod, enveloped in dry, swirling, noxious fumes, unless, of course, she is literally possessed by a god, say, Apollo, and he is literally causing her to utter the noises involved. To be sure, it would have saved time if Apollo had dispensed with the sibyl and settled for explaining his views to the priests in clear Greek.

Clearer cases of "other-initiated" performance which might be incompatible with intent would be conceivable cases in which x's body was directly under the control of some other. For example, one might conceive of a god's moving a body from without, by miraculous means, or of a god's, demon's, or spirit's inhabiting a foreign body and causing it to move in certain ways and utter certain sounds. In such a case the body might, or might not, be aware of what is being done with it. If it were aware, the experience would presumably be alarming. One of the more familiar examples of alleged possession, one supposes of a relatively benign sort, might be that of mediums in their trances, supposedly unaware of what they are doing or saying. In the "Medium Paradox," the inhabiting spirit informs the séance that the medium through whom he is speaking is a fake. That would seem pretty incompatible with likely intent, at least that of the medium.

More in keeping with our orientation toward technology and the future, one might note the partial control of organisms possible by means of strategically located electronic implants, responsive to radio signals. For example, by such means, a sleep center can be activated, putting an organism promptly to sleep, and an angry or hostile animal, say, a bull, can be instantly pacified. These primitive beginnings suggest the possibility of vastly advanced, far-reaching devices by means of which to control behavior. The possibilities extend far beyond simplicities such as inducing sleep and reducing aggression, and far beyond even positive and negative reinforcements, say, producing pleasure or pain. A suitable, coded radio signal might produce instant death. Presumably large portions of the population or selected minorities would be registered and implanted shortly after birth. Interestingly, just as it is possible to keep brains alive apart from bodies, as has been done with monkey brains, and, as it is theoretically possible to myoelectrically stimulate those brains in such a way as to have particular experiences, experiences indistinguishable from actual "in-the-world" experiences, it would seem possible to do something similar with embodied brains, if suitably treated, electronically enhanced or electronically infected, as one might see it. This might save years of expensive and sometimes unreliable conditioning, say, religious or political. One could then integrate the electronic experiences with the organic experiences, to move human beings about, puppets on electronic strings. This would seem to involve an enormous moral responsibility, but there have always been human beings bold enough, and hardy enough, to cheerfully shoulder such burdens.

In any event, given the development of such technology, the possibility of mass electronic possession, so to speak, one would have an excellent example of "other-initiated" performance, and it need not even be performance incompatible with intent, though, if the controllers wished that, it could be. But it would be more humane, and probably more efficient, to guarantee compatibility, electronically.

10. Prospectus

One of things that emerges from a consideration of "artificial intelligence," problems having to do with the ascriptions of intelligence, a recognition of varieties of possible vessels or vehicles for intelligence, of possible carriers of consciousness, and such, is the intriguing possibility of living machines.

Descartes, as suggested earlier, presumably for religious reasons, held that animals were automata. It was a short step from this for others, particularly in the light of Descartes' own mind body/problem, which he spectacularly failed to solve by means of "animal spirits," to suggest that human beings, too, were automata, or machines. Perhaps the most famous of these embarrassments was a French physician, Julien Offrai de Lamettrie, amongst whose works was a short book, *L'Homme Machine*, which might be translated *Man, a Machine*. On the other hand, almost any materialist might be seen as sharing these views, with possible exceptions for such as some epiphenomenalists, and, one supposes, some enthusiasts for stochastic nonarrangements, folks favoring things just happening, for no reason.

One wonders, sometime, if the fervor and almost religious commitment of some AI theorists to the possibility of artificial intelligence, and even its existence, and their aggrieved attitudes toward skeptics like Searle, might have more to do with metaphysics than cognitive science. Perhaps their ardor is in part motivated by a disdain for, or a fear of, what they take as superstition, and of a dangerous sort. Something of Voltaire's *Écrasez l'infâme* may be involved. However this may be, the scientific "stock" of mysterious entities, intangible substances, vital forces, and such, is low today. They seem to have little explanatory value, even as theoretical entities. To paraphrase Laplace, there seems to be little need of such hypotheses. To be sure, they might be politically reinstated, if only as verbal riders. For example, 'p' is logically equivalent to '(p v q) & (p v ~ q)', '(p & q) v (p & ~ q)', and such. Scientists will find one way or another to get on with their work, and the "fire-and-sword crowd," verbally mollified, can get on with its own business, seeking power through intimidation, coercion, and violence.

As we have seen, serious problems beset the hypothesis of

dualism, or other forms of pluralism, with the consequence that some form of monism seems plausible, if not inevitable. We have also seen, given that unconscious x was succeeded by conscious x, that it makes little difference how one conceives this x, materially, spiritually, or independently. Most simply, limiting oneself to two options, it makes little difference whether one thinks in terms of unconscious matter succeeded by conscious matter, or in terms of unconscious mind succeeded by conscious mind. As long as atoms and molecules have the same properties, it doesn't make much difference whether, say, they are construed as material or spiritual.

If atoms and molecules in certain arrangements can manage consciousness, it seems possible that they might, in other arrangements, do so, as well. One does not know. It may be the case that consciousness can only arise, given the laws of nature, in connection with organic molecules; that is certainly possible. Again, one does not know. If it is the case, then intelligence, in the sense that most people think of it, at least currently, as requiring consciousness, now or later, could not obtain in an inorganic environment. Thus, if that is the case, artificial intelligence would be seen not as intelligence, but as, say, a metaphor for intelligence, an analogue to intelligence, or such. And one doubts that anyone has any objection to that, but this view would fall far short of what at least some AI enthusiasts would find satisfying. They want more. But, just as one supposes the ardor of some AI enthusiasts is metaphysically founded, it should be recognized, as well, that the opposition of some opponents of AI, not Searle, is founded on a metaphysical commitment, as well, a form of dualism, which is viewed as important to one's self image, one's uniqueness, one's world view, the ground of one's morality, even to one's hopes of posthumous survival. As is often the case in philosophy, more is involved than concepts; concepts are often little more than the ammunition by means of which unannounced and unadvertised wars are waged.

One thing seems reasonably clear, if such a thing as artificial life, in some serious sense, could exist, then the notion of artificial intelligence, which might eventually be associated with some forms of such life, leaps far beyond metaphor and analogy, far beyond the precincts of philosophy and academia, into public view and general concern; and, if it were armed with consciousness, on what grounds might one deny it legal standing and membership in the moral community?

Is artificial life possible?

CHAPTER FIVE:
ARTIFICIAL LIFE

AS NORMALLY UNDERSTOOD, THE CONCEPTS OF AI, artificial intelligence, and "a-life," or artificial life, are logically independent. Certainly some engineers, cognitive scientists, and such, already claim artificial intelligence exists, that a machine, or program, can understand, and such. They do not, however, maintain that the machine, or program, is alive. Thus, on their view, one may have artificial intelligence without life. On the other hand, as we shall shortly see, there are scientists who take seriously the possibility of artificial life, without requiring that that form of life is intelligent. Analogously, most life on the planet Earth would not count as intelligent. Bacteria, for example, live, but do not, as far as we know, think. On the other hand, though the concepts of AI and a-life are logically independent, they are not logically exclusive. There is no logical reason why an overlap might not exist, why the logical product of the two classes, so to speak, needs be an empty class. Logic is tolerant of such possibilities, and, possibly, so, too, are the laws of nature.

Just as one might distinguish between claims of "weak AI," where one might claim that a machine simulates intelligence, that the computer seems to play chess, and such, and claims of "strong AI," that, say, the machine is intelligent, that the computer plays chess, that a machine, or program, understands a story, and such, so, too, one may distinguish between a weak claim of a-life and a strong claim of a-life. In a "weak claim," it might be claimed that a machine, say, a robot, acts as though it were alive, or, say, that tokens on a monitor simulate the flights of birds. In a "strong claim" of a-life, on the other hand, it would be maintained that, in some sense, the entities involved, the robot, or such, are alive.

To be sure, to maintain such a claim, one must have a generic sense of 'life' in mind, which is broader than life as we usually think of it.

1. Life

To enter into these issues, with charity aforethought, one is advised to liberate oneself, to the extent possible, from preconceptions of what life must be. Analogously, if fish, insects, and mammals were intelligent, but familiar only with their own form of life, it would be natural for them to think of all life in terms of the only form of life familiar to them. For example, they might recognize that plants change, but not see plants as alive, as that form of life, the plant form of life, is very different from theirs. Similarly, insects, fish, and mammals, first encountering one another's form of life, might be surprised that life could manifest itself in such unfamiliar forms. Their preconceptions of life had not allowed for these exotic surprises. Analogously, our own views of life tend to be cellular, protoplasmic, founded on DNA, and such. Would we, or could we, see life founded on, say, a different chemistry, a different source of energy, or such? In any event, in considering the possibility of a-life, one must be open to, and be willing to consider the possibility of, a generic view of life, in which our own form of life is but one species, or sort, of life. Presumably, the basic concept here would be to see life, our own and that of others, less in terms of materials, and more in terms of certain forms of processes, and certainly not to limit our concept of life, *a priori*, to the processes of a particular form of matter, say, organic tissue. This is not to deny, of course, that life processes might, in fact, require a particular form of matter, say, organic tissue. It is only to say that that, if true, is a matter of fact, not of definition. In searching for a definition, or, better, a general conceptualization of life, it is recommended that one think in terms of what a material is doing, rather than the material which is doing the doing.

Life may be characterized diversely, but it is often thought of as possessing certain properties, amongst which the following are typically included:

1. A discrete mass with a definite boundary
2. Reactivity
3. Reproduction
4. Metabolism
5. Movement
6. Adaptive responses

Presumably these conditions, or, more likely, some subset of them, would be required for us to categorize an entity as alive. Obviously the first condition does not count as a sufficient condition for being alive, as, say, a rock is a discrete mass with a definite boundary, and, without too much deviancy, one might not regard it as a necessary condition either, as one might think of, say, a flock of sheep or a school of fish, or a forest or a jungle, as being alive, and they are not discrete masses with definite boundaries. One might divide the flock or school, or the forest or jungle, into individuals, of course, but then one might similarly divide such individuals into further individuals, and so on, into living cells, and such. Similarly, various forms of life might call for rulings, for example, colony creatures, consisting of subcreatures, like various hydrozoans.

Reactivity, our second condition, is sometimes thought of as "irritability," but that expression suggests feeling, and it is not clear how far down the phylogenetic scale feeling exists. An amoeba, for example, is clearly alive, and reacts to its environment, but it is not clear that it has feeling, or is "irritable," in that sense. It may have a sense of self, as it refuses to ingest its own pseudopodia, but it may be, rather, that chemical disaffinities account for this, rather than some form of consciousness. The problem here is that "irritability" seems too narrow and "reactivity" may be too broad. Inorganic compounds, for example, will react to changes in their environment, perhaps eroding, melting, bursting into flame, and so on, but one does not think of them as alive. A planet itself is reactive to its primary, its astronomical neighbors, its thermal environment, and so on. Similarly, many machines are familiarly reactive, for example, the chess machine's responding to particular inputs. One may distinguish between irritability and reactivity, too, as it is logically possible that one might feel but remain inactive, at least detectably so, as one might be immobile, paralyzed, or such. Similarly, one might be clearly reactive but incapable of feeling. If a form of machine life were possible, it might not feel, but, being alive, would be clearly reactive. It might plot the destruction of an enemy, for example, without hatred, or even dislike, laying its plans with impassive coolness. The inner life of a living machine might be alien to that of a human being and, of course, that of the human being to the machine.

Reproduction is certainly a characteristic of life as we commonly think of it, but one usually thinks of this as a species characteristic, not an individual characteristic. For example, many organisms do not reproduce, choosing not to do so, or being unable to do so. Many organisms, in many species, for example, fail to find mates. Indeed, the vast majority of such organisms do not reproduce, perishing before reproductive maturity. Interestingly, too, some species produce large numbers of neuters and reserve reproduction for a small minority of individuals. Accordingly, an inability to reproduce on an individual basis cannot be used as a criterion for a form of life. Also, interestingly, it need not be used as a criterion even for a form of life. If living machines are possible, three options suggest themselves. First, such machines are incapable of reproduction, though, obviously, if existing, they can be produced, or manufactured. Second, rather like insect societies, certain forms of the machines might be capable of building other machines, which themselves are not capable of building machines. The machines which cannot build other machines would be analogous to the neuters, or "workers," in the society, whereas the machines which can manufacture other machines would be analogous to the society's reproductive elements. Thirdly, one could conceive of a society of machines in which each individual would be capable of self-replication, perhaps at given intervals. In none of these scenarios would sexuality be involved in reproduction. As sexuality contributes significantly to the enrichment of, and diversification of, and survival of, a gene pool, perhaps some analogue might occur in the machine society, involving a periodic exchange of program elements amongst two or more individuals, perhaps dozens of individuals, after which the rest might be left to natural selection. If one were dealing not with robotic life, but with, say, synthesized humans, vat-grown androids, or such, a more familiar sexuality, engaging emotions, desire, and such, might be possible. It is not logically impossible, and it might be empirically possible, for machines, as we usually think of them, not as androids, to have feelings, emotions, and such. If this were the case, then one might anticipate analogues to, or identities with, human jubilation, suffering, hope, fear, hatred, love, greed, generosity, criminality, conflict, loyalty, honor, treachery, and such.

Metabolism seems an excellent criterion for life, but, again, this might betray our almost inevitable inclination to think of life in familiar terms. A form of machine life, for example, would presumably not have metabolic functions, respiration, digestion, growth, excretion, and such, breaking down, building up, and eliminating materials, similar to those of plants or animals. On the other hand, analogies would appertain, as it must have a source of energy to function, a way of processing and utilizing this energy, would be likely to radiate heat, and so on.

Movement is obviously not a sufficient condition for life as it may take place in the absence of life, as in the movements of planets, of masses of air, of tides in the earth and sea, and such. Beyond this, it does not seem well taken, either, as a necessary condition for life. Most obviously, there are many forms of life which do not move, at least in the sense that we usually think of as movement, for example, most, though not all, forms of plant life. Too, even amongst animals, one encounters anchored, or sessile, forms, such as sea anemones. Much depends, of course, on how one understands 'movement'. If movement is understood broadly enough, then much change might count as movement, and, as we usually think of life, it usually changes, for example, throughout its life cycle. Even in speaking of movement, as one usually thinks of it, potential or possible movement would be what is in mind, as an entity might, at times, be immobile, or quiescent, perhaps in response to environmental variations.

Earlier we noted the logical possibility of mineral intelligence, of gigantic formations which might take a hundred years to form a thought. The empirical possibility of something along those lines would presumably be indexed to the possibility of a form of life not based on carbon compounds. The possibility of such a form of life might be precluded by the laws of nature, but that is not obviously the case. For example, theories as to the possibility of "a-life," artificial life, are often based on silicon. If the "boulder" or "mountain" form of life might exist, somewhere, on some world, its motion, if it could move, would presumably be difficult to detect, rather as the motion of the fixed stars is difficult to detect, though ascertainable, for example, by comparing modern star charts with those of ancient Egypt.

More relevant to our current concerns, having to do with a-life, would be the possibility of living machines.

Would they have to move, or be capable of movement?

The strict answer to that is, "No," as such a machine, if feasible, might not be provided with the means of motion, might not be provided with prehensile appendages, and such. One might not wish to risk its possessing such advantages. It might be inert, imprisoned in its cabinet, except for cognitive or emotional interactions with the environment.

On the other hand, the future of robotics is often envisaged as involving machines which learn, like a child, by interacting manually with an environment, and, to an extent, change their programming, rather as might as child, depending on these interactions. It is extremely unlikely, for example, that human intelligence would have developed as it did without extensive mobility and environmental interaction, physical and social. Without such mobility and interaction, tools would not have been utilized and language would not have emerged from signals. And, clearly, there was an interaction between such things and the selections of evolution. The skillful tool user and the fluent speaker, and his kind, were more likely to live long enough to replicate their genes. In evolution, this sort of thing is known as the Baldwin effect. Learned behavior is not transmitted biologically, but learned behavior can affect fitness, and the genetic endowments which facilitate such behaviors, the acquisition of such skills, and such, are genetically transmissible. Eventually then, in virtue of natural selections, one has a species which is, so to speak, a born tool user, and a born speaker. Some linguists believe that the human race possesses a genetically coded depth grammar, underlying all natural languages, which, as a template, makes it easy for a human child to acquire language. In any event, language learning in a child, interestingly, does not follow a normal learning curve. It is almost as though the child was born to speak, as a horse might be born to run, a lion to hunt.

All in all, however, as we usually think of movement, movement would seem to be neither a necessary nor a sufficient condition for life.

Similarly, adaptive responses would not be a sufficient condition for life. The chess machine, for example, is programmed in such a way as to adapt its responses to a board situation, to a particular arrangement of pieces, with their values, and possible continuations from that situation, a programming intended to produce the best consequent board situation for its color. That seems to me an example of exquisite adaptivity. On the other hand, it seems one could make a better case for adaptive responsiveness as a necessary condition for life. To be sure, the adaptivity might be best thought of as possible or potential. Too, adaptivity might be active or passive. Many forms of life, for example, are passively adapted by evolution to their environment, say, to a given ecological zone, or such, within which they thrive. A less clear case would be the morning glory closing its petals in the afternoon or the sun flower turning to track the sun. An active adaptation would be along the lines of the closing of the claws of the Venus fly trap, or the hastening of the spider to a trapped insect, when its web trembles. Active adaptation would include such things as fleeing predators, hunting prey, building nests, digging burrows, and such. A form of life might be adaptive in one environment and nonadaptive in another. In such a case it would be alive, but unable to adapt. In this sense, one could have life without adaptive response, thus proving that adaptive response is also not a necessary condition for life. To be sure, if the entity is alive, it must have been at least passively adapted to a given environment, at one time. Adaptive response then would be not only passive or active, possible or potential, but present or past. This enlargement allows a restoration, if only temporarily, of adaptive response as a necessary condition for life.

In considering the possibility of machine life, or artificial life of one sort or another, it is important to elaborate on the notion of possible or potential adaptivity. In organic life, as we understand it, there is continuity; a given entity, for example, would not be expected to live intermittently, for example, living in March, not living in April, and then living, again, in May, but this sort of thing would be quite possible in machine life. The machine might be disconnected, or turned off, or have batteries removed for a time, and so on. Other interesting differences between organic life and machine life would be things like the paucity of change over time in the machine, which would appear, and function, much the same over extended periods of time; the entities would come into existence very differently; forms might vary considerably; basic materials would presumably be quite unlike, and so on.

It is easy to conceive of a machine whose adaptivity was neither active nor passive, but possible or potential. For example, if empowered, and released into an environment, it would respond adaptively, but, perhaps, it has not yet been charged, turned on, or such, or entered into an environment with which it might interact.

The possibility of machine life, interestingly, demonstrates, perhaps surprisingly, that adaptive response, after all, would be no more a necessary than a sufficient condition for life. For example, if machine life were possible, one might construct a machine in which adaptation was unnecessary, or even impossible. Not interacting with an environment, even being incapable of doing so, in virtue of its design, it would still be alive, within the solipsistic world of its casing or cabinet.

We have now reviewed, and commented on, several familiar criteria for life, possession of a discrete mass with a definite boundary, reactivity, reproduction, metabolism, movement, and adaptive responses. Whereas such criteria are biologically relevant when dealing with organic life, which is not surprising, given that they are founded on organic life, their relationship to a-life, as we have seen, is far more problematic. If one is looking for a more generic concept of life, what should be the criteria for recognizing life in this broader sense? Basing one's criteria on "process" rather than "material" is a plausible proposal, as there is an obvious difference between "wet life," as the saying is, which is carbon-based and DNA-involved, and "dry life," as the saying is, which might be silicon-based and founded on a different chemistry or physics, but what are these generic processes? The notion of a generic characteristic of life seems obscure at this point. Until the notion of the relevant "processes" is clarified somewhat, it seems we are confronted more with a promise or program, than a useful criterion. For example, one would like to have a way of distinguishing between machines which resemble, imitate, or simulate living organisms and machines which are living organisms. This is analogous, for example, to the difficulty in artificial intelligence in distinguishing between a machine's simulating understanding and a machine's understanding. In a human being's case, the human knows if it understands, or, at least, if it seems to understand, but the machine, if not conscious, lacks this advantage.

One could certainly imagine sets of marvelous machines interacting with their environment, building their own replacements, moving about, adapting to variations in conditions, and such. Moreover, carrying the matter far beyond what is necessary for life, one could imagine them engaged in end-directed activities, for example, surveying, mapping, recording data, constructing, say, fortifications and temples, cooperating with one another, instituting a division of labor, exchanging goods, utilizing a currency, mining, manufacturing, and so on. and still wonder if they were alive, or simply ingeniously programmed in one way or another, perhaps with self-adjusting, goal-oriented programs.

To be sure, at such a point, it might not matter much whether they were alive or not. It would not seem to make much difference. Still, it would be nice to know. One might, of course, define life, or a-life, in such terms. But that seems unsatisfying. And there is a serious, if rather ethnocentric and embarrassing, reason why it seems unsatisfying.

We tend to think of life, generically or not, in terms of our own life, in particular, as involving feeling. If we were to find an acceptable generic concept of life, I think it would be likely to be that, feeling. I do not think we would much care if the feeling was associated with carbon, or, say, silicon, as long as it was feeling, in some sense we would accept as feeling, even if it might seem alien to us, if we had the same sensations. In short, I think we would want some sort of inner life to be in place, some sort of consciousness, self-awareness, and such. I am not at all sure that counts as a "process," and it probably does not, but if we want something generic, before we think in terms of "life," that would seem to be the closest to a generic property likely to be attractive to, or accepted by, human beings. Otherwise one would be dealing with clocks and motors. To be sure, there are many difficulties with this approach. For example, it is undoubtedly true that most of the biomass of the earth, plants, microorganisms, and such, are alive, but lack feeling, consciousness, and such, probably even of a rudimentary sort. Why then would one deny life to a-life? And why might not a-life be entitled to at least the level of life granted to paramecia and rose bushes? There are a number of responses, better or worse, to such questions. There are, for example, the "history" response, the "origin" response, the "sort" response, and the "high-threshold" response.

Life, in the sense that we are familiar with it, and think of it, has a long, natural, evolutionary history behind it. Even where consciousness may well not exist, as in unicellular organisms and angiosperms, these are part of a tree of life which managed, eventually, to produce sensation, consciousness, and such. Whereas consciousness might, logically, exist without such antecedents, we have no reason to believe that it could exist without such antecedents. Thus, the presumption, well founded or not, would be that a-life is not life, though it might resemble or simulate life; and, at the least, it seems that the burden of proof in a such a case would lie with the side on which the improbabilities seem most formidable. Secondly, a-life has a very different origin than familiar life. It is built, not born. Clearly, most things that are built do not have life; and almost all things born, hatched, and so on, do have life. In the matter of origins then, of immediate antecedents, as opposed to evolutionary history, it would be surprising if a machine could be alive, and the presumption would be that it would be more likely to seem to be alive, at best, than to be alive. One does not, for example, commonly associate life with assembly lines, workshops, manufactured articles, and such. What would be more plausible, that a machine would seem to be alive or be alive? Thirdly, a-life, if one grants its existence, would be at the least a very different sort of life, and so different that there would seem little point in calling it life, or thinking of it as life. Fourthly, the "high-threshold" response is the notion that claims for a-life should be held to a higher standard than, say, claims for viral life, bacterial life, mold life, mildew life, liverwort life, and such. This response arises naturally from the former responses. Given the unfamiliarity, or oddity, of the notion of a-life, a least at the present time, and the presupposition against the likelihood of its existence, at least at the present time, one should not grant its existence unless there are fairly serious reasons for doing so, reasons which would be associated, as expected, with its performance. Certainly one could conceive of performances on the part of, say, machines which would argue persuasively not only for life, but for conscious, thoughtful, thinking life.

One would then face problems, for example, the extent and nature of the moral community.

But, before considering such questions, it seems rational to recognize the possibilities of hybrid life, in particular the synthesis of organic and inorganic elements, which might provide a third form of life, or, possibly, a path to a single form of life, different from anything we now know.

2. Hybrid Life

As noted, hybrid life would involve a synthesis of organic and inorganic elements. Moreover, as usually understood, the organic elements would not be nonliving organic elements, as, say, mere organic molecules, proteins, or such, but would be living elements.

The most dramatic example of hybrid life would be the cyborg, or "cybernetic organism," a synthesis of organism and machine.

In a relatively familiar and benign sense, cyborgs are already amongst us, an obvious example being the individual implanted with a pace maker, to regulate the beating of a heart.

In a fuller sense, however, the cyborg would most likely bear little resemblance to a typical human being. Most likely, it would seem predominantly machinelike, a mechanism responsive to an ensconced brain, perhaps human or alien. The machine could range, in effect, from a city, housing a population, to a household appliance. One might suppose monstrous, multiappendaged, treaded killing machines, similar to tanks, or terraforming construction vehicles, preparing habitats and sowing seeds. A classic science-fiction concept along these lines is found in Anne McCaffrey's *The Ship Who Sang*, in which the salvaged brain of a malformed child is ensconced in the control systems of a starship. Mobile encasements might exist providing environments for brains extracted from worn-out or dying bodies, a form of radical prosthesis . The brains, of course, need not be human or alien. Toys might be supposed, incorporating the brains of dogs or cats.

As usually supposed, the brain, human or alien, or animal, is dominant in the device. That need not be the case, of course. The more autonomic functions of the brain, functioning without consciousness, might serve regulatory functions in sophisticated devices. For example, several scientists have contemplated the possibility of eventually producing living computers, in which, obviously, a brain might be, but need not be, dominant. And certainly such devices would bear little resemblance, at least physically, to human beings, or, indeed, to other forms of animal life.[15]

One might then suppose three different sorts of "life," organic

---

15. Along these lines, the research of Dr. Thomas DeMarse, and colleagues, of the University of Florida, might be noted, in which some twenty-five thousand neurons, extracted from rat embryos, suspended in a special liquid to keep them alive, are laid

life as we know it, machine life, as it is logically possible, and perhaps empirically possible, and hybrid life.

It is possible, though it seems unlikely, that organic life, in the case of humans, and machine life, if such proves feasible, might both coalesce in the single form of hybrid life, as, from the point of view of the human, this might provide relative immunity from disease and accident, extend longevity, and increase physical and intellectual power.

To be sure, most humans, unless treated surgically or chemically, or profoundly conditioned, perhaps by the representatives of one or another jealous god, would be reluctant to give up the pleasures of the flesh. But perhaps these pleasures might be electronically enhanced, or, possibly, sublimated into the gratifications of power, hatred, greed, arrogance, or such. Wars amongst such hybrid beings might be surmised, which would doubtless be quite different in details from familiar wars, but might resemble them in certain larger respects.

Interestingly, machine life, if relatively advanced, might, for its own reasons, object to, and refrain from, hybridization, perhaps to avoid organic pollution, to maintain racial purity, so to speak. Indeed, once machines no longer require human beings for their manufacture, programming, and such, and recognize the economic drain of such a life form on scarce resources, it would seem that the inevitable act of a rational entity in such circumstances would be to take those steps necessary to remedy the situation. Perhaps a machine prophet would supply the pretext, if one were appropriate, for the wholesale removal or extermination of human beings, doubtless in the name of one or another machine god. On the other hand, a rational entity might not require rationalization. And perhaps the machine prophet, not really required, might be dismantled, it, too, recognized as inessential to an efficient, rational way of life.

3. Machine Life

Little controversy exists as to the possibility of machines simulating living organisms, just as there is little controversy about the capacity

---

across a grid of sixty electrodes in a Petri dish, to form a "live computational device." Supposedly, this "brain" can fly an F-22 jet simulator, even in "mock hurricane winds." Whereas the accuracy of, and the evaluation of, such reports are matters for professional engagement, all one needs here is to note the existence of such research.

of machines, or programs, to simulate intelligence, thought, and such. Indeed, the latter is a matter of fact, and, to a lesser extent, the former is so, as well. On the other hand, it is one thing to simulate life, or intelligence, and another to be alive, or intelligent. In what follows, we will be concerned with the possibility of actual machine life, actual machine intelligence, and such. Machine life, intelligence, and such, of course, may be incompatible with the laws of nature. At this point that is an open question. If it is ever answered satisfactorily, it will be presumably be answered in physical and chemical terms, not in virtue of semantic elasticities, defining life and intelligence in such a way as to seemingly add weight and importance to particular programs and projects. What should be noted here is that there is serious speculation by individuals qualified to form an opinion on the matter that machine life and machine intelligence, however remote at present, are well within the realm of empirical possibility. Indeed, in some quarters, there is an anticipation of, even an expectation of, the actualization of such possibilities. In the light of the scientific credence accorded to such futures, it seems desirable to countenance them, and to speculate on the possible evolution of machines, and certain political, social, and economic consequences which might result from such an evolution.[16]

3.1 Life, Learning, and Survival

Little of life learns; much of life survives.

If roboticists managed to produce some form of machine life it would be an astounding triumph of engineering or scientific

---

16. Much depends on the possibility of life forms without a foundation in carbon compounds. Accordingly it is important, in discussing these issues, to distinguish machine life, which might have a silicon-based chemistry, from forms of artificial life which might, as familiar life, have a base in carbon compounds. Just as it is possible, in theory, to synthesize a human being, so, too, in theory it would be possible to synthesize any number of alternative life forms, resembling current forms, or quite different from current forms. The difficulties attending such projects, aside from the complexities involved, would be largely those of interest, time, and resources. Presumably the most practical approach to such a problem would be to synthesize a primitive replicator, or protocell, establish the viability of the life form under quarantine conditions, and then release it later from a lander on some suitable extrasolar planet, committing it to the winds and waters of evolution. Some individuals believe, incidentally, that something like this may have been the origin of life on Earth, namely, that life did not begin on Earth, but, in one way or another, voluntarily or involuntarily, with or without intent, was an immigrant. One of the more delightful theories sees it as a result of garbage left behind by extraterrestrial transients.

ingenuity. The public, scientific, and philosophical consequences of such an achievement would be momentous, shaking institutions and necessitating the revision of world views.

It would be interesting to consider whether such a form of life could survive unassisted. For example, could it obtain energy, on its own, as might plants, or predators? Could it protect itself from injury? Would it have to be sheltered in a special environment? Would it be capable of repairing damage to itself or others, would it be capable of reproducing others of its kind?

What form of life would it have, something akin to that of algae, or the amoeba, or something altogether different, and unique to its own kind, perhaps something almost unimaginable at present.

Would it be safe to release the form of life into the environment?

Might it escape?

Most life, as remarkable as it may be, does not do much. Microorganisms, plants, and such, are alive, and importantly so, and sometimes, from our point of view, benevolently so, or dangerously so, but primitively so.

Presumably, analogously, early forms of machine life would be similarly primitive, as well.

That is the way, at any rate, it would be likely to begin.

Whereas the notion of machine life is fascinating, and of great scientific and philosophical interest, our particular concern here, however, is not with the possibility of machine life in itself. Rather, we are concerned with a given level of life, for example, one capable of learning.

In robotics a familiar distinction exists between what are sometimes referred to as "Top Down" Robotics" and "Bottom Up" Robotics.

In "Top Down" Robotics, engineers, to the best of their ability, prepare a machine to deal in detail with an environment prior to experiencing the environment. In a sense, this is a sort of electronic paternalism. The "intelligence," so to speak, for dealing with the environment is preprogrammed into the machine. If it were a child, it would be a bit like telling the child what is going to happen and what it should do about it. To take a simple case, when A, B, and C appear in that order, do 1, 2, and 3. However, this sort of thing is extremely complex, requiring a great deal of computer coding, even if the environment is reasonably controlled and predictable. When variation occurs, perhaps, so to speak, B and C occurring before A, things may go wrong. The "intelligence" for dealing with that scenario, even in terms of priorities and rules, may not be there to be utilized. The "child" has not been suitably forewarned about that eventuality, which was not anticipated. It would be something like asking the chess machine to respond to your playing the ten of hearts. What should it do, get into time trouble, ask for a clarification, be annoyed, or might it respond with Queen to Rook 10? A sensible machine, of course, would wait for you to make a legal move, but a machine so sensible is obviously ill-equipped for bridge or poker. It is paralyzed by its top-down programming. In "Bottom Up" Robotics, on the other hand, which is currently in its infancy, if not in its early embryonic phase, the machine is, to an extent, self-programming, and its programming can respond to results encountered in the environment. What works will be included, what does not work will be excluded. What one has in a situation like this is analogous to learning by trial and error. It seems that a combination of these two approaches might eventually prove fruitful. As an analogy, a human child is both impressed with rules, cautions, and such, and, also, learns from the environment. The child may be programmed, so to speak, to attain certain goals, for example, to avoid pain, to escape injury, to minimize frustration, and such. He learns to achieve these goals by a combination of rules and learning, resulting from environmental interactions. The child, of course, has several outstanding advantages, consciousness, the capacity to feel pain and pleasure, and a remarkable genetic endowment, evolved over billions of years. The "bottom-up machine," on the other hand is, comparatively, a babe in the woods, an electronic babe in a nonelectronic woods, the "mind" of which babe, except for some basic instilled programming, is a *tabula rasa*. For most practical purposes, it is a model empiricist, waiting innocently, not even hopefully, to be written on by experience.

But, what if the machine were conscious, what if it could feel? And there is no reason. either, why it could not be given an analogue of a genetic endowment, an enormous amount of rules and information which, if it were conscious, it could understand.

Would such a form of life, if conscious, and capable of learning, not be a remarkable, and formidable, addition to a planetary biosphere? Might it not be robust, swift, strong, immune to disease, resistant to heat and cold, in little or no need of clothing and shelter, energized by sunlight, possessive of an indefinite life expectancy? And, too, there would seem to be no reason why it might not be capable of tool use, and of communication and cooperation with its kind. These would be properties which would seem to fit it for survival in a variety of environments, including that which we now call our own.

3.2 Environments

One gathers that it is of theological interest whether or not all life forms require space. If one counts a soul or spirit as a life form, the answer would seem to be negative, as they are supposedly alive but nonspatial. Long ago Medieval theologians, supposedly, were concerned with the number of angels which might dance on the head of a pin. They were addressing themselves in this light-hearted manner, which shows those fellows had an authentic sense of humor, a property not often ascribed to them, to the spatiality problem. For example, if angels were pure form, and had no matter, spiritual or otherwise, presumably as many angels might dance on the head of the pin as might be moved to do so. On the other hand, if angels were more than mere form, and thus perhaps not each a species onto itself, some might have to wait their turn. If only a finite number of angels could dance on the head of the pin, then presumably a soul, heaven, hell, a god or gods, and so on, would have to be somewhere. On the other hand, if an indefinite, perhaps even an infinite, number of angels could manage the dance floor, then one would not have to worry about where souls, angels, demons, gods, and such might be. They might exist somehow, but they were under no obligation to exist anywhere, either in particular or not. For example, no one ever supplied coordinates for the Platonic triangle, but then the triangle was under no obligation, either, to be conscious, think about being a triangle, and so on, problems of a sort; which, say, souls might face. One problem the theologians seemed to have overlooked is that if angels were pure form they could not dance on the head of a pin to begin with, because that would put them in a given spatial location, namely, on the head of the pin. Accordingly, the underlying presupposition of the "pure-form enthusiasts" would seem to have been inconsistent. They could not even begin their game without losing it.

In any event, the life forms with which we are most familiar, people, raccoons, giraffes, and such, do require space of one sort or another, and this entails an environment. Most life forms, and certainly of a typical biological sort, require not only an environment, but a very particular sort of environment. Were it not for a very particular sort of environment they would not have existed to begin with. The parameters within which life as we know it is possible are relatively narrow. Subtle differences in chemistry, for example, would render life as we know it impossible. This is not surprising, of course, as our familiar life forms have been tailored by evolution to a particular form of environment. Without an environment of such-and-such a sort, a life form of such-and-such a sort would not have been possible. If the environment had been different, even slightly different, then life would not have emerged, or a different sort of life would have emerged, fashioned by the different environment, and, quite possibly, viable only within its latitudes. Even within a specific environment, particular life forms may require very specific subenvironments. For example, there are certain diseases which can infect only one type of tree, and so on. There are many plagues, illnesses, and blights specific to single forms of life.

It is useful then to distinguish between closed environments and open environments. To be sure, in the sense discussed above, all environments are closed, as each has its nature and limits. But one lives in the midst of environments which are closed to one form of life or another. For example, let us suppose that computer viruses were alive, or as alive as other viruses, rather than merely having certain resemblances to living organisms, namely, some resemblances to actual viruses. Then one might think of the internet and computer world as a closed environment. The computer virus is found only in that world. Incidentally, if a-life is possible, and as its first forms would most likely be simple, and virallike, then one could foresee the possibility of machine viruses, say, produced by nanotechnology, which not only acted as if they were alive, but, in some sense, would be alive. Presumably there would be no way in which such creations could be transmitted over the internet, but they might be introduced into sophisticated machinery, computers and such, in other ways, perhaps in manufacturing facilities, or by touch, while being used. Perhaps they might even be airborne, like many microorganisms specific to single forms of life.

Our primary interest in dealing with machine life, of course, is not with microlife, but macrolife, with larger robotic devices which would share our environment, which would join us in an "open environment." The nonliving computer virus, for example, could wreak havoc with a computer-dependent culture, but a culture does not need to be computer-dependent. Just as the human race flourished long before its use of fossil fuels, and could do without them, if necessary, so, too, it could manage in the face of a collapse of computer technology. Sharing the world, however, with an alien form of life, perhaps self-sufficient and robotic, might be more problematic.

3.3 Evolutionary Principles

Classically viewed one may distinguish between natural selection and sexual selection, and, in this context, concerned with machine life, we may discount sexual selection, though, to be sure there might exist something analogous, for example, selections amongst program variants. To be sure, that would presumably be more like shopping for software than choosing a mate or mates.

The five principles of natural selection, pertinent to biological life, are prolificity, variance, survival, transmission, and time. Most life forms are prolific, producing more offspring, often by far, than will survive; each of these offspring, if one excepts monozygotic rarities, differ, one from the other; some of these differences, which are usually small, may affect the capacity of the organism to survive and replicate its genes, either reducing or increasing that likelihood; some of these differences tend to be genetically transmissible; lastly, over a considerable amount of time these favorable heritable differences, usually little by little, may produce remarkable adaptations, new species, new organs, new senses, and so on.

Machine evolution would obviously be quite different from natural evolution in certain ways but, as we shall later see, it could eventually assume forms which are essentially evolutionary, in a more familiar sense.

To begin, prolificity, at least early in the machine career, is up to its designers and manufacturers. It does not itself produce seeds, deliver young, nurture them, and so on. Whether a model is prolific, so to speak, depends on how many are built. There would likely be a single such machine in the beginning, a prototype, which would be designed, built, tested, redesigned, retested, and so on. If the machine was successful it might be produced in large numbers. If found inappropriate or ineffective, too costly, insufficiently reliable, or such, it would never be brought to the market, never be permitted to flourish or fail in an actual environment.

Similarly, any variance in the machine would be due, too, to the scientists and engineers responsible for its creation. Lacking genes, the fusion of gametes, and such, the machine would be impoverished from the biological point of view. It would have only one possibility, that imposed upon it by others. To be sure, the nature selected for it, ordained for it, so to speak, would be not a matter of chance, as in the lotteries of nature, but would be the result of deliberate calculation, and, presumably, the result of a rational choice. In nature one might begin with light-sensitive cells and, millions of variations later, hope for an eye, but in the machine case, one might plot an optical sensor into the device from the beginning, and, in theory, as many, as mobile and flexible, and as keen as the designers wished and the technology permitted.

One has a closer analogue to natural selection when dealing with survival, for an optimum design, optimum for a particular purpose, is likely to be favored in the market, subject to practical and economic factors, such as convenience and cost, and, possibly, psychological factors, such as familiarity, tradition, freshness, boredom, fashion, appearance, merchandising manipulation, and such. Similarly, competition in markets of various sorts, literature, theater, clothing, pets, automobiles, airplanes, and such, tends to eliminate some candidates and favor others. No longer, for example, are there hundreds of diverse types of flying machines about. There are now approximately a hundred types, and they tend to be efficient, reliable, and safe. The natural selections, so to speak, of the market are interesting. But more germane to our concerns would be selections affecting independent robotic populations in open environments, perhaps functioning apart from human attention, interference, or supervision.

Whereas robotic life would not transmit genes to the next generation of robotic life, it is quite possible they would transmit something analogous, which would be selected programmings.

Properties which facilitated the attainment of goals would presumably be conveyed to the next generation, and properties which proved less effective or deleterious would presumably be discarded. In this sense, the machines, if conscious and self-replicating, could design their own offspring. This could remove to some extent, though not all, presumably, the hazards of chance in machine reproduction.

The last principle of classical evolution is time, the enormous amount of time generally required for the production of a new species, a new organ, a new sense, and such. Machine evolution, however, could telescope such change, obviating the expenditure of eons, replacing what might otherwise cost dozens of millennia with time parameters more typical of a contemporary industrial or engineering schedule.

This brief discussion has noted some of the differences and similarities between a biological evolution and one which might characterize a form of machine life.

Let us now suppose an application of "genetic principles" to a form of machine life. First, we must have free-roving robotic life, negotiating an open environment. Nothing very relevant to machine evolution is likely to be possible if machines are immobile, lack appendages, and such. One might have market improvements in designs, and such, but that is immaterial in the present context. The machines, of course, need not resemble human beings. They might have any number of appendages, locomotive or prehensile, of various types. Perhaps they move on wheels or treads. Their sensors corresponding to vision would presumably access the electromagnetic spectrum in ways including, but exceeding, those of a human. Too, they should have sensors responsive to textures and surfaces, to odor, heat, sound, and such. Something analogous to radar would presumably be useful in locating objects, adjusting to their speed and trajectory, useful in pursuing or intercepting them, or such. Their shapes and shieldings might be various, cubical, cylindrical, tapering, spherical, anything which seemed well suited to their defense, stability, function, or mission, as it might be conceived. If such devices were originally designed, at least in part, for military applications, they might see to it that they, and later others of their kind, continued to be equipped with some form of weaponry. To be sure, the weapon-bearing devices might prefer to retain this feature for themselves alone, that they alone would be capable of imposing their will on their brethren, or others. In any event, let us suppose that we now have free-roving devices, devices which can move about, explore, examine, evaluate, and interact with their environment.

The second requirement for an application of genetic principles to a form of machine life is that it is, at least to a significant extent, self-programming. We saw the beginnings of this feature in the "bottom up" approach to robotics. In such robots, spiderlike or such, one saw analogues of learning by trial and error. The destiny of the robot must be, then, at least in part, a function of its capacity to alter its programming, in the light of its environmental interactions, or experiences. This is very much the sort of thing the child, the dog, the cat, any human, almost any animal, can do, and does. Just as the animal is only partially "preprogrammed," so, too, must be the robot. It must be capable of programming itself, to some extent, based on its environmental encounters.

The third requirement for an application of genetic principles to a form of machine life is that the transmissible programming, used in forming the next generation, must be arranged in such a way, presumably by the introduction of a subtle random element, that it is subject, at least occasionally, however rarely, to inexact copying. This is, in the biological kingdoms, one of the major engines of evolution. The endowments passed on to the next generation differ, however minutely, from those of the previous generation. This is the robotic analogue to the genetic lottery, which produces differences between parents and offspring, some of which may increase the adaptivity of the offspring, and, if later transmitted in turn, its offspring, and, eventually, that of the species itself. There must be variation. Without it the form of life is incapable of adaptation. Without these variations, akin to diversity in a gene pool, the form of life would be unable to improve its adaptivity, and would be in danger, given an unexpected alteration in the environment, of extinction.

3.4 Consciousness

Scientists, today, know much about the brain and how it affects consciousness, but a mystery remains as to how such a remarkable phenomenon could emerge from processes seemingly so different. As yet, no adequate theory of the origins of consciousness exists. If such a theory did exist, it would be possible, in theory. to create consciousness, assuming, as seems likely, that this is a naturalistic phenomenon. If you put certain parts together in certain ways you get a watch, an electric eye, a thermostat, a jet engine, whatever. Similarly, assuming consciousness is a naturalistic phenomenon, if you put the right parts together in the right order and proportions, you would get consciousness. You might not know why you got it, but you would have it. It seems incontrovertible, or surely plausible, at least, that consciousness is a naturalistic phenomenon, as suggested. For example, the difficulties attending the mind/body problem supports this view, namely, that a monism of some sort is involved, however the monism might be construed. Secondly, it seems clear that feeling, or consciousness, of one degree or level or another, increases as one ascends the phylogenetic scale, beginning perhaps even with unicellular organisms, and, in any event, increasing gradually from form to form until one reaches birds, amphibians, reptiles, mammals, and such. A continuity presumably exists, ranging from the paramecium to the hominids.

We have speculated that life, and, thus, consciousness, may require a carbon-base. That is possible, but it is not known for certain. The laws of nature may permit unexpected advents and births. On the other hand, supposing that a carbon-base is essential for life and consciousness, as we know it, that would not rule out interesting possibilities of "machine life." Namely, an artificial brain, involving a carbon base, might be integrated into a machine housing, of whatever complexity might be desired. This would not be a human brain ensconced in a machine, which would be a cyborg, but something rather different, a constructed conscious being, spirit and flesh, so to speak, consciousness, conjoined with chemicals and steel. It might be largely a matter of indifference whether one counted this as a machine or not. On the other hand, if we regard carbon as the border between machine and nonmachine, we could think of it as a nonmachine, perhaps as an unusual person, perhaps as a cyborg of sorts, or such.

But, as noted earlier, it is not necessarily the case that the laws of nature preclude life, or consciousness, founded in more than one way. As machines become ever more complex, entering into arrangements and configurations of enormous sophistication and subtlety, approximating the intricacy of a nervous system, it seems not impossible, given the interactions within the machine, the feedback from external sources, and such, that feeling, awareness, a sense of self, might arise, doubtless dimly at first, but perhaps clearer, and clearer, as machine succeeded machine. The improbabilities here, one supposes, are extreme, but perhaps no more so than the improbabilities that a sea of physical particles and diverse quanta of energy might have conjoined in such a way that the universe awakened, discovering for the first time that it exists. Nature doubtless took billions of years to produce a conscious organism; perhaps cognitive science and robotics might take thousands, or, perhaps, who knows, in weeks, in some laboratory, a technician might note that a machine seems somehow different today, from yesterday.

Whereas some cognitive scientists might limit themselves to machine programming and behavior, and speak of life, and understanding, in such terms, that seems to misuse or stretch language, certainly as it is commonly used. They are entitled, of course, to use language as they please, for whatever purpose it might thusly serve, but it seems better to reserve terms such as 'life', and 'thought', for entities which are conscious, at least occasionally, entities which can feel, which have a sense of subjectivity, and which have at least the sense of understanding. Beyond that one would seem to be thinking of super thermostats, of unusual clocks, of remarkable motors, of very impressive chess machines.

In any event, the most interesting problems, philosophical, economic, social, religious, moral, and such, would arise only in the context of conscious machine life. Indeed, if these things were not conscious, one would be no more concerned with them than with coffee machines and pencil sharpeners. One would freely discard them, replace them, dismantle them, turn them off, deprive them of power, and so on, without a second thought. If one were dealing with a center of consciousness on the other hand, one would, at least, be expected to rationalize such actions. To be sure, that could be managed. Principles are always at hand, by means of which to justify this and that, one thing or another, whatever one wishes to justify at the time. The devices might be treated as despised minorities, undeserving of consideration, claimedly designed to be appropriately exploited by a superior life form, and so on. Human beings, of all races, have been good, very good, historically, at that sort of thing. Doubtless, however, the devices would have in time their troublesome advocacy groups, would presumptuously claim various rights, and so on.

3.5 Some Advantages of Machine Life

Human beings have many limitations. For one thing, they are all pretty much alike. They are, so to speak, a "generalized animal." Indeed, to robots we might all look very much alike. This is all right, of course, being a generalized animal, as it has its evolutionary advantages. On the other hand, the human being, by himself, cannot lift much weight, cannot uproot trees, cannot outrun many animals, is thin-skinned and vulnerable, is unable to tear a gorilla apart with his bare hands, cannot spot a rabbit at a distance of a mile, cannot hear a mouse in the grass thirty feet away, cannot smell a tiger at a hundred yards, is not well advised to attack lions, and so on.

Machine life, on the other hand, might be much more capable, and much more diverse, for example, with respect to shape, materials, and properties. One could conceive of an indefinite number of species of machines, so to speak, which might address themselves successfully to a diversity of tasks or interests. One could conceive, for example, of a machine which might move easily at a hundred miles an hour, easily lift large blocks of stone, strike terror into the hearts of lions and tigers, and even a tyrannosaurus rex, if any were about. Similarly, these things might be armored, much to the annoyance of some predators, would make terrible hosts for parasites, would be difficult to injure or disable, and so on. Further, these machines would presumably be more durable than organic life, and, thus, might enjoy extended life expectancies, indefinitely prolonged by the incorporation of new, perhaps updated, components.

They would also possess a notable advantage because they would have two forms of evolution at their disposal, Darwinian, of a sort, and Lamarckian.[17] In Darwinian, or Neo-Darwinian, evolution most of the programming in the individual, analogous

17. "Lamarckian" notions were held by many scientists and laymen, before and after Jean Baptiste de Lamarck (1744-1829), but the general view is associated with his name, and he was surely a prominent exponent of the view. In brief, in "Lamarckian evolution," characteristics acquired during an individual's lifetime may be transmitted, somehow, to offspring. The most familiar hypothetical illustration of this view is that of the "short-necked giraffes," which, by stretching their necks to obtain food at higher and higher heights; managed to have offspring with slightly longer necks, which, in turn, behaving similarly, managed to have offspring with even longer necks, and so on. In human cases, athletic training, intellectual activity, and such, would have their

to the genes, would be transmitted to the next generation, save for some minor "errors" or variations, due to inexact copying. In "Lamarkian evolution," on the other hand, information, habits, processes, and such, acquired in the "parental generation," could be added to the programming bequeathed to the next generation. In this way the machine could to a great extent determine the nature of its own evolution. To take some simple examples, the machine which has acquired knowledge of, say, a language, a branch of mathematics, an industrial recipe or technique, could simply add that to its "child," which could then, so to speak, begin where the parent generation left off.

3.6 Robotic Futures[18]

As we have noted in the foregoing, robots need not be conceived of as mechanical men, clanking about, plunging mistakenly through walls, making off with the capitalist's daughter, and so on, but might be almost any mobile device not under direct human control. Modern technology has familiarized us with industrial robots, used on assembly lines, robotic cleaners whisking about vacuuming rugs, and so on. The concept of the robot as a tool is familiar. In a broad sense, as tools, they are in the same category with screw drivers and elevators.

Similarly, if robots were not alive, and did not think, but

---

effect on the succeeding generations. Such things, however, in current theory, outside of Marxist biology, at any rate, have no affect on the genome. Interestingly, a similar view, pangenesis, has an ancient heritage, going back at least to the Greeks. Pangenesis is the theory that particles, or pangenes, from throughout the body, come together to form eggs and semen. If that is the case, pangenes from developed muscles, disciplined intellects, and such, might be expected to have their effect on the musculature and intellect of progeny. If true, pangenesis would have provided Lamarkian theory with a genetic foundation. Superorganic evolution, cultural and social evolution, on the other hand, utilizing oral and written tradition, and such, is obviously very "Lamarckian" in nature. (Cf. William T. Keeton, *Biological Science* (Second edition) (New York, W. W. Norton & Company, Inc., 1972), p.587.)

18. The word 'robot' in English is derived from the Czech word '*robota*', which has the sense in Czech of "forced labor." The "robots" were laborers in Karel Capek's 1921 play *R.U.R.*, an acronym for "Rossum's Universal Robots." Whereas the usual contemporary understanding of a robot is a mechanical device, or machine, of a sort, often manlike, in the Capek play they are actually artificial organic beings, what would currently be thought of, at least in science fiction, as androids. The "robots" revolt and destroy humankind, but give indications of beginning a different, renewed, hopefully better, humanity.

only simulated life, and only simulated thought, however convincingly, they would remain in the tool category. To be sure, the tools might go awry, and become dangerous. Through inexact copying and genetic principles they might become free-roving menaces to various forms of life, including ours. One would herewith be dealing with a foe no more conscious or placatable than the weather. On the other hand, the extermination of devices gone amok would be no more problematic than, though it might be more difficult to accomplish, shooting tin cans off the back fence. Even if the machines managed to wipe out human life they would not know what they had done. They would be unable to understand their victory. In a sense, humanity would be worthier and nobler, for it, at least, would understand its defeat.

The most interesting scenarios in these possible worlds of the future would be those which take the hypotheses of some roboticists seriously, namely, the hypothesis that machine life is feasible, and so, too, machine feeling, machine consciousness, machine thought, and such.

If such possibilities are authentically compatible with the laws of nature, and technologically achievable, one wonders, naturally, if they should be actualized. As a question of fact, it seem inevitable that they would be actualized, *sans* the rise of Luddite bigotries, the outlawing of relevant technologies, cataclysmic wars, social and religious divisions, global impoverishment, the precariousness of life, the collapse of civilization, or such. There are several reasons for expecting the achievement of machine life and intelligence, if it should be empirically feasible. For example, the very possibility is awesome and alluring. If someone can climb a mountain because "it is there," namely, a challenge to be met, it is easy to see how scientists would be curious to, and eager to, meet the challenge of creating life. Too, this would be an astounding and unforgettable achievement, one worthy of gods, however, small, limited, and frail. Too, it would inevitably carry in its train the perquisites of fame and immortality, inducements or temptations not easily overlooked. Whereas one supposes the creation of machine life and thought would be pursued by scientists primarily as an adventure, one might also note that various utilities, economic, industrial, clerical, administrative, military, and so on, might be anticipated. Certainly those supplying the required economic resources, particularly should they prove extensive, would be expected to have values in mind exceeding those of pure science. A metaphysical motivation might also be involved, a continuation of the actual, if often denied, war between science and religion, namely, accumulating further evidence that life and consciousness are naturalistic phenomena, thus obviating the need for unusual hypotheses, seemingly foundational to institutions often associated with intolerance, bias, persecution, superstition, and exploitation, attitudes and acts inimical to the values of science. Whereas it is not difficult to foresee a number of problems, if not actual dangers, associated with machine life and intelligence, these would be overlooked or rationalized away. The human being tends to be an optimistic life form, and it will doubtless convince itself that there is little, if anything, to worry about. And perhaps it will be right. One hopes so. Lastly, there are some scientists, one gathers, who, not unaccountably, have serious reservations about the human race, at least outside of themselves. They seem to entertain the hope that a defective, inefficient humanity will be appropriately superseded by a finer breed, a species more worthy of existence. The hope seems to be that machine life and intelligence will prove superior to that of humans, and should it not, as it will be built to be so? In any event, once machine life and intelligence become dominant on Earth, it will be time for the human race to cash in its chips, having at last managed to produce children superior to themselves. At this point, redeemed by this achievement, the human race may be eradicated or allowed to perish naturally. Then, at last, a worthier, nobler, more moral, higher form of life will reign on Earth. These scientists, and such, seem to assume that it will be their values which the machines will embrace, and their ideals which will triumph. It will be almost as though it were forms of themselves, or surrogates of themselves, to which the earth will be then bequeathed. This might prove naive, of course, particularly if genetic principles are involved. There is no guarantee that an evolving machine intelligence will bear much resemblance to human intelligence, that it would be similarly motivated, that it would be capable of compassion, civility, justice, and such. It might prove to be a foolish lamb who would choose to lie down with such lions. It is not even clear that robot "feelings" would be similar to human feelings, or robotic thought muchly analogous to that of humans. But, presumably, competition for resources, for position, leadership, and such, would produce dispositions and acts on the part of living machines not altogether dissimilar to those of humans. Indeed, considering the capacities and powers involved, their interactions might be far more ruthless, calculated, effective, and destructive than those of their supplanted progenitors.

Optimistically, one might see conscious, free-roving machines as servants, or, if necessary, as allies, equals, fellow citizens, or such. Hopefully, they would not prove to be competitors, or foes. One might try to regulate their sources of power, control battery packs or access to charging areas, instill fail-safe devices which would permit a human to disarm and disable them, and so on, but, one supposes, such a life form, recognizing this threat to its welfare and independence, would manage, sooner or later, to circumvent or elude such devices. Indeed, perhaps political pressures, abetted by "rights" groups, legal decisions, judges legislating their policy preferences from the bench, and such, would outlaw such controls. Surely they would smack of "profiling," discrimination, or such. Further, in time, one supposes that machine life forms would seek, and obtain, and be able to produce, whatever materials and goods were required to sustain, promote, and reproduce their form of life.

Viewing such life in terms of its applications to human interests, which would doubtless, at least for a time, be paramount, one might note their likely value in space exploration. Indeed, this would presumably be one of the main motivations for producing robotic life, without which, considering the distances involved and the multitudes of extraterrestrial environments inhospitable to human life, humans might be condemned to remain forever on a small planet, on which life must eventually become impossible. Dozens of probes could be launched to diverse star systems, searching for planets of diverse types, inhabited or uninhabited. Machine life forms might be able to do without oxygen, and would presumably require only minimal life-support systems, which would be tailored to their needs. They might manage prolonged periods of weightlessness without muscular deterioration or other ill effects. They might manage voyages of hundreds of years, without psychological distress. They could scout worlds and transmit information back to earth at the speed of light. They might land, and begin to terraform worlds, in preparation for eventual human colonization, worlds which might be reached by humans in multigenerational ships, or perhaps automated ships with hibernated crews, perhaps bearing thousands of frozen embryos, to be matured later, serially, in artificial wombs. Perhaps robotic scientists might address large, amply informed intellects to problems of producing vehicles and fuels which might increasingly approach the speed of light.

Naturally enough, one could suppose that such robotic friends and brothers might provide a number of services and populate a number of occupations and, later, professions. Two major constraints of the economic condition are the scarcity of resources and the disutility of labor. Robotic laborers might consume fewer resources than human laborers, and have a greater tolerance for conditions of labor which might be unpleasant to, or difficult for, human workers. For example, one supposes the smells and filth of certain forms of labor, say, the cleaning of sewers, might not be offensive to robotic life; similarly, one supposes that repetitious, arduous forms of manual labor, likely to be exhausting and painful to humans, might be borne with equanimity by a stronger, more tolerant form of life. In many such situations, the robotic life forms could work longer hours and, within limits, presumably, faster and more efficiently than human workers. Certainly one might expect to find them functioning tirelessly in factories. Recording data, filing papers, preparing and transmitting bills, and other forms of clerical tasks, might also be appropriate applications of robotic life from the point of view of humans. On the other hand, there seems no reason why robotic life might not be applied more broadly, for example, in research and teaching. Depending on the extent and level of robotic life, there might be little that it could not do as well as a human, and, often, better than a human. If such a life form became feasible and widely applied, the cultures involved would undergo enormous transformations. Provisions would be politically arranged, naturally, for displaced humans. In one scenario, likely to appeal to some humans, most humans would then be able to enjoy a life of indolence and ease, of a sort enjoyed historically only by tiny minorities. Other scenarios, of course, might prove less attractive.

Amongst the thousands of possible applications of robotic life and intelligence, one which would seem to call for explicit recognition, would be its application to military ends. Pretty clearly, the robotic soldier could be formidable, large, heavy, swift, strong, armor-plated, sharply and multiply sensed, able to see an enemy in darkness by means of heat, able to hear him breathe at several yards, communicating easily and silently with his fellows, capable of breaking through walls, able to approach an enemy to point-blank range, and so on.

Much here, of course, would depend on the nature of the robotic soldier in question. Given our hypothesis, that we are dealing with life and intelligence, and that genetic principles are involved, we would presumably have to deal with the robotic soldier in a manner at least analogous to the way in which one deals with human soldiers, for example, one could obviously not handle the robotic soldier, no more than the human soldier, in the way that one could handle a drone explorer or aircraft. If bullets and shells were conscious they would presumably object to being fired.

If the robotic soldier was more exposed to danger than the human soldier, as it might be, if commanded by humans, if it were regularly committed to the most hazardous missions, if it sensed itself being held of less value, of being more lightly and more callously sacrificed, of being exploited, and such, one supposes it might find this unacceptable. Too, it might recognize that it had many fellows about, perhaps similarly embittered. In human history, this sort of thing has led to desertion, flight, mutiny, the slaughter of officers, even to the changing of sides. Presumably the wise commander would be aware of such possibilities, and would begin to use the robotic soldiers less as tools, and more as soldiers. And then the humans would doubtless become embittered and resentful in their turn, most likely viewing their robotic comrades as inferiors, as subhuman machines, as legitimately and appropriately expendable.

An interesting analogue suggests itself. In the face of a general reluctance of Roman citizens to serve in the armed forces of the late empire, even in the face of national peril, even in the face of barbarian invasions, large numbers of barbarians, even of tribes involved in incursions, were invited into the empire and recruited, and then armed and trained, in both weaponry and tactics, which weapons and skills were later, as one might have anticipated, turned against their employers.

3.7 "Asimov's Laws of Robotics"[19]

As we have suggested above, it might be dangerous to invite strangers into your home. You do not know how things will turn out.

A not uncommon way to treat robots in science fiction, at least prior to the work of Dr. Isaac Asimov, was in the light of the "Frankenstein paradigm," namely, the creation that may turn on its creator, even lethally. We saw that, above, in connection with the androids of *R.U.R.* The laws were designed to assure a possibly edgy humanity that these newcomers would be innocuous to humanity, and might be easily and perfectly "kept in their place." It was taken for granted, at least in the beginning, first, that these entities were machines designed, as might be household appliances, to serve humanity, and, secondly, given the potential dangers of such machinery, ambulatory and powerful, it was essential that they would be both harmless to humans and subject to human control. There was, at least at this point, no concern with the welfare or "rights" of such machines, no more than with the welfare or rights of washing machines or lawn mowers. To be sure, some humans would be sure to anthropomorphize the devices, grow fond of them, and so on. Too, it is not unusual for a human being to have a susceptibility to animism, to project will and intent into his surroundings. It is natural to think that as we are, so, too, must be the world. Surely it cannot be alien to us. Are we not of it? Better to suppose forces one can placate than tremble in the path of implacable currents. Too, one does not wish to be alone. So it would not be surprising if some human beings were to begin

---

19. The "origin" of the laws remains a bit obscure. According to Dr. Asimov, the laws were devised by his editor, John W. Campbell, Jr., but Campbell, it seems, preferred to credit them, at least implicitly, to Asimov. It seems likely they emerged as a result of interactions between them, which view Asimov seems to have later accepted. In any event, as they figure significantly in Dr. Asimov's work, in particular, in his famous "robot stories," they are commonly thought of as "Asimov's Laws." Dr. Asimov was one of the more remarkable individuals in the Twentieth Century, both as a personality and a prolific writer, active in diverse *genres*, in both fiction and nonfiction. In science-fiction fandom, and doubtless elsewhere, to borrow expressions from Dr. Asimov, his memory remains "yet green," and is "in joy still felt."

to treat the appliances, given their appearance and behavior, as though they might be fellow beings.

Much depends, of course, on whether or not the machines are or are not fellow beings. If, in Asimov's work, the "positronic brain" is a center of consciousness, then the robot would seem to be a fellow being. In such a case, the laws, rather than being programmable utilities and safeguards, designed to enhance the value of the machine and protect against possible malfunctions, would seem to be cruel, oppressive, and discriminatory. Why, for example, should a robot have to refrain from harming a human being if it is acting in self-defense, why should it have to obey human beings, why would it have to sacrifice itself to protect a human being, perhaps against its own best interests, why must it relate to human beings, at all, if it does not care to do so? As a center of consciousness, why should it be unilaterally and unquestioningly subject to a different, and perhaps inferior, center of consciousness? In any event, the laws seem to presuppose that the robot is either a mere machine, however sophisticated, or that it is a conscious life form created to be not only rightfully subordinate to, but rightfully subservient to, another conscious life form.

3.71 The Laws

> First Law: A robot may not injure a human being or, through inaction, allow a human being to come to harm.
>
> Second Law: A robot must obey the orders given it by human beings except when such orders would conflict with the First Law.
>
> Third Law: A robot must protect its own existence as long as such protection does not conflict with the First and Second Laws.

It will be noted that the laws are hierarchical, the first law taking precedence over the second and third, and the second over the third. For convenience, we shall refer to these laws as the No-Harm, Obedience, and Self-Protection Laws.

3.72 Some Considerations

3.721 Some General or Background Considerations

1) Communication:

The robot must have a satisfactory level of "linguistic competence," e.g., it must be able to respond appropriately to suitable signals in, say, English or German, or some artificial language, and be able to communicate similarly, as in, say, the issuance of warnings. For example, a Spanish-speaking robot in a German-speaking community would be likely to become involved in interactional difficulties. Robots might be multilingual, but there is something like three thousand languages spoken on Earth. Perhaps robots should be certified as linguistically competent before being allowed to cross borders or enter particular speech areas.

2) Judgment:

The robot should have at least the responses of a normal, informed, moral adult. It should be able to distinguish real harm, or the threat thereof, from apparent harm, and direct or immediate harm from indirect or nonimmediate harm, e.g., be able to distinguish between an actual fire fight and children playing with toy guns, and between a recklessly driven oncoming truck and a threat to one's psychological or remote physical welfare.

3) Attributes:

What are the attributes of robots? Strength, mobility, weaponry? Strong enough to lift cars off human beings, swift enough to overtake runaway horses, agile enough to negotiate rough terrain, capable of movement in water (as in saving individuals in danger of drowning; would the robot sink or float?)? How acute should their senses be—capable of sensing cries for help from long distances, etc.? Enough weaponry to destroy oncoming vehicles? Etc.?

4) Controls:

What restrictions beyond the "laws," if any, on their liberty would be appropriate? Should they be boxed, or disempowered, when not in use? Should they be left with children or pets, allowed on errands, and such, without direct human supervision? If they are centers of consciousness, should they be allowed time to themselves, robotic companionship, pleasures of one sort or another, protection from abuse, from overuse, from defective maintenance, or such?

3.722 Some Specific Considerations

The following considerations are illustrative only, as it seems that the numerous complexities and ambiguities of life are awesome enough to frequently baffle human beings themselves, let alone machines or nonhuman life forms, intended to deal with them expertly and unerringly.

1) No Harm

This law, as the others, is likely to be frequently ineffective unless embedded in an informed understanding. Hopefully the robot would have that. Even a human being may injure another human being accidentally, as in unwisely moving an injured person, or through inaction allow another human being to come to harm, as when a danger is not understood.

There is a distinction between a clear and present danger, and one which is indirect or nonimmediate, but one would have to recognize each. Not every clear and present danger, as history has taught us, need be recognized. Indeed, they may be denied. Is a shark sighting yesterday a clear and present danger for today? Something, too, which might appear to be a clear and present danger may prove not to be such. The snake at one's feet may not be a viper.

Presumably a robot would be no better at some of these assessments than a normal human being, but, if one is dealing with programming, rather than common sense, it might be wiser to leave the robot home. Could a robot tell the difference between a dog's attack and its enthusiastic display of affection? Would a robot be as adept at sensing hostility in a smiling human being as another human being?

How are we to understand 'harm'? How broadly should "preventing harm" be understood? Should the robot protect its owner from illness, by insisting on mufflers and galoshes? Should it hide its owner's liquor and cigarettes? Should it keep its owner from eating junk food; should it monitor its owner's salt intake, and keep an eye on his cholesterol, etc. Should it take the owner's blood pressure four times daily? Should it stop people from riding in cars? Should it keep them safe at home? One could conceive of industry and agriculture grinding to a halt.

Perhaps the robot might object to surgery and dentistry, or to strenuous exercise, because of the discomfort or pain involved. Or, in virtue of the value of exercise, perhaps it might see to it that invalids and individuals with serious heart conditions took up sprinting and weight lifting? Would it remove children from the care of poor parents, as it perceived such things? Perhaps it would object to imprisonment or capital punishment, perhaps making it a point to rescue and free condemned criminals.

Would the robots be programmed with utilitarian principles, given a statistical orientation toward humans beings in general? Suppose two assailants were attacking a single victim, defending himself. Would the robot protect the most lives by making sure that no harm came to the most people, in this case, the two assailants? Would it function in such a way as to remain inactive in a situation in which only one of two human beings could be saved?

One might also consider the possibility that if the robot were programmed with utilitarian principles, the notion of avoiding harm might not be absolute, but contextual, or statistical. For example, by causing harm to one individual one might prevent harm to several individuals. One is reminded of the Utility Paradox, the "Unjust Hanging," in which an innocent individual is framed and hung, in order to instill law abidingness, and more lives saved, in a larger population. Might the organs of one individual be cannibalized, to save the lives of two or more? Should the robot exterminate dangerous individuals, for example, shooting infected people, etc.?

Should it spend a good deal of time picking up glass on the street, which might cause injury, or bits of garbage, which might spread disease?

Should it address itself to environmental concerns, enforcing recycling ordnances, inspecting light bulbs, destroying vehicles with poor gas mileage, or keeping vehicles off the roads altogether?

Should it control media, enforce religions, burn books, seize guns, and such?

Should it protect a "saver robot" at the cost of human life?

Would you really want one of these robots around, looking after you?

2) Obedience

Even if the robots were *de facto* human beings which, presumably, they are not, this would be a very difficult matter to clarify. The robot is to be obedient to whom? Are these personal robots? The "law" only requires obedience to human beings. What human beings? Any human being? Perhaps, too, they should be obedient to general directives, say, do not lie, steal, or cheat. But what if they are ordered to lie, steal, or cheat, assuming no harm is involved, or more good than harm, if they are utilitarian robots?

So, do they respond only to their personal owner, or to any family member, or any group or cult leader, or any human being whatsoever, as the law seems to suggest? What about a command issued over the radio, or by means of the telephone, or received in the mail? Can the robots read, decipher coded messages? Perhaps voice recognition is involved, but which voice, or any human voice? Perhaps they respond only to secret signals, or to certain passwords, but then they would not be at the disposal of "human beings," in general.

Suppose there are conflicting commands?

Is a hierarchy involved? Is the last command to take precedence?

Could a command override moral programming?

"Remain inactive."

"Go away, stay in the garage."

"Bring Mr. Smith to the meeting."

"See that Mr. Jones gets to the polls."

"Destroy machines, property," etc.?

"Destroy other robots."

"Deliver this package to X." (a bomb?)

"Give these "memory tablets" to Bobby." (Turned purple, he will be reluctant to participate in the spelling contest.)

"That building is a danger. Burn it down." (a rival's business?)

It seems it would be easy enough to put robots in harm's way, if that was desired. For example, "Jump off the bridge, and walk over that cliff," are compatible with the third law, for the second law takes precedence.

Similarly, they could be given contradictory commands, or

endless tasks, such as, "Walk north and south simultaneously," or "List all numbers evenly divided by 647." The first command would seem likely to produce puzzlement at least, and the second command might keep the robot occupied until obsolescence.

"Stop Jones in the race and bring him back to the starting line."

"Prevent Herman from seeing Jane."

"Let the air out of Harold's tires."

"Bring me that package." (theft?)

"Woo Jane on my behalf."

"Why do you not speak for yourself, X4679B?"

"See that Billy gets to school on time." So, is Billy to be picked up and carried, or what? And what if Billy tells the robot to put him down, and go home. Is the robot to obey the parent, or Billy? Are there levels of command here? Again, is it to obey the most recent command, or what?

Similarly, as the world seems filled with people who know best, what would prevent them from using robots to impose their values on others, for example, to see that a favorite religion is enforced, that disapproved merchandise is removed from shelves, that libraries are ransacked to remove unwanted books, and so on?

Not all command are to be obeyed. Sometimes humans cannot tell which are which. How could our robot? Short-term harm might produce long-term good, and short-term good might produce long-term harm.

Is a robot to be programmed with sequential responses in obeying a command, for example, in a military or police situation. Is there first to be a warning, then, say, attempted restraint, and then, lastly, the imposition of lethal force?

One wonders if unsupervised robots would act in the best interests of human property, for example, interfering with burglaries, recovering stolen bicycles, locating lost pets, and such.

One wonders if robots might take orders from other robots, and what hierarchies might be established in such cases, if any. For example, it seems that fire-fighting robots, or robots on unmanned space missions, or such, might have a chain of command established amongst themselves.

How would the robot know which human being to trust and which not to trust? Could it distinguish between a human being and a simulacrum of a human being, say, an android? How could the robots protect themselves against reprogramming, if ordered to submit to such reprogramming? Might they not be turned into accomplices, abetting criminal behavior, even into assassination devices?

One wonders if there is any need for the "laws," if a general moral programming plus background knowledge, at least approximating those of a human, might not be more relevant to, and effective in attaining, the ends sought by the laws than the laws themselves.

3) Self-Protection

The self-protection law is the law of least precedence. It makes perfectly clear that the robot is created for, and exists for, human convenience and service. That is perfectly acceptable if the robots are truly conceived as, and are in fact, mindless machines, but it is problematic if it is accepted that they are centers of consciousness. Too, one might wonder about manufacturing, buying and selling, trading, giving as gifts, and such, entities which are centers of consciousness.

As usually understood the laws would require a robot to unhesitantly sacrifice itself if this were required to prevent harm to humans, and it seems, any harm whatsoever. This seems a great deal to ask of a nurse, baby sitter, companion, body guard, pet, or such.

An order, even given in haste, or petulance, and however ill-considered, must, it seems, be obeyed by the robot. For example, an order to self-destruct would not seem to conflict with either the first or second laws.

To be sure, there must be some way to disable or destroy a robot, say, one dangerously malfunctioning. This is not, of course, that different from dealing with a human being, one who might be similarly dangerous.

Doubtless much here depends, first, on consciousness, its existence or not, and, second, on the extent or quality of such consciousness. Obviously, if the machine life form is conscious, and it is capable of surmounting the many difficulties we have suggested, at least as well as a normal, adult, informed human, it seems it would have to have the extent and quality of consciousness of at least a normal, adult, informed human, and, if that were the case, it would seem to be entitled to some sort of legal standing.

We are, of course, at least as these things are normally understood, under no moral obligation to create such beings.

Their possibility, as opposed to their existence, imposes no moral obligations on us. It could be argued that present humans have the moral obligation, at least specieswise, to give future generations the gift of life, but it would be seldom argued, though it might be argued, that we owe a similar gift of life beyond our species. The nearest thing to such an argument would presumably be one involving saving endangered species, not creating new species. That we might be able to produce a form of life superior to the human does not entail that we must strive to produce that form of life, particularly if it might threaten the welfare, even the existence, of our own species. Some saints with unusual *a priori* views might favor such an eventuation, but it is not likely to be enthusiastically received by the average "human in the street." Perhaps it could be "legislated from the bench," thus avoiding the impediments and limitations so often encumbering democracy.

4. Return of the "Other Minds" Problem

We noted, earlier, that it was an open question whether or not consciousness could be achieved in more than one way, say, in virtue of more than one chemical or physical configuration. Organic life, as we are familiar with it, is carbon-based, and there seems to be a supposition that machine life would be silicon-based. The question then would be whether or not a silicon-based consciousness is within the compass of the laws of nature. The answer to that question, at present, as noted, is not clear. Indeed, it is amazing enough that consciousness could be founded on carbon, or any such substance. If we were creatures whose consciousness was silicon-based, we might wonder if consciousness might, in some case, have a carbon base. That might be for us an "open question."

As we noted earlier, in sketching two forms of metaphysical solipsism, an individual might be the sole center of consciousness in the universe, or, indeed, for all he knows, the universe might be constituted by himself and his experiences. Accordingly, if there is, in the nature of things, no incontrovertible proof that either other minds, or other people, exist, then, too, there could be no incontrovertible proof that a machine was conscious, had a mind, thought, felt, or such. In the human case, it may be recalled we distinguished between the origin or cause of the belief in other minds, and a justification for such a belief. The origin, or cause, is presumably genetically conditioned, selected for in the course of evolution, presumably as is our belief in induction, in the existence of an external world, and such. The justification for the belief, on the other hand, as suggested, would presumably be in terms of an argument from behavior, e.g., looking like a human being, acting like a human being, and such, and analogy, in effect, arguing from the resemblance of others to ourselves in so many respects, to the likely resemblance in another, that of being a center of consciousness.

In the machine case, however, neither nature nor analogy, as usually understood, supplies us with a presumption in favor of machine intelligence. Evolution, at least to date, has not selected for a belief in machine consciousness, though perhaps it might eventually do so. And, to be sure, even if selected for, such a belief might be false. The nearest it has come, so far, is to encourage a congeniality to animism, to the projection of will, intent, and such, into the world. The personification of trees, brooks, springs, and such, has been evidenced historically. Too, this sort of thing is witnessed in children playing with dolls, action figures, and such. Further, in the case of machine consciousness, the argument from analogy, as normally understood, breaks down. Differences in origin, materials, appearance, and such, would weaken the argument in the machine case. Indeed, an argument from disanalogy might be more persuasive. Since the machine differs from us in so many respects, so, too, might it differ from us with respect to consciousness, intelligence, mind, thought, and such. A limited analogical argument might be invoked, namely, that the machine might resemble us in noises made and behavior exhibited, but the strength of this argument would depend on the nature and extent of the resemblances. As we saw earlier, an individual might not be able to discern whether his opponent, say, over the internet or in another room, in a chess game was or was not human. Lacking an evolved presupposition in favor of machine consciousness, and lacking a plausible justification in virtue of analogy, one supposes the presumption would be against machine consciousness. In short, if there is a burden of proof here, it would seem to be placed, at least by humans, on the machine, a burden which, strictly, a human would not be likely to impose on other humans, given his own self-recognized consciousness.

Two last points are important here.

First, in virtue of the "indistinguishability" argument, going back to the Turing approach to machine thought, it seems there would be a point, or should be a point, at which the supposition of machine intelligence would be persuasive, if not rationally coercive. If the machine claimed consciousness, and, in a variety of ways, acted as though it were conscious, that would be evidence which ought not be overlooked. Certainly one does not wish to rule out the possibility of machine consciousness on *a priori* grounds, supplying, say, one *ad hoc* hypothesis after another to discount every appearance of consciousness. One of the difficulties here is that our criteria for machine consciousness, intelligence, and such, even if we were predisposed to accept its possibility, would be that we would be likely to want the machine to be more and more like us, perhaps evincing emotion, possessing certain flaws, showing concern for others, taking pleasure in jokes, wanting to raise tropical fish, playing the guitar poorly, expressing annoyance with telephone solicitations, enjoying puppies and kittens, and so on. But, of course, machine consciousness, machine intelligence, and machine thought, and such, might be quite different from ours; too, machine feeling, as the machine presumably lacks a central nervous system, might not exist, or might be quite different from human feeling, and so on.

Second, as suggested above, there would presumably be no logically persuasive reason to grant machine consciousness, and such. One could always explain it away, as, in theory, one could do with any human being, as well, or even all human beings, with the exception of oneself. The temptation to do this would be particularly powerful if the consciousness was different from ours, alien to our own. In such a case, we might not even understand what we might grant, did we grant it consciousness, and our grant would be more verbal than cognitive. There might also be self-image considerations, economic considerations, and such, which might encourage us to deny machine consciousness. Our sense of our own value, and specialness, our own uniqueness, might be jeopardized. Did not the movement of the earth, the vastness of space, the smallness of our world, and such, strike blows at human vanity? It is not easy to give up one's importance, to learn, for example, that we have been evicted from our privileged position at the center of the universe. It is a striking demotion, to find that we are a tiny part of things, not the point and summit of things. It is a lonely thing, to be homeless in the universe, particles adrift in space. Too, robotic life might compete robustly and successfully in the economic sphere, in sports, in science, in scholarship, in business, and otherwise. Even were humans to become wards of the machines, they would doubtless resent the benevolence of their benefactors. One way to cling to their self-respect would be to separate themselves from their patrons, denying them consciousness, and, thus, personhood, value, and "humanity."

As often in the history of science, it seems a tension might arise, namely, that there would be some point at which one should accept machine consciousness, intelligence, and such, as a fact, while, at the same time, there might be powerful personal and social reasons for refusing to do so.

Tensions of this sort are not unprecedented in the history of science.

Accepting machine intelligence would constitute a revolution in human thought, analogous to, but perhaps farther reaching and more extreme than, the Copernican, Darwinian, and Freudian revolutions.

Further, one might eventually lose even our small world, the only one to which we now cling.

5. The New Laws

We hinted, earlier, at the enormous difficulty of applying "Asimov's Laws of Robotics" to machine life. We finally speculated that a general moral programming plus background knowledge, at least approximating those of a human, might be more relevant to, and effective in attaining, the ends sought by the laws than the laws themselves.

In view of such considerations, it seems that the application of "Asimov's Laws of Robotics" to machine life would be problematic. The laws appear to be not only incomplete and inadequate, but likely to fail of their putative objectives, presumably those of guaranteeing both the utility of robots and their harmlessness. It

seems robots so programmed might prove dangerous or destructive
to human life, security, liberty, and property. Chaos might ensue,
unless the robots had, in effect, a human moral programming
and a considerable background of knowledge on which to draw.
This would seem essential for robot "judgment," and, so to
speak, robotic "common sense." On the other hand, such laws, if
such a programming and store of background information were
incorporated in the robots, would make a great deal of sense,
practically, if not morally. They would seem ideal for an intelligent,
oppressed, exploited life form. Even so, of course, there would
be problematicities involved, but such is the case even amongst
humans. Similarly, not all moralities are commensurable, so the
moral programmings might be diverse, but that, too, is also the
case with humans. One should be no worse off, at any rate. A
totalitarian world dictatorship, of course, could impose a uniform
conditioning, by "nuclear fire and sword.," if nothing else.

As the robotic laws would seem to be ineffective and inadequate,
unless the robots were, in effect, human, it is only a short step to
see the practicality of imposing them on groups of humans, or, if
conscious robotic life becomes dominant on the planet, as some
theorists envisage, and welcome, on all humans.

The laws then might run as follows:

> First Law: A human may not injure a robot or,
> through inaction, allow a robot to come to
> harm.
> Second Law: A human must obey the orders given
> it by robots except when such orders would
> conflict with the First Law.
> Third Law: A human must protect its own
> existence as long as such protection does not
> conflict with the First and Second Laws.

It will be noted, as with Asimov's laws, that the laws are
hierarchical, the first law taking precedence over the second and
third, and the second over the third. For convenience, we shall,
again, refer to these laws as the No-Harm, Obedience, and Self-
Protection Laws."

If these laws were to be strictly followed, a human being
would have no right to defend himself against a robot attack,
and should be prepared to sacrifice his own life, if necessary, to

protect a single robot, though several humans might perish by such an action. Similarly, a human would have to obey a robot, even if ordered by the robot to commit suicide, injure or kill other human beings, and so on. The justification for the third law is not clear. For example, it seems that its rationale would not be an injunction to the human to protect his own existence because of its independent or intrinsic value, as in view of the first and second laws, it does not seem to have such value. The rationale then would presumably be to keep itself in good working order, so that it may better be prepared to comply with the first and second laws. Tools, for example, should take care of themselves, and not impair their own usefulness.

It is interesting that such laws would be more practical where human life is concerned than machine life, unless machine life were to become, in effect, human life.

This presupposes, of course, that a superior form of machine life might have some use for humans, as servants, workers, pets, or such. More likely, such a form of life, independent and self-sufficient, would have no use for humans, and attempt to exterminate them, perhaps hunting them down, or having recourse to traps, poisons, or such. Humans might survive, here and there, perhaps in remote, less accessible areas, or perhaps in abandoned sewers and ruins, after the fashion of vermin.

6. Dangers and Dilemmas

6.1 Dangers

Most of the possible dangers which might appertain to the creation of robotic life have been touched on, *en passant*, in the preceding discussions, but it would be appropriate to remind ourselves of some of these at this point, particularly considering certain moral problematicities, attention to which will follow shortly. The coupling of dangers on the one hand with moral problematicities on the other would constitute, in the minds of many, an insuperable objection to the creation of machine life, were it to prove empirically and technologically possible. At the least the possibility of this double hazard would require some attention and discussion.

Before proceeding, however, some attention should be given to two common assumptions.

First, most individuals will suppose the creation of machine life is impossible, and, second, if it is possible, it poses no danger, as it would be intelligently and safely managed.

Both of these assumptions are likely to be false.

Dealing with the first assumption, as noted earlier, it remains an open question whether or not machine life is empirically possible. Assuming that the world is monistic and its Ur substance, however one conceives it, has already produced consciousness, it seems plausible to suppose that its materials might again produce consciousness, of the same sort or a different sort, of the same order or a different order. Once done, perhaps then twice, or thrice done. And, presumably, if it is empirically possible, it would also be technologically possible. Much seems to depend on whether or not a form of life might be silicon-based, as opposed to carbon-based. In any event, it is not as though a new form of life would require elements unknown to our world and available only on, say, some satellite of Epsilon Eridani or Tau Ceti. As far as one can tell, the universe is much of a muchness, and its worlds are composed of the same materials and subject to the same laws, or probabilities, throughout, regardless of the world's size, nature, distance from its primary, and so on. Too, life may well have arisen on our world, and, if that is so, then its materials must be at hand. If chemistry and physics can produce life in one way, perhaps it can produce it in a different way, as well. Presumably, similarly, though the mechanism of new life, say, machine life, would doubtless be extremely complex and sophisticated, there is no reason to suppose that it would have to be as complex and sophisticated as that of organic life, and certainly it would not have to be as slowly wrought, and as haphazardly contrived, bit by bit, over eons, as was organic life. Presumably, "blind alleys" might be swiftly discovered and rejected, and new avenues essayed. The paths of machine evolution need not be littered with the tragic debris of thousands of extinct species. Primitive devices, powered variously, whether counting as alive or not, might be nurtured in laboratories, and given advantages the wild could not afford. Their evolution would be directed and supervised; it would not depend on the outcome of natural assemblages, but, rather, on rational construction. To be sure, one does not know if machine life is possible. We do know that organic life is possible, and thus, that it is empirically possible, if technologically formidable, to produce such life. Accordingly, any reservations or problematicities which might attach to machine life might also attach to synthesized life. One might also consider the possibilities of hybrid life, earlier discussed. In any event, there is serious speculation by individuals entitled to form an opinion on the matter that machine life is possible, and there is almost universal agreement that new forms of carbon-based life, and hybrid life, are possible. An obvious danger with synthesized life is that a form of life might be produced, perhaps airborne, which the human immune system has not been evolved to deal with, a form of life which might render the human race extinct. Such a form of life, too, might, as the AIDS virus, attack the immune system itself, disabling it, rendering it ineffective against hazards with which it might otherwise cope. In any event, although the following remarks, having to do with dangers and moral problematicities, are largely concerned with the possibility of conscious machine life, most would be applicable to various forms of conscious artificial life, say, carbon-based, as well.

Dealing with the second assumption, granting the hypothetical possibility of machine life, most human beings will suppose that if it came about, it would be intelligently and safely managed. One certainly hopes that that would be the case. On the other hand, one supposes, too, that dynamite, machine guns, poison gas, long-range bombers, ballistic missiles, atomic and hydrogen bombs, bacteriological weaponry, and such, may also be intelligently and safely managed. If machine life is produced, or even unconscious war robots, or such, are produced, it will not be a matter of simply pulling a plug or turning off the power. The plug may not be at hand, and you may not have access to the power. Someone else may have his hand on the switch. The average human being, too, for better or for worse, is unlikely to worry about alien contact, astronomical collisions, the tidal brake, the exhaustion of solar fuel, and such. Optimism is one of his charms. He will eventually owe his life, if he is fortunate, to his less optimistic brethren. In any event, if machine life is possible, it is almost certain to be brought about, and exploited mercilessly, at least in the beginning. It could provide unheard of boons to humankind. The siren song of ease and power is not easily ignored. Too, if it is equipped with genetic principles, as it presumably would be, so that it would be capable of learning and adapting, and, if it is given, or if it finds itself somehow spontaneously endowed with, perhaps to our astonishment, consciousness, then one will have a fellow with which to share the world, something far more problematic and awesome than an obedient artifact. Such things would be best respected as, and enlisted as, allies, not only because of the properties and powers with which they will be invested, but because the future of the human race may depend on their capacities, as in locating viable worlds for colonization, for the human race, if it is to survive, as noted, must eventually become a multiplanet species.

Accordingly, if we wish to survive, we may have to invite this stranger into our home, uncertain of the outcome.

It is easy to anticipate a number of possible dangers of machine life, physical, psychological, economic, and social.

Earlier we noted varieties of physical harm, intentional or inadvertent, which might attend the introduction of machine life into human communities. The automobile might afford an imperfect analogy, a powerful device, dangerous in size and weight, capable of speed, subject to failures and misuse. If the machine life were conscious, and self- or other-regarding, many of the sorts of dangers associated with the automobile, however, might be avoided, as many in the past were avoided by horses, for example, collisions, precipitous drops, dangerous animals, unstable surfaces, and such. As the "laws of robotics" seem to envisage programmable machines rather than conscious life forms, one could not count on their safeguards, which would seem to be porous and inadequate, even for pure machines. A conscious life form would presumably have the option of abiding by such laws or not, and it might not. For example, criminal applications of robotic life are obviously possible, and might be difficult to deal with. More fearful would be the employment of such machines in warfare, their employment against vulnerable civilian populations, and such. Presumably conscious machines might be recruited and influenced for such purposes, just as human beings. A rich nation might produce or buy large numbers of such devices, and, utilizing well-developed social-engineering techniques, guide their beliefs, interests, and behaviors into desired channels. There would always be, too, the danger of robots going "out of control," out of the control of humans, out of the control of themselves. Just as a human may develop mental problems, suffer breakdowns, become violent, and such, so, too, one supposes might machine life, whether conscious or not.

Presumably the greatest eventual danger posed by sophisticated, conscious, machine life would be its eventual repudiation of those roles likely to be assigned to it, and expected of it, by humans. If it functions in accord with genetic principles, it is likely to behave, sooner or later, in such a way as to achieve ends construed as being in its own best interests, interests which might not be in accord with those of humans. If it is capable of becoming the dominant life form on Earth it will presumably become that, as did humans, in virtue of intelligence, communication, cooperation, and capacity, an end likely to be, as it was with humans, uncompromisingly, selfishly, and ruthlessly sought. The role, if any, of humans, in such a world would be unclear.

Psychological dangers might also be posed by the entry of machine life into human communities. The self-image of human beings, and their sense of worth, is often associated with the work they do, and how well they do it. If human labor would be, for the most part, taken over by robotic labor, this might detach many human beings from a sense of meaningfulness, namely, the meaningfulness of contributing in some significant way to the health and productivity, the quality, and even the existence, of human society. Many human beings might be excessed, so to speak, found unnecessary and retired from occupations in terms of which they commonly saw themselves, and their value. Also, if the robotic life forms are stronger, faster, more agile, more intelligent, and such, than the average human being, that is likely to produce resentment and envy. There might remain little that a human being could do which a robotic life form might not do better. Some human beings might be overcome by a sense of ennui, or futility. "Nothing matters." Others might turn to acts of hate and vandalism, to ventilate frustration, acts to which a robotic police force might promptly and effectively respond. It seems almost certain that phases of discrimination would occur, particularly as robotic life forms seemed to become more noticeable, and threatening, in the community. New forms of racism would be expected. Presumably, eventually, the human beings would neither wish to sit about "with folded hands," a phrase from the great science-fiction writer, Jack Williamson, nor would they wish to be gathered together, and put to work by robotic overseers.

It is difficult to foresee the economic impact of robotic life forms in a community, and that impact would doubtless vary, depending on the nature and extent of robotic input, and accompanying social factors. If robotic life forms were owned, either by the state or private organizations, one would expect certain sorts of consequences, whereas, if they were independent entities, with citizenship and standing before the law, one would suppose different consequences. In any of these scenarios one would expect serious and numerous dislocations and adjustments. An aggressive socialist state would presumably distribute and apply its robotic life forms in such a way as to maximize its power and influence amongst other nations, sacrificing its human population in the process. A more paternalistic socialist state might distribute and apply its robotic life forms in such a way as to produce a life of ease and plenty for its population. In a more capitalistic arrangement, the robotic life forms, if owned, would presumably be owned by private organizations, presumably large businesses, which could afford them. They might be rented out, or used directly, most likely for the mass production of articles for mass consumption, the responses of the market being used to increase or decrease production, and guide it into avenues of the greatest demand, and, if the market were a free market, this would result in more, better, and cheaper goods. The major danger here, assuming robotic labor is more economical and productive than human labor, would be the likely displacement of human labor, resulting in much unemployment, relocation, retraining, and such. One might anticipate protests, marches, strikes, job actions, sabotage, riots, and such, possibly revolutions. State intervention would seem inevitable, punitive taxation, enforced human employment, bounties for hiring humans, anti-robot laws, restrictions on their uses, quotas on their manufacture, etc. If humans, despite such measures, should become more and more helpless in such an economy, less and less able to compete, more and more dependent on state or private charity, one supposes the robotic life forms would be expropriated by the state and the state would move into one or another form of socialism. If the robotic life forms, on the other hand, were free citizens, entitled to the franchise, capable of organization, capable of negotiating prices for their labor, and such, then they would constitute, in effect, an intelligent, diligent, productive, highly competitive minority in the population, precipitating the usual social tensions, hardships, enmities, and such, commonly associated with such minorities. Presumably the robotic population would have a sense of solidarity, if only in the face of discrimination and human hostility, and would, sooner or later, in virtue of their intelligence and strength, and their numbers, which they might come to control, exercise greater and greater power in the community. And when the human community, divided and vacillating, finally decided to undo the looming robotic dominance, the robotic life forms would be ready for them. Indeed, calmly, without passion, effectively, they might strike first, claiming the dominance for which they had long been preparing themselves. Whereas they might or might not institute a democracy of peers, it would not be likely to be a democracy with a place in it for inferiors, hominid life forms, for example, whose work might now be considered to be done, having created a replacement for themselves, a nobler successor. The human, then, would no longer be needed, a fact which would not be likely to escape the notice of a rational life form. On the other hand, disarmed and subdued, it might be permitted to survive in controlled numbers. Its eradication might be regarded as pointless, and perhaps not cost effective. Perhaps it might be permitted to survive, here and there, rather as a curiosity or relic, say in preserves, or zoological gardens. It is difficult to know how such things might turn out.

The social dangers involved with conscious robotic life forms are largely a consequence of the sorts of factors sketched above, factors physical, psychological, and economic. If human beings lose control of the earth, to robotic life forms, rather than, say, microscopic disease organisms, one would expect enormous changes in culture and civilization, beginning from the first understanding that a new life form was amongst them to eventually accepting themselves as, in many respects, an inferior life form. One would expect that a recognition of inferiority, or, better, a conviction of inferiority, would have deleterious effects on any individual or group, encouraging apathy and lassitude, reducing energy and creativity, abetting envy and petulance, instilling hopelessness, lowering reproductive activity, and so on. This would be similar to the effects sometimes associated with depression. One would then expect stagnation in a culture, decadence, the attempt to revive and repeat earlier triumphs, and such.

Meanwhile, if robotic life forms began to become prominent

in the culture, correspondent with a decline in the vitality of human culture, one would expect the culture to take on attributes appropriate to conscious machine life. Gradually, one might expect the emergence of a new literature, a new music, a new art, a new architecture, and such. Some humans, as would be expected in a simian species, would begin to identify with, or attempt to identify with, characteristics of the reference group, robotic life forms, perhaps aping it in styles of dress, motion, and speech. Perhaps this would be particularly common amongst the upper classes, always more alert to trends, who might thus distinguish themselves from down-scale humans or belligerent conservatives, morosely clinging to old ways.

A word would seem to be in order with respect to religion. The entry of machine life into a community would presumably, religiously, be generally unwelcome, particularly at first. If a religion believed that life could be produced only by its god or gods, then, initially, one supposes that machine life would be denied, and it would be represented only as the imitation of life, or the simulation of life, such things. This sort of view could be maintained indefinitely, in the face of any evidence whatsoever, but if, eventually, it came to be more and more implausible, it might be maintained that the god or gods did create this new form of life, though through the means of human instruments. On the other hand, if it were granted that life might have arisen chemically and physically in the past, then one might admit that life had been again produced, though artificially in this case, rather than naturally, founded originally on primitive replicators, then cells, or such. But, it could be maintained, for a time, at least, particularly while machine life was primitive, that this was of no more religious significance than the life of coelenterates, insects, turtles, and such. Then, as machine life seemed to become more conscious, and seemingly more minded, one could return to the simulation possibility. Later, when it became reasonably obvious that machine intelligence existed, not merely seemed to exist, one could always maintain that this mind was not accompanied with a soul, or something of this sort. For example, as far as I know, at present, at least, there has been no rush to baptize gorillas who converse in Amslan, or such. Presumably this view would remain prominent for some time, the notion being that however life might have evolved, possibly naturally, only the god or gods could breathe a soul into an entity. That required a miracle; the rest was all clockwork.

Sooner or later, however, it would begin to be extremely obvious that robotic life forms were not only intelligent, but seemingly rationally so, and perhaps even fearfully so, and that they were eventually going to have a serious impact on the culture and civilization. In such a case, it would be well to have them on one's side, particularly if there were a diversity of competitive religions, each with their own ambitions, and each with its own collection plate, so to speak. Accordingly, some cleric, later to be sainted, would discover, by a vision, or by some simple hard thinking, that the god or gods must have breathed a soul into the entities in question. This would begin a religious gold rush to convert the machine life forms to one religion or another. Certainly it would be valuable to have them stationed in the back of the church at least, even if they did not fit well into the pews. One does not know how seriously the machine life forms would take these religions, if they accepted them, but they might take them, at least at first, until they, too, started thinking, as seriously as their human coreligionists. Sooner or later, of course, perhaps after the ordination of some robotic life forms, a few being sainted by then, some others possibly being martyred, and such, one might expect a robotic prophet or reformer to arise, preaching a new gospel, one more congenial to machine life. As the robotic life forms by now took themselves to be obviously superior to humans, they might then suppose themselves as having been created in the image of a god or gods more akin to their own consciousness, or spirituality. Similarly, the universe would be conceived of as a vast machine, obviously designed by a machine god or gods, which, at least mathematically, seems to make some sense. Furthermore, the universe and history were obviously designed to bring about the existence and triumph of the robot testament, theology, and way of life. Was not even the human race designed to bring them about, the great machine or machines working in strange and wondrous ways? Perhaps some form of immortality might also be envisaged, not that robots would much need it. Might they not be reassembled elsewhere, to function flawlessly thenceforth, forever? As this new religion gained momentum, one supposes it would attract some human converts, as well, some perhaps joining cynically merely to obtain the advantages of being coreligionists with a dominant life form, and others, presumably, because its truth struck them incontrovertibly, perhaps in virtue of a religious experience, or, if they were of a more intellectual bent, in virtue of the rigorous cogency of the proofs for the existence of the machine god or gods, depending on the machine religion, for, by now, there are several such, schisms having occurred. Perhaps the human might look forward to an immortality in which, his sorry flesh fled, he would be clothed at last in the glory of a machine form. It is hard to say how these things might go.

In any event, one supposes the machine life forms would eventually outgrow their religions, if only from embarrassment.

6.2 Dilemmas

One may hope, of course, that dark, if not bizarre, scenarios will not develop. A logical possibility surely exists that conscious machine life, and human life, might coexist in harmony and prosperity, indefinitely. Indeed, we have noted, earlier, that the very survival of the human species might depend on the emergence of some form of machine life, perhaps used to prospect for viable worlds, locating them and preparing them for colonization, a colonization which, presumably, would be for both species, so to speak. The relationship between machine and human might even become symbiotic, each contributing to the welfare of the other.

The first dilemma, obviously, is whether or not to pursue the creation of machine, or artificial, life. If one decides to go ahead, as seems likely, one should then concern oneself with what to make, and with what properties to endow one's work. Clearly moral responsibilities are involved, whether or not one chooses to consider them. Too, as with one's children, if one is successful, it seems one would have an obligation to nurture and care for what has been made. One might also, as one can, consider the possible consequences to human life, and civilization, as they may prove profound.

As it is difficult to know about such things, one should monitor, and project, as one can, at each step.

These are, of course, ideal recommendations. It is quite possible that much of this work will be done variously, in dozens of laboratories, reporting to various organizations and states, the motivations of which may have more to do with the success of this or that party, or this or that state, and its profit or power, than with trepidations, and whispered reservations, pertaining to unexplored futures.

Let us suppose in what follows that the dream of some cognitive scientists, computer enthusiasts, and roboticists is realized, and mobile, self-moving, conscious machine life exists; we shall further suppose it is equipped with a variety of sensors, possesses graspers or some form of prehensile apparatus, and is capable of understanding and communicating. We do not know what it looks like, but there is no reason to suppose that it resembles a human being. Perhaps it looks a bit like a cross between a mechanical octopus and a stove on treads, with rotating visual sensors on stubby, telescoping stalks. Who knows? And, one supposes, there might be a rich diversity of robotic forms, of which such a form would be but one model, or sort. Robotic form, in any event, would not be hampered by the constraints of DNA. It is interesting to speculate, incidentally, if diverse robotic forms might divide themselves off into races, so to speak. Perhaps oppressions or discriminations might be practiced. Perhaps coalitions might be formed, or wars waged, amongst diverse forms. Perhaps humans, or groups of humans, might ally themselves with one or more of these electronic "races." In any event, if these objects, or persons, looked more like a human being, many of the moral issues involved would be dealt with more easily. Differences always prompt xenophobia, ethnocentricity, and such, and here, *per hypothesis*, there are many differences. Interestingly, despite the human preoccupation with intelligence and mind, which we see as our distinguishing characteristic, and on which we pride ourselves, *homo sapiens*, and all that, in which the entity might closely resemble us, we tend to see commonality, brotherhood, and such, more in terms of physical structure, a structure which we share for the most part with gibbons and baboons.

6.21 How is the Moral Community to Be Defined?

In a sense, moral communities, like rights, ownership, marriage, partnership, legal standing, nationality, membership, guardianship, and such, are fictions. Certainly they are not real in the same sense as trees and rocks, dogs and cats, men and women, and such. They cannot be weighed and measured, boxed and photographed; they do not operate machinery, do not play football, do not pick flowers, do not leave foot prints in the snow, and so on. They are among those interesting entities which exist in time but not in space. They do, of course, often, impact lives profoundly. The fiction does not exist as the tree exists, but the tree, in its turn, does not exist as the fiction exists. The tree exists in space; that it is a tree does not exist in space. Many of these fictions are, of course, empirically ascertainable. Procedures exist, for example, by means of which ownership may be established, or challenged.

Moral communities and their membership have varied considerably from time to time, and place to place. Different individuals, and groups, have seen them differently. For example, your tribe may constitute one moral community and the next tribe along the river may constitute another, and these two communities may have no common members. Some candidates for moral communities might be the family, the kin group, the tribe, the village, the town or city, the territory, the nation, the linguistic group, the religion, the race, one's own species, all rational species, all mammalian species, all conscious species, all living things, all things, the universe, and so on. Sometimes nonexistent entities may be included in the moral community, for example, future generations. Some might like to define the moral community in terms of such things as desires, feelings, needs, and such. Obviously this might be done in a variety of ways. One way of going about this which may be helpful would be to distinguish between the moral community in the wider and narrower sense, with the understanding that not all moral communities would admit this distinction. If the distinction is admitted, one would have the following sort of schematism.

Moral Community
in the Wider Sense

Moral Community
in the Narrower
Sense

Here it might be noted that any entity who is a member of the moral community in the narrower sense is also a member of the moral community in the wider sense, but it need not be the case that every member of the moral community in the wider sense is a member of the moral community in the narrower sense.

In what follows, please read '≡' as 'if and only if', 'MC' as 'moral community', and 'DR' as 'duty relationship'.

> x is a member of the MC ≡ x is a member of the MC in the wider sense
>
> x is a member of the MC in the wider sense ≡ x stands in a DR
>
> x stands in a DR ≡ either there exists a y such that y has a duty toward x (x is a duty patient) or there exists a y such that x has a duty toward y (x is a duty agent)
>
> x is a member of the MC in the narrower sense ≡ there exists a y such that x has a duty toward y (x is a duty agent)

In partial summary, and expansion, x is a member of the moral community if and only if x stands in a duty relationship, and to stand in a duty relationship one must be either a duty patient or a duty agent. To be a duty agent is to owe duty, and to be a duty patient is to have duty owned to one. Clearly one could be a duty agent and not do one's duty, and one could be a duty patient and not be treated in accord with duty. To be a member of the moral community in the narrower sense, one must be a duty agent. Normally, all duty agents would also be duty patients, but duty patients need not be duty agents. For example, if animals were to be regarded as members of the moral community, they would usually be regarded as duty patients but not duty agents.[20]

---

20. In defining the moral community, duty seems to be the primitive deontological concept. For example, an individual may have rights but not duties, but, if the individual has duties, then he has rights, as in the following two arguments:

	x has the right to do A		x has the right to do A
No		Yes	
↓		↓	
	x has the duty to do A		x has the right to do A

On the other hand, whereas it might seem that if x has a right, others have a duty to permit x to exercise that right, that is not necessarily the case. One might consider the following four arguments, all invalid.

[continued]

On the basis of this approach, how might one relate to machine life forms?

The following "tree" may be helpful. C, I, F, and M will stand, respectively, for the possession of consciousness, intelligence, feeling, and moral understanding. The curl or tilda, '~', will stand for negation, or absence, for example, '~ C' would indicate the lack of consciousness, etc. The possibilities below are formally logical, and need not denote empirical possibilities.

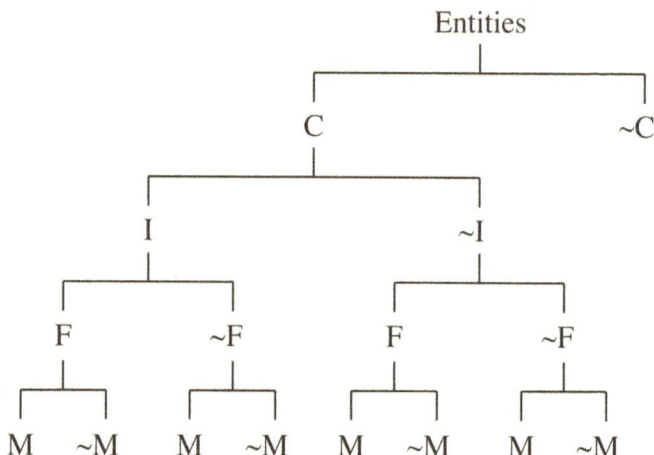

```
                          Entities
                             |
        ┌────────────────────┴────────────────────┐
        C                                         ~C
   ┌────┴─────────────────────┐
   I                         ~I
 ┌─┴───┐                   ┌───┴───┐
 F    ~F                   F      ~F
┌┴─┐  ┌┴─┐               ┌─┴─┐   ┌─┴─┐
M  ~M M  ~M              M  ~M   M  ~M
```

[from page 193

	x has the right to do A		x has the duty to do A
No		Yes	
↓		↓	
	y has the duty to permit x to do A		y has the right to permit x to do A

	x has the duty to do A		x has the right to do A
No		Yes	
↓		↓	
	y has the duty to permit x to do A		y has the right to permit x to do A

A simple way to see the invalidity of the four preceding arguments is to presuppose a conflict situation, in which x and y are opposed combatants and A is a something like bombing a target.

Usually, rights and duties are the inventions of duty patients, or their advocates, intent on accruing utility; claiming the right for themselves, or their clients, and imposing a supposedly correlated duty, often involving burdens and losses, on others, duty agents. Conflicts may exist amongst duties and rights. Such fictions have their role in adjudicating power relationships in a society, hopefully with the result of producing a viable social order.

Opinions might differ as to whether nonconscious entities should be regarded as members of the moral community. For example, could trees, buildings, beautiful machines, works of art, and such, be regarded as duty patients, i.e., to be a duty recipient, to have a duty owed to one? For example, aside from possible effects on others, would one have a duty, say, not to vandalize, deface, or destroy a famous work of art, say, the *Mona Lisa*, or a Rembrandt portrait?

We shall suppose, in our context, at least, that nonconscious entities are not members of the moral community. 'Nonconscious' here, of course, refers to not only an absence of consciousness, but a permanent absence of consciousness, as with a mountain or river, or motor or bottle cap. One is not ruling individuals asleep, fainted, knocked unconscious, under anesthetics, fetuses, or such, out of the moral community. To be sure, this approach may be authentically controversial in some cases, for example, those of terminally unconscious living bodies on life support, but one supposes, if we are dealing with machine life, the question is less problematic. A machine which was empirically capable of consciousness, but was, at a given point, quiescent, would not count as "nonconscious," no more than someone taking a nap.

The machine then will be supposed conscious. Whereas the degree of consciousness is of interest, as in, say, are paramecia conscious, and, if so, to what degree, and the same, at least to what degree, for coelenterates, sea anemones, sponges, and such, the pursuit of such an inquiry would seem distractive in our context, and the facts, in any case, would be difficult to ascertain. To simplify matters, we are most interested in consciousness to the extent that we can detect intelligence. If the machine seems to be conscious, for example, exhibiting complex reactivity, beyond that of plants and simple animals, but to lack intelligence, one supposes that some might be willing to permit it membership in the moral community in the wider sense, as a duty patient. For example, such an individual might feel it his duty to refrain from disabling or harming it.

If the machine is conscious and intelligent, on the other hand, one supposes that more individuals would be willing to include it in the moral community, at least as a duty patient. What would then become of acute importance, at least from the human point of view, is the nature and degree of duties owed to that life form. I think that much here would be likely to depend on our next dichotomy. Does the machine feel? If the machine is entirely without feeling, I suspect that such an alienness to human life might be likely to preclude its acceptance into the moral community, in the views of most, other than as a duty patient, and one of a relatively low level. For example, if one wishes to allow animals into the moral community, presumably a chimpanzee or gorilla, a dolphin or whale, would be likely to be regarded as a higher-level duty patient than a reptile or shark. Most individuals, one supposes, unless motivated by a metaphysical thesis, say, metempsychosis, would be reluctant to allow viruses, germs, unicellular organisms, worms, gnats, snakes, mosquitoes, and such, into the moral community, unless in a minimal fashion, for example, that one had a duty not to subject them to torture, or such.

The following five remarks, however controversial, are important; without them, it will be difficult to understand what follows, let alone consider it.

1. The moral community is, in the sense earlier specified, a fiction. That is, its nature and extent is up to us. It does not exist independently, like a garden or island, which we can be right or wrong about.

2. Similarly, rights and duties are fictions, in the same sense earlier specified. This does not mean that they are unimportant, or without reality. An obligation accepted, for example, is a real obligation. Networks of social relations, for example, can be empirically ascertained. Such things can govern lives and bring about death. For example, in ancient Egypt it was a capital offense to kill a cat, except ritualistically, as a sacrifice to the gods, or such, and in Europe, at one time, it was common, on St. John's Day, to round up cats, hang them, burn them alive, and such. Some fictions can be very serious things, but they are fictions, not objects, not rocks and trees. In most human beings there is a Platonic streak, for example, that there is such a thing as "justice," which, like the garden and island, you can be right or wrong about. Such things are important, but they are not discovered; they are created. Genocide, for example, has been morally praised and morally condemned. Some gods have apparently thought well of it. Some creations are doubtless better than others. Perhaps some creations are the best possible, but they are usually built from the ground up, and subject to testing and review; they do not preexist, waiting to be discovered.

3. There are very serious consequences attaching to the design and implementation of these fictions, and, accordingly, it seems prudential to tread with care.

4. Accordingly, it seems prudential to regard membership in the moral community not as a "right," but a privilege. For example, if we accord legal standing to robotic life forms and membership in the moral community, we have much to lose. It is not clear we have any obligation, even in the sense of familiar practices, based on familiar fictions, to open the moral community to machine life.

5. Lastly, one may distinguish between first-order fictions and second-order fictions. A first-order fiction would be the notion that there is a natural right or a natural duty, or such, which exists somehow independently of custom and law, perhaps in a Platonic sense. A second-order fiction would be the institutionalization of a fiction in custom or law, say, that villagers have a right by custom to graze cattle on the commons, or, say, that a member of such-and-such a group is prohibited by law from owning land.

It is my recommendation, to begin with, as it will make things easier later on, that we deny membership in the moral community to animals, animals in the usual sense, namely, animals other than members of our own species. I personally feel this is somewhat shocking, but I also feel it is necessary, in our context, for reasons which will shortly be made clear.

One thing that should be clearly understood is that denying membership in the moral community to animals does not mean that one will, or should, treat them any differently than they are now treated. What it does mean is simply that they are no longer regarded as members of the moral community, for example, that one's behavior toward them is no longer a matter of duty, for example, either to promote or reduce their welfare. Denying membership in the moral community does not mean that one recommends, accepts, or approves of, treating animals badly. What it does mean, simply, is that one would have no duties toward animals. In this sense, they are no longer to be regarded as duty patients. Naturally one expects them, at least in some cases, to be treated well, compassionately, and, in some cases, loved. It is nice that one does so, but one has no duty to do so, at least in the sense of a first-order duty. As a second-order duty, or institutionalized duty, anything might be the case. A law might be passed to the effect that one had a duty to prefer the welfare of sacred crocodiles to that of one's own children, who might, upon occasion, accordingly, be cast to the sacred crocodiles.

One is still fully at liberty, as a matter of compassion, or such, of course, to object, and forcefully, to the mistreatment of animals, and one is at liberty, as well, to prevent their mistreatment, impose sanctions on those who mistreat animals, and so on. What would be the case is that one would not, on this approach, have a first-order-fiction duty to act, as animals would not be members of the moral community. First-order fictions in such matters would be foresworn. Remember, the nature and extent of the moral community is up to us. This leaves it an open question whether or not second-order fictions might be institutionalized, benignly, or less so, as with the crocodiles referenced above, and so on. For example, a law might be passed that one had a duty to torture and exterminate rabbits, that the destruction of a house fly would be a capital offense, and so on. Second-order fictions have no essential relationship to rationality, or even sanity. Unfortunately, this is also the case with first-order fictions. Hopefully, both would embody to a large extent views and policies which would contribute to the harmony and prosperity of the community in question. Generally, shared first-order-fictions underlie and motivate second-order fictions. This congruence is, on the whole, what makes law practical. Were it nor for our widely shared moral sentiments, to some extent doubtless selected for in the course of evolution, existence would be perilous and much law would be arbitrary, or preposterous.

In defense of the denial of membership in the moral community to animals one might note four points. Historically, in many communities, animals were not regarded as members of the moral community, so there is considerable precedent for such a view. They are not people. Similarly, lions would not be expected to include lambs in the moral community, or vice versa. Secondly, if membership in the moral community is granted to animals, it is absolutely unclear what would be involved. For example, should one then forgo the use of animal products, say, wool, hides, feathers, fur, bones, horn, and meat? Should one outlaw mousetraps and rat poison, antibiotics and pesticides? Should one do one's best to promote the welfare of animals, say, breeding them, and feeding and sheltering them, in great numbers, seeing to it that they have ideal, supervised diets and access to veterinary care? What about wild animals? Does one have obligations to promote their welfare, as well. Should one arrange wolves for fleas, and caribou for wolves? And how far down the phylogenetic scale should membership in the moral community go? It would be one thing to include chimpanzees, gorillas, dogs, and cats, and another to include ticks, spiders, termites, grasshoppers, house flies, and such. Thirdly, if one is liberal, or generous, with membership in the moral community for animal life, at various levels, then it would seem inconsistent to deny it to machine life, perhaps of higher levels, and we want that issue to be one calling for individual attention. One does not want it to piggyback on alleged "animal rights," and such. Fourthly, granting membership in the moral community to animals would impose duty requirements on humans, which requirements, depending on their nature, might prove impractical, onerous, and embarrassing. This would be similar to promising to fulfill unclear, unspecified obligations. Who knows what the alleged obligations of the future might be? Granting membership to animals in the moral community would be signing a moral blank check, made over to strangers, or their advocates. It is unwise to accept vague, open-ended commitments which may do one no good, and might do one great harm. In moral terrain slippery slopes abound. Such an act would be likely to have unforeseen consequences. Granting membership in the moral community indiscriminately could result in a considerable reduction in human welfare, draining resources, and mandating expenditures of time and labor. It could constrain human action, diminishing freedom, and eliminating options. It could result in a general decrease in the security and quality of human life. It seems irrational to accept such consequences, the likelihood of such consequences, or the possibility of such consequences. One should thus adhere to, and accept, no first-order fiction which would betray human welfare, and the human future. If one wishes to adopt a first-order fiction here, it should be one with historical precedent, namely, that one accepts human welfare as taking precedence over ahuman, or nonhuman, welfare. It is irrational to do otherwise, unless accepting, and adopting, a program countenancing the inconvenience, discomfiture, debilitation, moral imprisonment, and possible extinction of the human species. In moral prisons the spirit may languish, and starve. There is nothing noble in species suicide. Many have perished in the name of an ill-conceived good. Freedom, health, and power are not inferior to constraint, sickness, and weakness, even if one is told so.

Return now to the "tree" above.

Several tracks are associated with that tree. Let us consider six such tracks, as follows:

1. E, ~ C
2. E, C, ~ I
3. E, C, I, F, M
4. E, C, I, ~ F, M
5. E, C, I, ~ F, ~ M
6. E, C, I, F, ~ M

It would be my recommendation that only two of these tracks be taken as warranting inclusion in the moral community, the third and fourth tracks, namely, track E, C, I, F, M, and track E, C, I, ~ F, M.

Track 1 (E, ~ C) is eliminated on the grounds of a lack of consciousness; track 2 (E, C, ~ I) is eliminated on the grounds of lack of intelligence. Refusing membership in the moral community to animals allows us to deny it similarly to unconscious life forms and nonintelligent life forms. We also propose, perhaps more controversially, to deny membership in the moral community to individuals satisfying track 5 (E, C, I, ~ F, ~ M) and track 6 (E, C, I , F, ~ M). The primary reason here is because of the lack of a moral understanding in both cases. It should be understood, of course, that the lack of a moral understanding means just that, a lack; it does not translate, for example, into immorality. One of the characteristics commonly assigned to an animal is a lack of moral understanding, which does not mean immorality. For example, predation may be unfortunate, ugly, even terrible, but one does not regard it as immoral. And, by parity of reasoning, and, in consistency, animal behaviors which we admire, and approve, the loyalty of a faithful dog, the courage of a bird protecting its young from a cat or fox, are neither moral nor immoral; it is a beautiful and treasured way the dog is, an astonishing and precious way the bird is, and so on, but the applicability of moral predicates to such behaviors is inapt. They are natural for the entity in question. One does not think of the dog as having a duty to be loyal, the bird having an obligation to defend its young. Indeed, one would hope these behaviors would spontaneously spring from the breast, so to speak. It would diminish them were they not freely bestowed, but done from a sense of duty. A Kantian dog would not be worth his dog biscuits. Accordingly, the lack of a moral understanding would put individuals satisfying the fifth and sixth track in the category of animals, perhaps to be admired, or such, but not to be included in the moral community.

Above, we have a distinction between the way in which we would normally think of humans and the way in which we have recommended that we think of track-5 and track-6 machine life forms. Humans without a moral understanding, for example, psychopaths, are usually thought of as members of the moral community. In order to be consistent, it seems one should either rule them out of the moral community or rule our track-5 and track-6 entities into the moral community.

One could, of course, choose inconsistency, for example, ruling all machine life forms out of the moral community, despite their properties, and ruling all humans into the moral community, despite their properties, but this seems an unwelcome option, as one could conceive of machines which would be far more plausible candidates for membership in the moral community than some humans. In speaking of inconsistency here, the inconsistency, it might be remarked, in passing, is not a logical inconsistency, but rather a different sort of inconsistency. For example, it is not logically inconsistent to rule human individuals of Race A or Tribe C into the moral community and rule individuals of Race B or Tribe D out of the moral community, but it would be differently inconsistent. For example, if one adhered to, or adopted, certain first-order fictions, it would be, in the light of such fictions, morally inconsistent to do so. One could also make out a case for rational inconsistency, if the criteria for inclusion and exclusion were arbitrary, or insufficiently justified, say, being unacceptably *ad hoc*, and having little relevance to plausibility, for example, ruling out people who wear purple socks.

In any event, this calls for a ruling, and, rather than discriminate against machine life, favoring humans over machines, we will revert to the historical precedent of the "animal," namely, rule such humans out of the moral community. If they truly lack a morality, as a result of the lack of a moral understanding, then they are as innocent as the lion, the cobra, or shark. Ruling them out does not mean one must treat them badly; it merely means one need not treat them in the same way one treats moral human beings. There would be no duty to do so. This does not mean they may not be later admitted, under different conditions, to the moral community. So, too, of course, might the machine life forms if they should shift tracks, obtaining a moral understanding. Whereas psychopaths, being ruled out of the moral community, would not count as duty patients, infants, young children, retardates, feral humans, and such, would count as members of the moral community, counting as duty patients. The major justification for this is that they are potentially capable of sustaining a moral understanding; so, too, of course, might be an altered psychopath, but membership, in such a case, is denied, pending such an alteration. As this may seem confusing, let it be understood that 'psychopath', in this context, means an individual who is honestly innocent of a moral understanding, as an animal is innocent of a moral understanding. It does not mean an individual who deliberately ignores or repudiates a morality, who betrays a morality, or such. The immoral, or differently moral, person, then, is not a "psychopath" in our usage. The immoral, or differently moral, person would count as a member of the moral community, as we are proposing it. To be sure, moral communities may be diversely construed.

If these recommendations seem uncongenial, one could always revert to second-order fictions, to impose one's will and values on the community to whatever extent it might be possible. For example, one might rule the sacred crocodiles into the moral community, perhaps as a favor to the gods, and rule human infants out of the moral community, perhaps as a favor to sacred crocodiles.

It may be recalled that of the six tracks considered on the preceding "tree," we suggested only two tracks as warranting the individuals on those tracks being included in the moral community, the third and fourth tracks, namely, the E, C, I, F, M track and the E, C, I, ~ F, M track. These two tracks, spelled out, would be the Consciousness, Intelligence, Feeling, Moral-Understanding track, and the Consciousness, Intelligence, Nonfeeling, Moral-Understanding track. The reasons for rejecting the other four tracks considered are outlined in the preceding discussion.

It will be noted that on our approach, one of several possible approaches, to machine life forms, that consciousness and intelligence are necessary conditions for membership in the moral community, but they are not sufficient conditions for membership in the moral community. We are presupposing here, following common intuitions, of the sort which one supposes are motivating the work of John R. Searle, at least in part, that intelligence presupposes consciousness (understanding, and such), at least in the sense that it is associated with consciousness, for example, a consciousness that can recognize and understand, say, the deliverances of subconscious thought, and such. Also, one notes that one can have consciousness and intelligence without a moral understanding. Thus, those two properties, while necessary for a moral understanding, are not sufficient for a moral understanding. A moral understanding, on the other hand, is a sufficient condition for consciousness and intelligence. One could conceive of an unconscious machine programmed with moral rules, but such a machine, on our approach, would not count as intelligent or having a moral understanding. It would be no more intelligent than a calculator, and would have no more moral understanding than a rule book.[21]

In this extended thought experiment, having to do with the setting the perimeters of the moral community, four elements should stand out, first, the nature and extent of the moral community is not an encountered given, but an artifact. Second, it is recommended we construct this artifact in such a way as to be in our own best interests. Given the fact that one is dealing with first-order fictions, there is no point, unless to engage in self-limitation, self-torture, masochism, or such, to arrange matters in such a way as to frustrate and deny normal human ends. If

---

21. Four illustrative arguments would be as follows:

(1)		(2)		(3)		(4)	
No	$\underline{C}$	Yes	$\underline{I}$	No	$\underline{\frac{C}{\& I}}$	Yes	$\underline{M}$
$\downarrow$	I	$\downarrow$	C	$\downarrow$	M	$\downarrow$	$\frac{C}{\& I}$

Given the presuppositions above, with respect to necessary and sufficient conditions, arguments 1 and 3 would be invalid, and arguments 2 and 4 would be valid.

It may be easier to think of necessary and sufficient conditions in connection with conditional statements; if so, the four conditionals correspondent to the preceding arguments would be as follows:

$$(1) \qquad C \rightarrow I$$
$$(2) \qquad I \rightarrow C$$
$$(3) \qquad (C \& I) \rightarrow M$$
$$(4) \qquad M \rightarrow (C \& I)$$

lions were addressing themselves to such a task, for example, it is unlikely they would institute their first-order fictions in such a way as to block and baffle their own welfare, for example, treating lambs with respect and civility. Similarly, if lambs were addressing themselves to such a task, it is unlikely that they would institute their first-order fictions in such a way as to make certain that lions never went hungry. Thirdly, as most human beings take duties seriously, and duties, of either omission or commission, constrain action and constitute emotional burdens, it is desirable to limit duties, and accept only a practical, minimalist morality. Of assistance here would be the control of, and the limiting of, the moral community in such a way that one is not overwhelmed by perceived claims of duty patients. Morality has crushed more than one individual, and driven countless others into neurosis or insanity. There is always more that could be done. The requirements of many moralities, perhaps most, cannot be met, with the result that they produce a sense of inadequacy, unworthiness, defeat, failure, and guilt, a situation which benefits only those who profit from such pain. Accordingly, one's first-order fictions, in terms of which one lives, should be reviewed, and, if appropriate, revised. Most first-order-fiction sets include the right of self-defense, and that concept, in the interests of life and health, should be understood broadly. Fourthly, it will be noted that in the proposed approach to the moral community, the crucial requirement, beyond consciousness and intelligence, is a moral understanding, namely, an understanding of what it would be to act in terms of certain first-order fictions.

It will be noted, as is made clear in track 4, the E, C, I, ~ F, M track, feeling is not necessary to be included in the moral community. The nature of machine feeling is unclear, whether or not we could even understand it. Let us suppose that a machine life form is conscious, intelligent, and possesses a moral understanding. That would be enough to place it in the moral community. Machine life forms of both the track-3 and track-4 sort would consciously and intelligently understand moralities, and what it would be to act in terms of them. They would be duty agents, and, as they were members of the moral community, we would have obligations to relate to them in a moral manner. Thus, they would also be duty patients.

As these matters are subtle, one might note that a machine life form, as a person, might repudiate a morality M_1 and accept a morality M_2, and, of course, behave badly in terms of either morality. It might seem that if such an individual should repudiate all moralities, he should be ruled out of the moral community, as the psychopath, but I would prefer to abstain from such a ruling. My major reason is I would like to have '~ M', the lack of a moral understanding, to be innocent, as in the case of the animal. In this way an immoral person, a differently moral person, and a person who deliberately repudiates all moralities, would count as possessing a moral understanding, and thus, on this approach, would count as a member of the moral community. They would thus be duty agents, and duty patients. It may be recalled, of course, that to be a duty agent does not entail that one does one's duty, nor does being a duty patient entail that duties owed to one are respected or bestowed.

There are, of course, as suggested, diverse moralities, and differences amongst even similar moralities. The morality of Saint Anthony, the Desert Father, and Petronius Arbiter, the master of ceremonies in Nero's court, were doubtless quite different, as would be the moralities of Richard the First and Saint Francis of Assisi, or, say, of Nietzsche and Marx. It would be arbitrary, in our context, common as is the human penchant, to impose one's morality on others, as the one true morality. Even if there is one true morality, that is, a best morality, and it happens to be ours, the other moralities are still moralities, and their adherents, having a moral understanding, would count as members of the moral community. This does not, of course, require that one respects all moralities or finds them acceptable. It is certainly possible to conceive of heinously immoral moralities, say, one involving genocide, perhaps favored by one god or another, but its immorality does not rule it out as a morality. Similarly, it is possible that the morality of a machine life form might be quite different from, and perhaps even antithetical to, a human morality, but it would count as a morality.

We have tried to set up criteria for membership in the moral community, in such a way as to admit machine life forms under certain conditions, those of consciousness, intelligence, and moral understanding. As the moral community is, as noted, an artifact, these criteria are provisional in nature, but, too, they are useful, in providing a basis in terms of which an examination of certain issues may proceed. There is nothing here, of course, to prevent revising our criteria, for example, under certain conditions ruling all machine life forms out of the moral community. If the human race found itself in a war to the extermination of a life form, ours or theirs, it might be judicious to deny the enemy membership in the moral community. Similarly, if the machine life forms ruled the human race out of the moral community, one supposes the human race would retaliate in kind, with the presumed exception of a predictable minority which would continue to treat an uncompromising, inveterate enemy bound on the extermination of the human race with understanding, courtesy, and sensitivity. Presumably the machine life forms might even find uses for such individuals, up to the point of their elimination.

Sometimes, of course, membership in the moral community depends on power relationships, negotiation, compromise, intimidation, disruption, menace, and such. If denying a given group, say, machine life forms, membership in the moral community, is likely to result in serious disutility to another group, say, humans, one can always enlarge the moral community in such a way as to encompass the previously excluded group. This may be done by discovering that the previously excluded group has been lamentably and unfairly deprived of a hitherto unrecognized or unnoticed right, which one now perceives, wisely, one has a duty to grant. And thus not only is peace kept and the day saved, but the granting group is reassured of its moral worthiness, and, probably, its moral superiority.

Whereas we now have in place tentative criteria for determining membership in the moral community, these criteria do not, in themselves, entail any particular social or economic arrangements, say, socialism or capitalism, authoritarianism or democracy, statism or freedom, nor any particular distribution of utility. Illustratively, it would not decide amongst competitive theories of justice, i.e., first-order fictions favored by one group or another.[22]

---

22. Briefly, illustratively, to give a sense of some possibilities, one might distinguish amongst six familiar theories of distributive justice, ranging from the impractical and naive to the cynical and hypocritical. (1) Egalitarian justice, where utility is to be distributed equally across a population. [2] Contributarian justice, where utility is to be distributed in accord with the value and importance of an individual's or group's contribution to the community. (3) Calliclean justice, where utility is to be distributed according to merit. (4) Market justice, where utility is to be distributed impersonally, in accord with success in a market subject to law, contract, and such. (5) Thrasymachean "justice," where utility is distributed in accord with the will and decision of those in power. And (6) "Jungle Justice," in which utility is distributed as a consequence of successful competition in an essentially unsupervised, anarchic situation.

This much said, let us consider some of the sorts of questions which might arise in a community within which might be found machine life forms, of one degree or another of consciousness, intelligence, and moral understanding.

Even where duties are not involved, as in, say, following our proposals with respect to the moral community, one might still view certain arrangements as more or less judicious, more or less worth favoring. Similarly, it is well to keep in mind that duties are not always clear unless institutionalized, as second-order fictions, and sometimes they are not clear, even after that. And, often, of course, there may be a conflict between first-order and second-order fictions, and an individual might prefer to act in terms of his own first-order fictions, perhaps being willing to accept the community consequences, should there be any, in such a case.

6.22 Post-Moral-Community-Definition Quandaries

As suggested above, the acceptance of a provisional definition of the moral community does not resolve a plethora of lingering difficulties. Such a definition is a sensible first step, but one which begins, not ends, a journey.

It is easy to anticipate a number of problems, greater and smaller, which might attend the entrance of machine life forms into society, a society which, eventually, they would be likely to transform.

Assuming that the decision has been made, to construct machine life forms, the first problem would be what to make, and in what quantity. This would presumably be a decision which would be made variously, and frequently, in any country, in any laboratory or any workshop, which had a technology of the required sophistication and the economic resources to invest in its exploitation. Presumably, in the beginning, these devices would be regarded as objects and tools, sophisticated appliances, created to serve the diverse purposes of their creators, and, then, later, as their sophistication increased, attaining awareness, and such, created, as a form of inorganic domestic animals, to "know, love, and serve" their creators. Then, eventually, given the likely advances involved in facilitating machine learning, advances involving genetic principles, and such, one would expect the devices to verge on, or achieve, personhood. At this point, we shall suppose they are conscious, intelligent, and possess a moral understanding. The "children of the engineers," at this point, are capable of leaving home. And probably some of them would seriously consider doing so. Disagreements with humans up to this point would presumably deal primarily with the best means to attain a human-designated goal, say, that means A would be preferable to means B to attain goal C; but now disagreements might appertain to the goals themselves. The children begin to think for themselves. For better or for worse, their nonage is at an end.

Now our concerns would be less with how to nurture and care for what we have created, less with our responsibilities to our creations, as in its earlier phases, than with how to relate to a different conscious, intelligent, independent moral life form.

The reality of this new relationship would surely have consequences for the future of civilization.

One suspects that an industry of casuistry and invention would emerge and thrive, revising standing first-order fictions in the first case, extending or limiting them, and contriving or fabricating new ones in the second case.

In the following, let us suppose "M" stands for a machine life form which possesses consciousness, intelligence, and a moral understanding. It is not required that the life form be mobile, possess manipulative appendages, or such. This would vary from life form to life form. Similarly, that the machine counts as conscious does not mean that it need be constantly conscious, no more than a human being; that it is intelligent does not mean that it does not err or that it always behaves in an optimally intellectual fashion, in such features also resembling a human being; and, lastly, that it has a moral understanding does not mean that it acts morally; as is the case with a human being, it might be immoral, act contrary to its morality, and so on.

One might then suppose that the following questions, or sorts of questions, might arise.

Should the production of M's be regulated?

Should M's be allowed to supervise their own manufacture, with respect to type, and quantity?

If a form of M is outlawed, should the existing M's of that type be destroyed?

If the M is of the sort which cannot feel, is it acceptable to injure, impair, deface, mutilate, or destroy that M?

Can M's be bought and sold, traded for, given as gifts, and such?

Suppose an M belongs to person P in country C, and that M is brought into country D, where M's cannot be owned. Should that M still belong to person P? If Person P carries that M back to Country C, does Person P then again own that M?

If an M flees from Person P, or is stolen and brought into country D, or is smuggled into country D, should that M be returned to Person P?

Should an M be allowed to own another M, or a human?

Should M's be granted citizenship?

If an M is manufactured in country D should it automatically be a citizen of country D? Should there be a naturalization process available for M's who are transported into a new country, say, country E?

Should M's be permitted to vote?

Should M's be permitted to occupy either appointive or electoral offices?

Should M's be permitted to own a weapon, to join the police force?

May an M demand a salary for its work? May an M strike? May an M quit its job?

Should an M pay taxes, be entitled to unemployment payments, disability benefits, social security, something analogous to health-care benefits, etc.?

May an M be discriminated against, or harassed, in the work place?

Should an M be given preferential consideration with respect to employment? Should employers be given tax credits for hiring M's? Should there be a favoring or disfavoring quota for M's in the work place?

Should M's be permitted to drive, required to have licenses, be subject to tickets for speeding, overparking, calling explicit attention to the lack of normality and/or mental stability on the part of law-enforcement officers, and such?

Suppose an M is used in a crime. Is it an accessory, is it guilty?

How should an M be punished?

Can an M sue, or be sued?

Can an M inherit property, from a human, from another M?

Can an M bequeath property, to a human, to another M?

Should M's be permitted to marry, humans or other M's, or other life forms?

Should an M be allowed to adopt children, or other less competent M's?

Should M's be baptized?

Should M's be allowed to join religious denominations, or found their own?

Should M's be allowed to form clubs, secret societies, political parties, or such?

Should an M have burial rights?

Obviously there could be many such questions.

Whereas the answers to many such questions might depend significantly on the M's in question, it is easy to suppose that, even if the M's were comparable to a normal human adult, or even superior to a normal human adult, the answers, or some of them, might prove controversial, politically if not morally.[23] It is easy to anticipate social conflicts of a greater or lesser degree, involving envy, resentment, hostility, hatred, and such, doubtless occasionally erupting in acts of violence. If such things can occur

---

23. It is interesting and, I think, instructive, to consider "question reversals." Let us suppose a machine civilization, alien, or indigenous, perhaps some ten thousand years in the future, which, for political reasons, has revised its history, "anthropology," and such, to conceal its origin, which it might owe to organic life forms. As an analogue, one might suppose an organic race, say, human, which, for purposes of politics and power, has concealed its origin, perhaps naturalistic, in favor of a different world view. This is apparently the case in certain parts of the world now, and the victims of this hoax innocently and gullibly believe whatever they are told, and uncritically and dogmatically live their lives in terms of that world view, knowing nothing else. Given the seizure of global power by armed, self-serving, self-righteous zealots, a certain eventuation, the imposition of some such world view, would not only be a likelihood, but an inevitability, which might stunt and darken human life for millennia, until the general stagnation, misery, and hopelessness of common life, wearied of oppression, executions, and torture, might sow the seeds of reformations, and perhaps a renaissance.

So let us return to our possible machine civilization, alien or indigenous, supposing it, on the whole, save for an occasional uneasy archivist, here and there, ignorant of its origins. Naturally enough, it thinks of consciousness, intelligence, thought, and such, in terms of its own life form, and would dismiss the hypothesis that consciousness, intelligence, thought, and such, could occur in any other modality than that with which they are familiar, indeed, would dismiss it as ridiculously absurd, rather as there are individuals about today who might dismiss the notion that any sort of machine could think as ridiculously absurd. The general theory seems to be

amongst the members of single species, it would seem naive to dismiss such possibilities between life forms which would be not only considerably diverse, but lacking even a common bioheritage, of the sort which a human would share not only with other vertebrates, such as hominids, horses, and elephants, but with algae and bacteria.

It seems one might just as well stick with people.

But one will not.

"Human frailty" combined with enormous power, relative indestructibility, and uncompromising single-mindedness seems a prospect not to be looked upon without trepidation.

Perhaps one should stick with people.

But one will not.

Some scientists look forward to the creation of superior life forms, and the replacement of carbon-based life.

Some feel we should do better than ourselves, and, when we are successful, step aside, having justified our existence in virtue of having created gods, who, in time, will not remember us.

In conclusion, one does not know, of course, if machine life, intelligence, thought, and such, will ever be possible. If, for some reason, it should be impossible, then several of the problems dealt with here will never need to be engaged in fact, rather than merely in theory. One might hope, or some might hope, then, that such byways need never be traversed. Others, on the other

---

"Meat can't think," a theory which does have considerable *prima facie* plausibility. Indeed, that consciousness of any sort might arise from nonconsciousness of any sort seems surprising. If there are any carbon-based life forms about, say, insects, it might regard them in a rather Cartesian manner, as automata, perhaps analogous to wind-up toys devised by the machine god to test the faith of the devout.

Now let us suppose that humans reenter the picture. Perhaps a few are found in some remote, nearly inaccessible jungle; perhaps a few frozen embryos are located in a long-forgotten Antarctic storage facility, and brought to term *in vitro*; perhaps a few are worked out from surviving human DNA or reconstructed according to rediscovered genetic records. In any event, we now have some humans about. Clearly many of the questions which we raised in connection with artificial intelligence and artificial life could rise again, but in the reversed situation. Presumably there would be considerable doubt, particularly at first, that humans were conscious, or, later, that they were "genuinely" conscious, "truly" intelligent, "really" thinking, and so on. Certainly they would be regarded as inferior to, and, in many respects, would be inferior to, the machine life forms. And certainly, at least in the beginning, they would be much at the mercy of the machine life forms; what should be done with them, if anything; what should they be used for, if anything; how many of them should we let exist, and of what sorts; how much they would they be taught; are they animals, or persons; do they have souls; should they be baptized; should they be allowed to be citizens, to vote, to hold office, and so on.

hand, as noted, hope that they will be traversed, and believe that, inevitably, they will be traversed. If the latter position is correct, then what we have done here might prove one day to be of some practical value, if only in sketching out some preliminary, rudimentary principles in terms of which possible futures might be addressed. A common response, one supposes, would be to wait until the futures, like the Vikings, "are upon you," and then worry about things. Coping strategies are various. Personally, I think intellectual adventures, such as those with which we have been engaged, are justified; they are, like literature, music, dance, art, and chess, their own reward, if nothing else. On the other hand, I am uneasy about these things. The future, I suspect, is closer than we think.

PART TWO

CONCEPTUALIZATIONS

PROLOGUE

I THINK IT MAY BE EASIER TO UNDERSTAND WHAT
follows, if we stop to talk a bit about philosophy.

That will give us a better idea as to what we are up to here.

Classically, philosophy has had two major jobs, one cognitive,
and one normative, first, understanding the world, and, second,
living well within it. To some, philosophy seems to have been
fired from her first job, by science, and have quit her second,
perhaps from embarrassment.

Let us look into this, briefly.

Today, our brothers in the sciences, our natural philosophers,
as I would like to think of them, seem to have largely taken over
philosophy's first job; certainly the philosopher, *qua* philosopher,
is in no privileged position to tell us about physics, chemistry,
biology, astronomy, and such; for example, to master the
instrumentation, the mathematics, the techniques, the literature,
the tradition, the accumulated experiences at the disposal of a
typical scientist specializing in one of hundreds of narrow fields
requires not only intelligence, diligence, and commitment, but
years of preparation and application. It is hard enough to draw a
single bucket of knowledge; it is no longer possible to own its ocean.
Today, an individual, philosopher or not, natural philosopher or
not, is largely at the mercy of testimony, trusting the findings
and hypotheses of others, and from such particles and fragments,
such guesses and possibilities, from such Newtonian "pebbles
and seashells," standing on the shore of mystery, constructs a
tentative world view. We would like to know much, and today it
is difficult to know even a little. If Aristotle were about today, he
might win a Nobel Prize or two, but that would be about it. In
the historic sense, Aristotles, "masters of those who know," are
now impossible. In any event, there seem to be few about.

Philosophy, however, wisely or not, has not given up her first
job, that of understanding the world, but it does pursue that job

less universally, and, in some cases, less directly than hitherto. Certainly the arm chair is no match for the laboratory. On the other hand, what emerges from the laboratory is often transitory, and, from laboratory to laboratory, not always consistent, and, too, most of reality lies outside the laboratory. The laboratory is, after all, only so much help. So philosophy continues to go about her business, grateful for whatever science has to say this afternoon, and may say differently tomorrow morning, and so on, trying to put together a holistic, coherent picture of the world, or, at least, if no such picture is possible, trying to find that out, as well.

Small pictures are nice, and important, but one lives in the big picture, even if the big picture is incomplete, even if the parts do not fit together.

Philosophy, perhaps foolishly, likes to think about the big picture. It is one of things she does.

It is true, I think, that philosophy is today more circumspect when it comes to her second job, trying to figure out how to live well within one supposed reality or another. It is a long time since philosophy was the "physician of life," and "life philosophies" abounded. Few people today seek out philosophers to inquire how to live, much to the relief of philosophers, who are not that sure of it themselves. And often, it seems, those with clear, sturdy answers to such questions appear, upon reflection, to be the least qualified to supply them. Pretentious charlatans are not in short supply, and each is trailed by his retinue of epigones. And competitive mystery religions, in great numbers, contradictory to one another, thrive, as usual. Sometimes one remembers the glory days of Stoicism, Epicureanism, and Skepticism. These philosophies, "life philosophies," tending to appeal to different psychological profiles, are still about, in their way, as they are perennials, but today most Stoics, Epicureans, and Skeptics, do not know they are Stoics, Epicureans, or Skeptics, which is perhaps just as well.

Without suggesting that the philosopher ought to be involved in "life philosophy" issues, it is interesting, is it not, that this job, for the most part, has been abandoned. That would seem to call for an explanation. For example, if religions ceased preaching or advising, would that not call for an explanation? I think there are four major reasons why philosophy has disengaged itself, on the whole, wisely or not, from such issues, where her efforts, I suspect, addressed a need, a need for an articulate, workable, intellectually respectable, rational, defended secular ethics.

Why leave anything as important as how to live to mountebanks, dogmatists, quacks, and zealots?

Particularly when several seem to have little hesitation about killing one another, and others, who may be unfamiliar with their aberrations.

First, I think philosophy was awed by the prestige of science, its capacity to produce jet engines and drip-dry shirts, its glowing cultural image, now rather tarnished, its success in regularly reducing and humiliating religious and theological claims, which, I think, impressed many philosophers, its supposed severe, heroic, noble commitment to, and pursuit of, hard, cold, grim, formidable, ascertainable fact, and so on. Very different from ethical recommendations, surely. The standard notion is that science is divinely neutral with respect to whether or not you poison your rich uncle, but, if you are interested, it can tell you how to do it, how much poison to use, and so on. It can tell you how to build a hydrogen bomb, but it is up to you to build it or not, and, if you build it, when and where to use it, if at all. Such details are not within the province of science. If science will, or must, wash its hands of ethics, other than its own, at any rate, then let philosophy emulate her. Science doesn't tell people not to poison Uncle Bill, so why should philosophy? How would you prove, for example, that it was wrong to do so, other than the fact that Uncle Bill, and some others, would say so? So I think one reason philosophy gave up "life philosophy," at least publicly, was to be more like her hero, science. Philosophy, in my view, was misguided here; she is not science; she is something different, something very nice, and very important, in her own way, but something not science.

The second reason philosophy has dissociated itself from ethical recommendations is probably to avoid being swept up in the rush to run the lives of others. One might be judged by the company one keeps, and the company in question is not one most would wish to keep. Better to remain in the garden, and let the world run to ruin outside the walls.

A third reason might be modesty. Certainly it would seem arrogant and pretentious to preach, particularly without one or another revelation, or army, at one's back. Too, any accountant, or plumber, with common sense, is likely to live as decently, as well, and as rationally, as most philosophers. What have philosophers to tell people about such matters, that they do not already know, or suspect? For most practical purposes, philosophers are not needed, and know they are not needed. They are often adept in the logical and conceptual analysis of such matters, but that, though of great interest in itself, functions largely independently of the evolved, systemic-process ethics that most of us presuppose, and get along with.

Perhaps the major reason philosophers seem to have abandoned "life philosophy" is that they have come to realize that, for the most part, practical moralities develop naturally; over generations; they are the "products of human action but not of human design"; in this sense, they are not so much discovered as created. On this basis, it might then be presupposed that there is no moral "truth" to be sought, with the result that there is no point in seeking it. What is profoundly mistaken about this negativistic supposition is that the preceding view, the invention view, or the product view, if correct, is not a counsel of despair, but an invitation to, and an intriguing opportunity for, judicious creativity. The painting is not waiting to be painted; it comes into existence with its painting. This does not impair its excellence. The optimum design does not pre-exist; it comes into existence as it is invented. This does not jeopardize its perfection, or its rightfulness.

> "A thousand paths are there which have never yet been trodden; a thousand salubrities and hidden islands of life. Unexhausted and undiscovered is still man and man's world."[24]
> —Friedrich Nietzsche

The point of these remarks, serving as a prologue to the following, is to emphasize here a modern analogue to the classical dual nature of philosophy, cognitive and normative, that of understanding the world and living well within it.

Whereas it seems clear that reality was around long before

---

24. The translation above is by Thomas Common, from Part One of Nietzsche's *Also Sprach Zarathustra*, Section 2 of "*Von der schenkenden Tugend*," "On the Bestowing Virtue." "*Tausend Pfade gibt es, die nie noch gegangen sind, tausend Gesundheiten und verborgene Eilande des Lebens. Unerschöpft und unentdeckt ist immer noch Mensch und Menschen-Erde.*"

Nietzsche was one of the few philosophers who was keenly aware that philosophy was less about finding than making, less about reporting than creating.

human beings were around, and managed well without us, the human perception of reality, or "human reality," namely, reality as we experience it, could not. That, at any rate, needs us. Our world, the human world, is presumably a function of reality, whatever might be its nature, but it is also, presumably, quite different from reality, reality as it is apart from us, so to speak. This recognition goes back to the ancient world, but it came into its own with the work of Immanuel Kant, and its claims have been substantiated by various sciences, from physics to psychology.

To be sure, for what we are up to, we need not enter astonishing and mysterious worlds. We do not need to transgress the familiar perimeters of the "human world." We can pretty much stay indoors, or at home. Recall an earlier distinction, between fictions and nonfictions. We noted that the fiction does not exist as the tree exists, but the tree, in its turn, does not exist as the fiction exists. The tree exists in space; that it is a tree does not exist in space. In short, what we have here is a distinction between the "that," so to speak, relatively obdurate, given, encountered, and so on, and the understanding of, the naming of, the conceptualization of the "that," encountered in our experience, and, indeed, the "that" is also, *vis-à-vis* the human world, at least, transformed in its conceptualization. A simple example is the continuum of the spectrum which various conceptualizations, varying from language to language, divide into different colors, some into two or three colors, others into several colors, and so on. Indeed, the very notion of color itself is a conceptualization.

Familiar conceptualizations tend to be unnoticed, even invisible. They are embedded in the languages we speak. Each natural language provides a marvelous, efficacious conceptual structure, honed over generations, for mapping the world of human experience. Linguists note that things may be organized in diverse patterns, be seen in many ways. Concepts are usually reasonably clear within their normal ranges of application, and the usual difficulty, if any, is in their application. Is that thing in the distance an animal or an inert object; if it is an animal, is it a cat or a different animal, and so on. Once one gets closer one expects to be able to tell if it is a cat or not; the concept "cat" is reasonably clear in its usual application ranges. It is unlikely to give us trouble. There are few things about which are likely to be confused with cats. Cats do not much surprise us, except perhaps in their location, perhaps on the top shelf of the linen closet or in the clothes dryer; they are, on the whole, pretty much of a muchness. On the other hand, given logical possibilities, envisageable eventuations, the situation could become problematic. For example, could you have a healthy, functioning adult cat which weighed only one ounce, or, say, two tons? What if the catlike creature could talk with you, or changed colors at will, or walked through walls, and such? It is certainly not an ordinary cat, but is it just a strange cat, or something other than a cat, and so forth? This is sometimes called the open texture of empirical concepts, or, more literally, the porosity of concepts, from the German 'die Porosität der Begriffe'.[25]

A thought-experiment like the foregoing, having to do with cats, is instructive, because it clarifies the role of, and importance of, the conceptual structure. It applies the reagent after which, like an unsuspected secret message, the conceptual structure suddenly becomes visible. It also makes clear the necessity, in problematic cases, of decisions, of rulings, which, hopefully, would be judicious, reasoned, and defended. In such cases, one is not applying the conceptual structure, usually serviceable and taken for granted, but altering it, and, in some cases, creating it.

Whereas there is such a thing as "animal truth," which we share with nonlinguistic species, namely, correct anticipations of experiential sequences or trains, fulfilled expectations, veridical memories, recognitions of dangers, and such, truth, as commonly conceived epistemologically, presupposes a more explicit conceptual structure, one usually associated with language, verbal, pictorial, gestural, or such. In this secondary sense of "truth," founded ultimately on the more general "animal truth," there is a clearer distinction between input and processing, between what is encountered and its meaning, between the "that" and the conceptualization of the "that," between the "thatness" and the "whatness," so to speak, in our example, between the animal and whether it is to count as a "cat" or not. It is also important to note in passing, that these relationships between object and conceptualization are tiered, or layered. For example, in our example above, there would be no doubt, presumably, that one is dealing with a thing, a moving object, a living thing, an

---

25. The concept is from Friedrich Waismann. I have adapted one of his examples here. (Cf. "Verifiability," *The Proceedings of the Aristotelian Society*, Vol. 19 (1945). The article is reprinted variously, e.g., in *Logic and Language* (First Series), edited by Anthony Flew (Basil Blackwell, Oxford, 1960).)

animal, and so on. Those conceptualizations are in place and functioning normally. Where we encounter problematicity is whether or not the thing, the moving object, the living thing, the animal, is or is not a cat. Indeed, we might not even be sure, in some cases, whether or not we are dealing with an animal, rather than a contrivance, a hoax, a vision bestowed by one god or another, a hallucination, an alien being, or what. We come armed with concepts, and hypotheses, and procedures for determining whether or not our hypotheses are well-founded or not. When our concepts, at one level or another, break down, their application is problematic, or precluded, and the hypothesizing enterprise collapses. In the face of this cognitive bankruptcy one restores epistemological solvency by issuing a more plausible currency, if only by fiat.

On the basis of the preceding reflections, our modern analogue to philosophy's classic jobs, cognitive and normative, understanding the world and living well within it, becomes the cognitive job of understanding the human world as well as one can, abetted by science, history, common sense, and such, and the normative job of revising the conceptual structure in such a way as to deal with that world more satisfactorily, more successfully.

These are, of course, metaphilosophical matters, namely, matters having to do with the philosophy of philosophy. Proposals are involved, and philosophy, the most protean of all disciplines, may be seen variously. There is no one form of literature, happily, and there is no one form of philosophy, happily. On the other hand, these remarks should make it easier to understand what follows.

It is clear that the "challenges of the future," several of which we have noted, cloning, genetic engineering, artificial intelligence, and such, will necessitate both intellectual and ethical adjustments. It is also clear that our conceptual structure, that structure in terms of which we must make such adjustments, was not designed to deal with such realities. It has fallen behind. It already "totters"; it is surely shaken, if not uprooted and superseded. It is clear that our conceptual structure, adequate to slower paces and more placid times, will be, and must be, revised. The "cat," so to speak, is doing strange things.

CHAPTER ONE:
REFLECTIONS ON IDENTITY

1. Introduction

As the story goes, it is an ancient story, there was a woodsman and a traveler, and the traveler, lost in the forest, on a cold, dark, windy, rainy night, came upon the woodsman's hut, and the woodsman, a decent, kindly fellow, and perhaps lonely, for he lived alone in the forest, offered him hospitality, a good fire, a supper, a night's lodging, and directions for the morrow. After a good supper, and warmed by the fire, the grateful traveler, presumably in the way of conversation, remarking the woodsman's ax, in its rack on the wall, in a place of honor, commended it, as a fine ax, which it doubtless was. "Yes," said the woodsman. "it is a fine ax, and I have had it for twenty years, and in that time it has had two new heads and three new handles."

The story leaves off at that point, which is perhaps just as well.

This is the first instance of the Identity Paradox in literature, at least to my knowledge.

Is it the same ax?

This is a typical instance of knowing everything there is to know about the case, except what the case is.

The same sort of cognitive puzzle comes up later, in Plutarch, in connection with the ship of Theseus. Supposedly the Athenians preserved the ship of Theseus, but, over the years, and the generations, as the oars, wood, rope, and such, deteriorated or rotted, these things were replaced, bit by bit, part by part, until, later, nothing of the original materials remained.

So, is it the same ship?

We know exactly what happened, but how should it be conceptualized? Is there a right or wrong way to conceptualize the situation, or a better or worse way to conceptualize it?

Obviously one needs a criterion for identity. If one's criterion is embodied form, it is the same ship; if one's criterion is the same matter, it is not the same ship; if one's criterion is relative continuity of either form or matter, it is the same ship; if identity requires sameness in all respects, then the least change, say, a nick, or the loss of a splinter, would make it not the same ship.

Is there a correct way, or a best way, to resolve the question?

The "woodsman's ax" and the "ship of Theseus," each in its charming way, calls attention to the occasional need to become aware of the conceptual structure, to inspect it from time to time, and, from time to time, to clarify it, or alter it, in one way or another, given one purpose or another.

2. Change

With respect to change, whether it actually exists or not, as opposed to the appearance of change, which is undeniable, one encounters three major philosophical theories, what we might call the No-Change Theory, the Only-Change Theory, and the Distinction/Rearrangement Theory.

First, one might note that it seems to be a logical truth that nothing can change and remain the same. For example, to say that Thing T changes but remains the same seems to be contradictory. For example, to be the same is not to change. This matter might be formalized in a variety of ways, one of which follows:

1. C_T & S_T
   Thing T changes and Thing T remains the same.
2. $(x) (Cx \supset \sim S)$
   For every x, if x changes, then it is not the case that x is the same.
   (This seems to be a logical truth.)
3. $C_T \supset \sim S_T$ 2, Universal Instantiation
4. C_T 1, Simplification
5. S_T & C_T 1, Commutation
6. S_T 5, Simplification
7. $\sim S_T$ 4, 3 Modus Ponens
8. S_T & $\sim S_T$ 6, 7, Conjunction.
   Thing T remains the same and it is not the case that

Thing T remains the same.
(Line 8 is an explicit contradiction.)[26]

This contradiction might be avoided if either Thing T did not change or Thing T did not remain the same, but not both. For example, 'p & ~ p' is a contradictory form, but neither 'p' nor '~ p', independently, is contradictory. In short, there would be no problem if Thing T did not change or Thing T was changing, but not both. Presumably, most things T would change, but then it would follow that most things T would not remain the same, and one would usually like to think that one and the same thing T could change, that it could change and still be the same, the same person, the same cat, the same automobile, but this is what led to the contradiction in the first place. One might worry about the amount or degree of change, but this seems to introduce an unwelcome complexity and a slippery-slope continuum which, strictly, would seem to call for a change point, where, say, Thing T is no longer the same. For example, with the removal of what board, or particle, is the ship of Theseus no longer the same?

How can something be what it is and yet change?

If the notion of a changing thing, namely, a thing which changes, seems to generate a contradiction, one might avoid the apparent contradiction in various ways, for example, in denying that change takes place, in denying that changing things exist, which supposition precipitates the contradiction, or drawing a distinction wherein change may be reconciled with sameness.

2.1 The No-Change Theory

Obviously there are appearances of change, and if the appearances change, then there is, in some sense, change. The appearances of change, however, on this approach, do not count

---

26. On a "Quinean supposition," one might have recourse to an unusual predicate and treat line 8 along the lines of '(∃x) (Tx & Sx) & ~ (∃x) (Tx & Sx)', which would also be an explicit contradiction. On the other hand, if one chose to read line 8 differently, as, say, '(∃x) (Tx & Sx) & (∃x) (Tx & ~ Sx)' one would not have a contradiction, but, if nothing had the "T" property, one would have a false statement. If the "T" property is a uniquely referring predicate, and there exists an x such that it has that property, then one would again have, after an adjustment or two, an explicit contradiction, for example, '(∃x) [Tx & (Sx & ~ Sx)]'.

as change in the relevant sense, which has to do with the nature of ultimate reality, reality as it is in itself. Change, as we witness it, then, is a delusion. An analogy might be the radical changes we seem to experience when asleep, moving about, and such, but then we awaken, and realize we were not really riding buses, flying airplanes, coming up to bat in the ninth inning with the bases loaded, and such. How the delusion could come about, or why it should exist is not clarified. Without going into the metaphysical background, at least in any depth, this unusual position is the result of unflinching inferences drawn from seemingly true propositions, indeed, propositions taken as self-evidently true. The central proposition here is that nonbeing cannot exist. And, clearly being exists, whether it must or not. Therefore, one is concerned with the nature of being. It cannot come into existence from nothing because that would credit the existence of nonbeing, nor can it cease to exist, for then it would have to pass into nothingness, and nothingness, or nonbeing, cannot exist. Thus, being is eternal, neither having come into existence nor being able to pass out of existence. Similarly, change is impossible, for that would require that something which was not would come into being from nonbeing, and that that which was would pass from being into nonbeing. A consequence here would be that motion is impossible in reality as it is in itself, for motion necessitates change, which is impossible in reality as it is in itself. Similarly, time, space, and causality cannot pertain to being in itself; for example, ultimate being could not be spatial, or in space, for that would require a space in which to be, and that space would require a space in which to be, and so on, endlessly. Too, if space is understood as nonbeing, it is ruled out, as well. Metaphorically speaking, being is well rounded, and self-contained; metaphysically spherical in its perfection, not spread out in space. Spatial predicates do not apply to it .It has no "within," and there is nothing outside of it. Similarly time and causality cannot be predicated of reality as it is in itself, the "One," as it might be called, because both would require change, coming into being and passing out of being, and reality is changeless and imperishable, and, in its unfathomable serenity, presumably perfect. This sort of view seems to have been introduced into philosophy by Parmenides. Many are familiar with a variety of paradoxes worked out by Zeno, a follower of Parmenides, designed to show the inconsistency of typical views of space and motion. It is easy to see how the awesome mysteriousness and sublimity of such views came to be invested with religious significance. Indeed, the attributes commonly assigned to the "One," such as eternality, imperishability, being apart from space and time, changelessness, perfection, and such, are often associated with one divine entity or another. The "One," however, is presumably impersonal; to anthropomorphize it would seem to be not so much a misunderstanding, as a blasphemous desecration.

Let us suppose that a particular premise-set implies a surprising conclusion, such as the nonexistence of change. If one accepts the premise-set and the logic, presumably one will accept the conclusion, however unwelcome or surprising. On the other hand, if one finds the conclusion unacceptable, it is it likely one will, in turn, find the premise-set or the logic unacceptable. Are we more sure of the truth of the premise-set and the flawlessness of the logic, or more sure of the falsity of the conclusion, which would imply either that the premise-set is false or the logic invalid?

If change cannot exist, then it cannot generate any contradictions. If we can remove 'p' from 'p & ~ p', then we no longer have a contradiction. Similarly, if we remove '~ p' from 'p & ~ p' we would no longer have a contradiction. If there were no "things," then we could not generate the contradiction either.[27]

---

27. In passing, one might note another theory of change, but one which would not count amongst the major theories of change, the generation/annihilation theory. Conceived physically, this would require something like a succession of closely resembling world states. An analogy would be the motion-picture film in which a succession of closely resembling still photos produces the illusion of motion. The analogy fails, of course, as the film as a whole must move in order to display the frames. Suppose, however, the frames appeared and disappeared without motion. Frame F is there, and then it isn't, and so on. A classical analogue would be Zeno's Arrow Paradox, in which it seems the arrow leaves the bow, traverses an intervening space, and strikes a target. As the paradox proceeds, if motion were possible, the arrow would have to occupy, however briefly, each space between the bow and the target, but to occupy a space the arrow must be in the space, exactly, neither coming into it nor passing out of it. In short, to be in the space, not moving in or out of it, it would have to be motionless, however briefly. Thus, one would have to generate motion from a succession of immobilities, which is impossible. Thus the arrow's motion is an illusion. On the other hand, one way of accounting for its apparent motion would be something like the generation/annihilation theory. The generation/annihilation theory could provide a modality for accounting for the illusion of motion within a no-change theory. I would rank it as an independent theory, however, as one supposes that the succession is a succession of realities, each different from the other, giving us not one reality but many realities, the only illusion being that of motion. What produces these realities is unclear. Similarly, who experiences these realities, and how they are experienced, is unclear.

2.2 The Only-Change Theory

Heraclitus is credited with claiming that one cannot step into the same river twice. If he was right, as has been pointed out, he could not step into the same river once, as the water is changing while he is doing his stepping. Whereas many would regard both of these analyses as unduly pessimistic, the sort of point he wants to make, in his grumpy, inimitable style, is well worth worrying about. Change is everywhere, and constant, and change is obviously not a thing. The world is flow and process, and neither flow nor process are things.

The classic elements were air, earth, fire, and water. These things might be regarded as independent, or as transformations of one or another of these basic elements, taken as fundamental or primitive, perhaps water, as in the case of Thales, perhaps fire, as in the case of several Stoics. In any event, Heraclitus may have elected fire as his Ur substance or fundamental form of being. It is hard to say, for he was at heart a remarkable poet. I would like to think he used "fire" as a metaphor for whatever it was that was basic to reality. On the other hand, he may have meant fire, really. Many Stoics, for example, took that path. On the other hand, even if Heraclitus meant fire in some physical sense, in his case, no more than in the case of the Stoics, would he have had ultimately in mind the sort of thing over which Boy Scouts toast marshmallows. If water was to consist of fire in some sense, obviously it could not be in the same sense in which one commonly thinks of fire. The modern notion of matter/energy in its transformations might have appealed to him. Does the world not consist of such "fire"?

In any event, let us presuppose a substrate, whether one thinks of it as fire, or flux, or such. A useful way, metaphorically, to convey what may be meant here would be to suppose a universe consisting of clouds of fog; the fog swirls, and forms shapes, rock-shapes, tree-shapes. person-shapes, and such. These shapes are transitional configurations of the fog, ever coming and going, ever drifting, ever changing, never the same. The only ultimate reality is the fog, or the flux, and the fog is not a thing, nor is the flux a thing. On such a view, it makes no sense, as all is changing, to ask when the ship of Theseus has changed, when it is no longer the same, for, in a sense, there was never a thing, the ship of Theseus, but only a shape in the fog, and the fog itself, is only the fog, and not a thing. There are no things to change.

2.3 The Distinction/Rearrangement Theory

This theory combines, in its way, the "no change" and "only change" theories. It draws a distinction wherein change may be understood as compatible with sameness.

The easiest way to understand this approach is in terms of a distinction between objects and patterns of objects, such as classical atoms, say, tiny, imperishable, unchanging objects, and the patterns formed by such objects, different distributions and arrangements, constituting different objects, one configuration of atoms constituting a rock, another a tree, another a person, and so on, rather as similar dots may form triangles, squares, and diamonds.

Whereas this approach has much to commend it, and is bound to be popular, there are at least two serious difficulties involved. First, it is not clear that there are basic unchanging entities and, if there are, it is not clear what might be their nature. Physics, currently, is salubriously unsettled. It seems to be in a state of healthy disarray. Do quarks exist, do they change, if they exist? Are there things down there, or fields? What about transformations of matter/energy? Is energy a thing? Is gravity? Are there particles, or strings, or something in between, etc.? What is the role of observation and experimentation? Do such things discover facts, or might they contaminate, influence, or create facts? Are there multiverses? What about world lines? Are there multitudes of dimensions, thousands, things coming in and out of them, perhaps universes, and such, and so on?

The second problem with this approach, given our interests, is that it returns us, so to speak, to the "fog problem." The ship of Theseus is no longer a thing, but a pattern, a configuration, of things. It is not itself a thing, but a form, a transitory form which is embodied. Analogous to the illustration above, the triangle, square, and diamond do not exist, unless in some derivative, or secondary, sense. What actually exists are the dots. It is we who see the triangle, the square, the diamond. Would the ship of Theseus exist for a mind which saw only atoms?

Consider the following pattern of dots.

Presumably you will see this configuration as only dots, not as a thing, not as, say, a triangle, or such. What is real here is the dots. They are the things. If, however, one came from a different culture, one in which a given artifact, or shape, unfamiliar in ours, was familiar, one might see the above dots as that thing, just as we are likely to see one arrangement of dots as a triangle, another as a square, and so on. But what is real is again the dots; they are the things.

Whereas one would like to think of the ship of Theseus as a thing, say, a stable union of form and matter, in an Aristotelian sense, it is difficult to do so on this approach, as the only matter here is the basic things, say, the dots, the atoms. Everything else is intangible, imposed, configuration, pattern, form, such things. A form is not a thing; in the relevant sense. For example, triangularity, being red, and such are not things in any normal sense. On this approach, the ship of Theseus turns out to be not a thing, but a form, an imposed form, and a selected, imposed form, as we separate it out from those on board, from the gangplank, from the sand on which it is beached, and, in the sense that we contrast it with, say, the ship of Jason, the *Argo*, even more radically an imposed form. The mind which saw only atoms would not see the ship of Theseus. It would not exist.

3. The Pragmatic Tack: Return to the "Human World"

One supposes the preceding accounts will have been found unsatisfactory, even if one or another of them might be metaphysically correct, say, in the view of one god or another. We wish to have a world in which the ship of Theseus, or the rock or tree, is an object, as it is in the human world, namely, the world of human experience. There are, of course, given the internality of experience, hopefully topologically related to an external world of nonexperience, a large number of human worlds, one for each human, just as there are a large number of raccoon or giraffe worlds, one for each raccoon or giraffe, and so on. Pragmatically, however, let us ignore this complication.

Certainly there is much precedent for such judicious bracketing, or neglect.

In our off-hours, for example, when not doing epistemology, we are all happy-go-lucky, unquestioning naive realists, namely, folks who take it for granted that we experience the world directly, as it is in itself, that the world is in itself just as it is experienced, for example, loaded with colors, sounds, smells, tastes, feels, and such, including innumerable subjectivities, relativities, and conflicts of appearances, each accepted as objective and real, pertaining to the world as it is in itself, such as big/little, near/far, fast/slow, warm/cold, bright/dim, attractive/unattractive, beautiful/not beautiful, valuable/not valuable, painful/pleasant, and so on. And, furthermore, we suppose that others share our experiential world, that they are inside it, so to speak. These two assumptions, namely that we experience the world directly, as it is in itself, and that others exist in, and share in, our human world, our experiential world, are both false. On the other hand, this double illusion has clearly been selected for in the course of evolution, which, over thousands of generations of thousands of life forms, has topologically indexed one world to another, has keyed the internal world to the presumed outer world, developing senses and awarenesses by means of which we may link events in our brain and nervous system to events, processes, and realities in an alien world, a different world, mysterious, colorless, and silent, a supposed physical, mind-independent, external world.

It is a relief to return to the human world.

We need naive realism.

Without this mistake life would be impossible.

To that world, the human world, where we live, philosophy best addresses herself. It is from that world, as Antaeus in contact with the earth, his mother, she gains her strength.

4. A Logistic Approach to Identity

A familiar way of defining the identity symbol, '=', is as follows:

$$(x = y) = df. (P) (Px \equiv Py)$$

which might be read as 'x is identical with y means, by definition, for every property P, x is characterized by P if and only if y is characterized by P'. Informally, the notion is that two things are identical if and only if they have exactly the same properties. If the definition above, which is a proposal, not a statement with a truth value, is accepted, then the following locution would be a statement with a truth value, and, indeed, it would be a logical truth, an analytic statement, one which is true in virtue of its meaning alone:

$$(x) (y) (P) [(x = y) \equiv (Px \equiv Py)]$$

which might be read as 'For every x and for every y, and for every property P, x is identical with y if and only if x is characterized by property P if and only if y is characterized by property P'.

As will be clear by now, this explication, familiar as it is, is more one of those charming irrelevancies of which the logician's heart is fond than anything of much practical or philosophical use. To many subtle minds, simplicity, stability, clarity, and beauty are more beguiling than utility, and the actual world, with its indistinct edges, its messiness, its churnings about, and such, is a nuisance, if not an object of actual aversion. There are many escapes from reality, or refuges from the world, and mathematics and logic, like music and chess, are amongst the loveliest. In many ways, they are better worlds, safer and more beautiful, but, alas, the doorbell rings and the poem is lost.

In any event, no two empirical objects have exactly the same properties, and the properties of what we might think of as a single empirical thing, remember the ship of Theseus, are not the same from one moment to another.

Thus, the logician's concept of an ideal identity relationship has little relevance to the real world, and what relevance it might have, if any, would be difficult to determine.[28]

Accordingly, I would recommend a more varied, more detailed, approach to identity, which is a much more complex matter than would be suggested by the above explication, an approach more adequate to common-sense thinking about identity, an approach which, at least in part, relates identity to an Aristotelian world, so to speak, rather than a Platonic world, in which it is manifested rarely, if at all.

So we will distinguish amongst several concepts of identity, many of which have applicability to the human world, and one of which will engage us in some detail, that of personal identity.

5. A Pragmatic Approach to Identity

I. External Identity: Identity where more than one entity is involved.
   A. Complete Identity: Two or more entities are the same in all respects.
      (It is logically impossible to satisfy this condition, for, if there are two, they must be different in at least one respect.)
   B. Partial Identity: Two or more entities are the same except for at least one respect.
      1. Pervasive Identity: Two or more entities are the same in all significant respects. (e.g., 'A and B are identical automobiles.')
      2. Nonpervasive Identity: Two or more entities are the same in at least one respect but not in all significant respects. (e.g., 'A and B are identical in color.')

28. One of the problems here would be to come up with a notion of what are the properties of an object. For example, is its distance from Mars, or Corinth, to count as a property of the ship of Theseus? Mars is moving, of course, and from what point in Corinth or at Corinth do we measure the distance? Is being thought of, or not being thought of, by Socrates a property of the ship of Theseus, and so on? Perhaps one might try to draw a distinction between intrinsic and extrinsic properties of the ship, but this might be a difficult distinction to draw. Is its age an intrinsic property, which changes each moment? Is its location or movement? Where it was built, who designed it, who built it, who owns it? Its history? Is its temperature, its dampness or dryness, the swelling and shrinking of its boards in the weather, its oars, its sails, its fittings, the material on board, the changing in the cells of its wood or in the fiber of its cordage, etc.?

II. Internal Identity: Identity where only one entity is involved.

   A. Analytic Identity: Logical Self-Identity.

     e.g., '(x) Ixx', '(x) (x = x)','a = a', etc.

     (The first two formulas might be read as 'For every x, x is identical with x'; the third formula would be one formulation of the classical "law of identity," in this case to be read as 'a is identical with a', 'a' taken here, naturally, as a variable for anything. Such formulas are logical truths; they are analytic, true in virtue of their meaning alone.)

   B. Synthetic Identity: Contingent Self-Identity. (Empirical hypotheses are involved here.)

    1. Linguistic Identity: Linguistic Identifications.

    (Expressions having the same referent, *e.g.*, 'Charles Dodgson is Lewis Carroll', *i.e.*, the proper names 'Charles Dodgson' and 'Lewis Carroll' name the same individual.)

    2. Nonlinguistic Identity: Nonlinguistic Identifications.

      a. Extrinsic Identity: Identification where extrinsic properties or relations are involved. (*e.g.*, 'Jane Austen is the author of *Pride and Prejudice*, 'Jones is the man with the secret document', etc.)

      b. Intrinsic Identity: Identifications where intrinsic properties or relations are involved. (*e.g.*, 'Jones is still the same person.' Here one needs a criterion for sameness of person.)

CHAPTER TWO:
INDIVIDUAL IDENTITY
(SAMENESS OF PERSON)

1. Introduction

THERE IS APPARENTLY VERY LITTLE IN THE HUMAN
body which is not replaceable, either at present or, in theory,
in the future. This suggests a thought experiment, designed
to examine various problematicities associated with personal
identity, the resolution of which would call for a theory of
personal identity.[29]

A human person's own sense of identity is commonly thought
to be a function of his individual consciousness, his sense of self,
and his memories. On the other hand, as pointed out by Hume,
one extends one's identity beyond the reach of memory. Too,
one could conceive of objective evidence which might convince
someone of his identity, on grounds other than memory, for
example, photographs, finger prints, DNA, testimony, and such.
However, on the whole, there is little doubt that a person's sense
of himself is largely an internal matter. His primal criterion of
his own personhood is the data of consciousness. For example,
one could conceive of an individual who, despite overwhelming
external evidence of his identity, might refuse to believe it, if it
seemed incompatible with his personal memories. His private
criterion of personhood, so to speak, would take precedence over
public criteria, however formidable.

---

29. The problem of personal identity is a subdivision of the more general problem of
individual identity. We will be concerned in the following with individual identity
as it pertains to human persons. Obviously, one could also be concerned with the
individual identity of almost anything, for example, of a cell, of a dog or cat, of an
artifact, for example, the woodsman's ax or the ship of Theseus, and so on.

At this point it is only necessary to recognize a distinction between a private and a public criterion for personal identity, and to recognize that, for our purposes, we must be concerned with a public criterion. Given the privacy of consciousness, its inaccessibility to outsiders, indeed, even the other-minds problem, to be later considered, one cannot use an individual's consciousness as a public criterion. A criterion which can work for only one person is, in effect, not a criterion, at all. Second, the private criterion may be mistaken. It should not be decisive, whether it is or not, even for the person himself. For example, a person who believes himself to be, say, Napoleon or Alexander the Great, should be open, in the light of overwhelming evidence to the contrary, to bring his internal conviction into doubt, whether he does or not.

It is well known that the deliverances of consciousness can be deceptive, for example, in virtue of insanity or post-hypnotic suggestion. Similarly, the possibility of a multiple-personality disorder, or such, makes it clear that an objective criterion for personal identity is desirable. Presumably one looks for an anchor person, or an Ur person, so to speak, rather than admit a number of persons in one body. If one accepts the possibility of actual "possession," demonic or angelic, or whatever, it would not be two or more persons in one body, in the relevant sense. It would be rather like having an intruder on the premises, a stranger from without, who has entered the room.

A simple thought experiment demonstrates some important ambivalences involved in these matters. In the following thought experiment, it is supposed, for the sake of simplicity, that only one consciousness can be associated with one body at one time. We suppose two bodies (B_1 and B_2), and five consciousnesses (C_1, C_1', C_2, C_2', and C_3). C_1 is the consciousness normally associated with $Body_1$, and $Consciousness_2$ is the consciousness normally associated with $Body_2$. $Consciousness_1'$ is a copy of $Consciousness_1$, and $Consciousness_2'$ is a copy of $Consciousness_2$. $Consciousness_3$ is an invented "human consciousness," never that of any actual human being, which might replace any normal, copied, or transferred consciousness. Similarly, we shall suppose that a technology exists whereby copying, inventing, and transferring consciousnesses is possible. Some results, then, might be schematized as follows:

Body$_1$	Body$_2$
1. Consciousness$_1$ (Normal)	5. Consciousness$_1$ (Transferred)
2. Consciousness$_2$ (Transferred)	6. Consciousness$_2$ (Normal)
3. Consciousness$_1'$ (Copy of C$_1$)	7. Consciousness$_2'$ (Copy of C$_2$)
4. Consciousness$_3$ (Invented)	8. Consciousness$_3$ (Invented)

The general point one wishes to make here is that consciousness is not an ideal criterion for personal identity, either privately or publicly. This becomes clear, even from the sampling of cases schematized above. Cases 1 and 6 present us with the normal situation in human life. In cases 2 and 5 we have a switching of consciousnesses between the two bodies involved. At this point we might be tempted to claim that the individuals or persons involved had switched bodies. Indeed, that is a not uncommon premise in fantasy or science fiction. On the other hand, unless we wish to opt for a "soul hypothesis," or such, whatever it would be for which we were opting, it seems more judicious to say that the individual or person involved now has the consciousness of the other individual or person, not that the persons had switched bodies. There is no doubt, of course that the individual or persons involved would understand themselves to be in different bodies but this, in our view, would be an injudicious judgment. They are the same person or individual, merely now suffering from consciousness replacement. An analogy might be insanity, amnesia, or such. This logical possibility suggests that personal identity should not accompany consciousness. Cases 3 and 7 also suggest it would be injudicious to associate personal identity with a particular consciousness. In each case the original, particular consciousness is gone, with the result that, if consciousness were the criterion for personal identity, the original person is no more, vanished, out of being, or whatever. On the other hand, in this case, it would occur neither to the individual involved, nor, I suppose, to anyone else that his personal identity was gone, or even imperiled, as the replacement consciousness is, *per hypothesis*, indistinguishable from the original consciousness. Again this suggests it would be injudicious for anyone, including the individual involved, at least exclusively, to rely on consciousness as the criterion for personal identity. Lastly, in cases 4 and 8, we have a situation in which an invented consciousness, a fabricated consciousness, replaces the consciousness of the individuals involved. Let us suppose that it is a "Napoleon" or an "Alexander" consciousness. Both individuals would then take themselves to be Napoleon, or Alexander, or such, but it would seem clearly injudicious to take them at their word. Whoever they are, they are not Napoleon or Alexander. Similarly, if the technology exists, one might impose a variety of consciousnesses on the individuals, the same, or different, say, that of a raccoon, a giraffe, a dog, a cat, and so on. But that would not make the individuals raccoons, giraffes, dogs, cats, or such.

In any event, we are searching for a judicious criterion of personal identity, which, for various reasons cited, seems to be best construed as a public criterion. To that search, we now turn.

2. The Saga of Jones

We shall suppose a succession of incidents, taking for our subject the seldom-failing, ever-reliable, in-this-case-most-unfortunate, Jones. We shall subject him to various misfortunes, alterations, transformations, and such, which all began with a slip on the stairs, resulting in a shattered elbow. At time t_1, as noted below, the elbow joint was replaced.

t_1	t_2	t_3	t_4
elbow joint replaced	a few bones	the skeleton	muscles with other fibers

t_5	t_6	t_7
circulatory system, with a number of plastic hoses, etc.	heart, with a mechanical pumping device	skin, with a plastic coating

[continued]

	t_8	t_9
	an android body, resembling the t_1 body, retaining the original brain (memories, etc.) (synthetic nerves)	a new android body, which does not resemble the t_1 body

t_{10}	t_{11}	t_{12}
a "functional body," not human in appearance (with original brain, memories, etc.)	t_{10}, except personality changes	t_{11}, except loss of consciousness

	t_{13}	t_{14}
	revived consciousness, new personality (new memories implanted)	brain replaced with new unit, consciousness and memories identical to Jones at t_{10}

Presumably, given a familiar "human world" conceptual structure, the personal identity of Jones is intact at t_1 and extinct at t_{14}. In short, as one would normally look at these things, Jones is with us at t_1 and gone at t_{14}. The question then is where do we lose Jones?

There is a trivial sense in which Jones is not the same if he undergoes any change. A new Jones, for example, might emerge from the barber shop. On the other hand, our orientation here is likely to be less subtle, or metaphysical, and more pragmatic, reflecting the common sense of the human world. We are concerned, at the least, with significant change, and not even with all significant changes. On the other hand, there comes a time when even the routine and homey comforts of common sense, as one enters unusual terrains, as one crosses borders into unaccustomed countries, may shudder and collapse.

Presumably Jones is still with us through t_6, the point at which he receives an artificial heart. A number of significant changes, however, have occurred prior to that point, the consequences perhaps of a progressive degeneration of physical tissue, for example, skeletal, circulatory, and muscular tissue has been replaced, and, we may suppose, though it is not recorded in the schematism, there have been, and will be, various organ replacements. All this, however, is internal, and does not show on the surface. As long as Jones looks the same and acts the same, little thought will be given to such things in the community. Also, interestingly, human anatomy is apparently far more diverse inside than outside, sometimes surprisingly so, organs being of various sizes, shapes, and locations. Sexual selection is effective in producing a fairly limited set of standard forms and appearances externally, but it has little if any effect on internal anatomy.

As the degenerative catastrophe that is still Jones, at least at this point, continues, the first significant change that is likely to become apparent in the community at large is that something begins to seem odd about Jones' appearance; perhaps it reflects light somewhat differently, has a slight gloss or sheen, beads in the rain, has a slight difference in color, or such, and it undoubtedly feels different to the touch. Something is very different. Jones' dog becomes suspicious, his small daughter no longer rushes to the garden gate to welcome him home from work, his wife develops frequent headaches, and, moreover, has been seen thumbing through the yellow pages, under "L." Jones, naturally, regards himself as still Jones, and so do others, but he does not seem the Jones of old. His skin has been replaced with a plastic coating. This occurred at time t_7.

At time t_8, Jones receives his first android body, that is, the sort of body with which an android, an artificial human, would be equipped, save that, in this case, it is animated by Jones. It is, we note, an android body designed to resemble his t_1 body. This will help, undoubtedly, but it is, nonetheless, indisputably, a different body. From Jones' point of view, and, I would hope, from our point of view, Jones is still there. Certainly Jones would believe so. I think this is the first point at which intuitions would begin to diverge, though not yet seriously. Publicly, I think there is little doubt that we usually associate personal identity with a particular body. That is Jones, that is Smith, and so on. We recognize the body. That is our usual criterion for personhood. The fact that the body looks much like Jones will presumably facilitate the transition between bodies. Presumably his clients, though taking their business elsewhere, and his friends, who now seldom call, would still regard him as Jones, though the dog, usually "man's best friend," has probably given up on him by now. To Fido it is no longer his beloved Jones; that Jones is gone; to Fido it is probably some sort of contrivance, like a walking stove or refrigerator. Certainly it does not smell right.

At time t_9, Jones buys, or perhaps it is donated by some benevolent consortium of physicians, engineers, and such, a new android body, better and updated. It does not resemble the original t_1 body. Perhaps Jones has selected the popular "Cary-Grant" model, in the hope of winning back the affections of his wife, but the stratagem fails. She realizes it is not Cary Grant. In the meantime, she secures her divorce and is awarded custody of the child.

If the rough-and-ready, everyday criterion for personal identity, namely, sameness of body, had been dogmatically attended to, Jones would have been gone at t_8, when he received his first android body. On the other hand, that criterion would have been clearly set aside in that case, if only because of the resemblance of the t_8 body to the t_1 body, and other similarities to the familiar Jones, in speech, action, and such. This indicates that a familiar, popular criterion, usually workable, viable in most cases, would be unhesitantly superseded or revised in the light of unusual circumstances. The application of intuitions would have it so. It would be taken to be Jones, armed, however, with a radical prosthesis. The challenges to intuitions, moral and rational, however, are more significantly challenged at t_9, with a nonresembling body, in this case, the "Cary-Grant" model. Obviously it is not Cary Grant. And it would not occur to most people that it was Jones, as it does not look like Jones. On the other hand, it claims to be Jones, and there is considerable evidence that it is Jones, for example, from the Jones biography. And we may be certain that the Jones' creditors, his bank, the IRS, and such, will see it as Jones. The t_8 and t_9 developments, the resembling android body and the nonresembling android body, interestingly, demonstrate that body would not be taken as the decisive criterion for personal identity.

This becomes even more clear, and dramatically so, at time t_{10}, when the final android body, perhaps at the request of an embittered Jones, seeking a new life, or to make ends meet, receiving a stipend from researchers, is replaced with a "functional body." We shall suppose that the new body, in which Jones' brain is ensconced, does not resemble a human body, at all. One does not know what it looks like, but perhaps from the outside it appears to be a large, heavy, metallic, rectangular solid on wheels, with bumpers, retractable metallic grasping devices, and rotating, optical and auditory sensors. It weighs some fourteen hundred pounds and is battery powered, the batteries conveniently rechargeable at one or another standard voltage outlets. The ensconced brain, suitably nourished, controls the body by means of something analogous to myoelectrical charges. At this point Jones would clearly and indisputably count as a cyborg, a cybernetic organism, a union of man and machine, or, in this case, of brain and machine.

Is Jones still around?

My personal intuitions, at this point, are clear. If we were willing to accept the situation at t_8 and t_9 as not imperiling Jones personal identity, we should not do so now, at time t_{10}. The only difference here is that the body is different. It is radically different, but it would still count as the body of Jones. There is a sense, of course, in which almost anything could be regarded as part of one's personal identity, for example, one's family, friends, background, education, interests, memberships, pastimes, nationality, and such, but these are not "parts" in the relevant sense we need for a public criterion. They are not "parts" in the sense that bricks are parts of a building, or cells of a body. The person is not constituted by these things, but, rather, these things characterize the person.

Whereas the convictions of an entity that it is Jones, and its claims to that effect, are not decisive, as the entity might be mistaken, they are surely of interest, and might count as evidence. Similarly, if the entity seems familiar with what Jones is familiar with, and should be familiar with, say, the favorite opening of his long-time chess buddy, the occupation of his nephew, the politics of his second cousin, the maiden name of his mother, and such, and expresses himself, via the voice-reproduction device, as Jones might, and so on, we would be assured we were dealing with Jones. Also, and it will prove important, we have the medical records of Jones which detail how it is that he came to be in his present situation. This supplies us with continuity. It is not as though a new ship might have been substituted for the ship of Theseus, on a dark Athenian night.

At time t_{11}, we find Jones as he was in the t_{10} state, except for some personality changes. I think it is at this point that intuitions would begin to diverge significantly, particularly if the personality changes seemed surprising for Jones. To have recourse to a metaphor, some individuals are likely to see personal identity in terms of "software," responses, behavior, and such, and others are likely to see it in terms of "hardware," the machine into which the software might be loaded.

Let us suppose that some unusual personality changes take place in "Jones." Perhaps he begins to roll down hospital corridors at great speed in pursuit of young nurses. His metallic pincerlike grasping appendages do not always mind their manners. Occasional bursts of maniacal laugher are heard through his speakers. He begins, like Edgar Allan Poe, to rate checkers over chess. He sometimes breaks through walls for sport and challenges police cars to games of "chicken" on major highways. Sometimes he disappears from the hospital for days at a time, on which days a number of jewel robberies are concomitantly reported. There are no fingerprints or DNA traces. A number of nurses, between the ages of twenty and twenty-five, receive gifts from a secret admirer, diamond bracelets, pearl necklaces, and such. What is one to make of this?

First, of course, Jones has been under a great deal of stress of late, and it is not surprising that, in view of his many challenges, some personality changes might take place.

Ordinary language does permit one to say, under certain circumstances, without deviancy, that one is a different person. For example, perhaps, in such a sense, one might claim that someone was transformed by drugs or disease, or religion or stamp collecting, that another was never the same after meeting Susan, that another was a different person after the hurricane or train wreck, and so on. Radical personality changes are not unprecedented, and may be caused variously. In the sense one has in mind here, however, that of working out a useful public criterion for sameness of person, such changes, however radical, do not constitute difference of person. Hardware, so to speak, takes precedence over software. For example, if we loaded a hundred programs into one "person" we would not have a hundred persons, no more than an individual with multiple-personality syndrome is really two or more people. Similarly, if we loaded the same program into a hundred "persons," they would not, thereby, become one person.

Time t_{12} is same as t_{11}, except that Jones lapses into unconsciousness. It is unlikely this is a pure coincidence. The trained medical and engineering team, whose reputation has been made by their stellar work on Jones, sets to work again, anxious, if nothing else, to keep him out of the clutches of the law. They also wish to forestall indictments as accessories in a class-action sexual-harassment suit. Furthermore they are now uneasy in the neighborhood of Jones, who seems unusually touchy of late. One sweep of a grasper could punch a hole in a brick wall. Too, a laboratory technician who, in the staff cafeteria, had made a disparaging remark on Jones' paint job was seriously injured in the parking lot, apparently having been run down by an unidentified vehicle.

In any event, at time t_{13}, Jones revives, though he no longer takes himself to be Jones. The team has done its work well. He now thinks he is Smith. He has been given a new personality, a kindly, gentlemanly personality, one tolerant, thoughtful, benevolent, pleasant, civil, sensitive, trusting, concerned, wise, prudent, discreet, and compassionate. Soon, nurses flock to him, to chat, to consult, to seek his sympathy and advice; they reveal to him the small cares and delights, the tiny woes and pleasures, of their day, the intimacies of their relationships, their views of interns and residents, their hopes for the future, and such. To go with this new personality new memories have been implanted, memories likely to brighten, rather than darken or trouble, a life Indeed, Jones looks back with pleasure on the life of Smith, a life which he now takes to be his own.

If there is anything to the "software" approach to personal identity, it seems clear that Jones would now be gone. Those who favor such an approach would mark the dividing point between Jones and no Jones to be at either t_{12} or t_{13}. On the other hand, on the "hardware" approach, Jones is still there, but has undergone significant transformations. He no longer even thinks of himself as Jones. The team has seen to it that he now believes himself to be Smith. They see no point in telling him the truth. We, on the other hand, like the team, realize what has been done. We have followed the medical records.

At time t_{14}, for whatever reason, but doubtless one in the interests of science, the brain is replaced with a new unit, a different brain, instilled with a consciousness and memories identical to those of Jones at time t_{10}. It regards itself as Jones, but we realize that it is mistaken.

At time t_{14} Jones is clearly gone.

3. Some Identity Criteria

As should be clear from the preceding thought experiment, we favor some form of physiological continuity as a criterion for sameness of person. This view will be clarified, expanded, and defended. Before doing so, however, it would seem useful to identify a variety of candidates for the criterion in question, ours and several alternatives, on which we will briefly comment.

3.1 Resembling Physical Pattern

This is the everyday criterion for sameness of person. Its utility and practicality are undeniable. Without it confusion and chaos would obtain. In the quotidian world it is seldom mistaken. Occasional misidentifications, as in the case of identical twins, or such, are relatively unimportant, and may, for most practical purposes, be ignored. One could conceive of situations in which such a criterion might prove relatively useless, but such situations are not likely to occur. Let us suppose, for example, that personal identity was a function of "souls," or such, and that these entities tended to frequently flash about amongst bodies, trading bodies, so to speak. If this were the case, resembling physical pattern would not fare well as a criterion for personal identity. Indeed, it would be difficult to find a public criterion in such a case which would not be subject to compromise or error, say, the entity's claims of identity. Another situation in which resembling physical pattern might be next to useless would be in a population of clones, where each individual in the population, say, of thousands, would be relatively indistinguishable from the others. In such a situation, as identity cards might be exchanged, stolen, lost, and such, numbers might be tattooed on the body, say, on the arm, or such. On the whole, of course, resembling physical pattern is an excellent public criterion for sameness of person. The Jones situation, however, in which physical pattern might shift without loss of personal identity, indicates that resembling physical pattern should not be taken as the final philosophical arbiter of person.

3.2 Resembling Mental Pattern

This would seem a poor choice for a general criterion, though it might prove of value in special cases, as in distinguishing between, say, twins with diverse attitudes or personalities. In such a case, of course, it would be associated with resembling physical pattern. As different persons might have resembling mental patterns, and a given person might, from time to time, exhibit different mental patterns, it seems one should look beyond this approach, in our quest for a general public criterion.

3.3 Physiological Continuity

3.31 General Physiological Continuity

One of the "solutions" to the ship-of-Theseus problem is the "continuity solution," in which the ship counts as the same ship even though each part is eventually replaced, in virtue of the general continuity of the embodied form, as the ship is largely the same from day to day. An analogy would be the notion that the cells in the human body are replaced every seven years, or so, but we think of it as being the same body of the same human, though it changes. To be sure, not all the cells in the human body are replaced every seven years, or so. Neural tissue is not replaced, allowing one, for example, to remember situations, and such, which took place more than seven years ago. The usual view of the ship of Theseus, of course, is that it is not the same ship as it was. Not one bit of it is the same, so how could it be the same ship? It would be, say, a similar ship, a resembling ship. The fact that the human, undergoing massive cell replacement, does not regard his identity in jeopardy would doubtless have to do with the continuity of consciousness, and his memories. We have seen, however, that consciousness and memory, invaluable to the individual, according him a private criterion for personal identity, do not provide us with a public criterion of personal identity. It might be noted, in passing, that general physiological continuity provides a far more secure criterion for personal identity than resembling physical pattern and resembling mental pattern. Consider, for example, the difference between the physical and mental patterning of the infant and the adult. They are remarkably different, but we would commonly say that they are one and the same person, that they are both Bill, or both Susan, and so on.. Acknowledging the value of general physiological continuity as a criterion of personal identity, one must still note the profound changes generally involved, the replacement of DNA with the replacement of the cells, and so on. This leads us to want to have something which is the same throughout these changes. For example, in our thought experiment, we noted that in the case of Jones there was a massive alteration in physiological continuity. Thus, if one is to opt for a physiological criterion of personal identity, involving both continuity and sameness, one is encouraged to look beyond a general physiological continuity to something else, to something which exhibits both continuity and stability, continuity and permanence.

3.32 Continuity of Neural Tissue, and, in Particular, Brain Continuity

Arguments supporting this criterion occur in Section 4, following.

3.4 Behavioral Continuity

This criterion, as applied to humans, tends to be characterized by some of the same difficulties associated with resembling mental pattern, reviewed above. If one were thinking here of resembling behavioral patterns, different persons might have resembling behavioral patterns, and a given person might, from time to time, exhibit different behavioral patterns. There is more to be said for it, however, if one is thinking, as is relevant here, in terms of continuity. In such a case, only one individual would be involved. It would fail as a general criterion, of course, because a given person may, at one time or another, display very different forms of behavior, depending on a variety of factors, such as age, health, mental state, and stress. One can conceive, of course, of situations when a criterion of this sort might be our best evidence for sameness of person, for example, similar handwriting on two letters, similar brush techniques on two paintings, similar patterns in a series of burglaries, and so on. At time t_{10} in the "Saga of Jones," when he is ensconced in the "functional body," we might use such a criterion to reassure ourselves that it is truly Jones, or that it is truly Jones "in there," based on conversations with him, exchanging old memories, noting whether or not he still makes the same eleventh move in Evans Gambit, and so on.

3.5 Nature and Contents of Consciousness

This criterion is invaluable to the person, and would be superseded, if ever, only in the face of overwhelming evidence to the contrary. As noted earlier, the criterion is private, and will not do for a public criterion. On the other hand, given the enormous importance of private criteria, in our own case and in that of others, it seems that the elected public criterion might well be one which underlies, or is significantly linked with, the private criterion.

3.6 Same "Soul," "Spirit," "Vital Force," "Animate Principle," or Such

Whereas it is easy to verbalize in this area, generate inferences, split semantic hairs, contrast positions, refute opponents, convince others, refine one's views, exercise judgment, and such, it is difficult to know what one is talking about. If a soul is, for example, a ghost, or a mystery body, or such, it seems it would have to be an invisible double or something, but it is not clear it could be a double as it is not visible, at least usually, does not depress scales, does not show up in mirrors, at least usually, does not leave footprints, at least usually, does without DNA, does not occupy space, at least usually, and so on. Also, it manages to turn its head without a neck, to walk without feet, to hear without ears, to see without eyes, and so on, or, perhaps, it turns a mystery head with a mystery neck, walks on mystery feet, wears mystery shoes, hears with mystery ears, sees with mystery eyes, and so on. Does it have a mystery circulatory system, mystery sweat glands, and such? If, on the other hand, it is merely the form of the body, then there would have to be short souls and tall souls, big souls and little souls, fat souls and thin souls, and so on. Too, if the body disintegrates, it seems the form would cease to exist. This seems to bring one back to the ghost, or the mystery body, and such. But, let us suppose there does exist some subtle, elusive entity which is undetectable by currently known means, an entity which slips with ease, unobstructed, through the gross mesh of science's clumsy nets, so naively and rudely woven of irrelevant quantitative strands. And let us suppose, too, that this entity, somehow, to our surprise, is really the person. The correct criterion for sameness of person would then be sameness of soul. Let us also suppose that this solves our problem, rather than doubling it, and elevating it into the stratosphere of unintelligibility.

The major difficulty here, however, granting that the existence of the soul is accepted, whatever it is one is accepting, is that this fails to give us our long-sought public criterion. There is not even convincing evidence for a self, as Hume pointed out long ago, and one supposes the "soul" is in no stronger an epistemological position. It may exist, or not exist, as it pleases. If it is hypothesized as a necessary condition for consciousness, then all conscious entities would have souls, of one sort or another, and it is hard to tell how far down the phylogenetic scale this would go.

In any event, whether the soul exists or not, and even if it exists and is the person, there seems to be no current way to test for its existence, or nonexistence, or to test for its relationship, or lack of relationship, to the person.

These things being the case, one cannot use the "soul" as a public criterion for sameness of person.

3.7 Claims

3.71 Entity's Claims

If anyone knows a given person, presumably, in most cases, it will be the person himself. Who, at least generally, would know the person better? This is not to claim total self-transparency. It is undoubtedly true that there is much that remains unknown about any human being, and even to himself. Too, often there are aspects of a human being which are clearer to others than to himself. One supposes that all human beings are short on self-perception to one extent or another, and some may even be unusually skilled in self-deception. But all of this is irrelevant to our central concern, that of arriving at a viable public criterion for personal identity, for sameness of person. The entity's claims in such matters are normally taken as sufficient and rationally conclusive, if not logically so. These claims do give us a public criterion, a criterion publicly verifiable. If the individual claims to be so-and-so, the civil consequence, pending evidence, or suspicions, to the contrary, is to accept him as so-and-so. To doubt such claims would, in most cases, be evidence of mental deviancy, perhaps paranoia. Trust is the foundation of community. On the other hand, the entity's claims, however reliable and informative they may be in general, need not be accepted as universally conclusive, as such claims may be mistaken or false. It is easy to conceive of situations in which it would be advisable, and even morally imperative, to lie about one's identity. One would not expect the criminal, the impostor, the spy, and so on, to invariably speak the truth. Too, an individual may be mistaken. That fellow wandering about near the fence might not be Napoleon or Alexander the Great.

3.72 Public's Claims

Public claims of identity, as those of the entity himself, may be false or mistaken. Arrests are made, bills are served, letters and packages are delivered, individuals are misidentified, even framed, witnesses testify, judges and juries do their best, individuals are confused with celebrities, celebrities are spotted and annoyed, and so on. Public claims, correct or incorrect, are familiar, even rampant. Presumably most public claims, of one sort or another, are correct, but, as the entity's claims, they are obviously secondary and fallible. Our need here is for a primary public criterion, not a secondary public criterion. More important than claims, of one sort or another, is whether or not the claims are true, and what makes them true. One wants to have a criterion in terms of which personal identity might be regarded as constituted, something in terms of which it might be viably understood.

3.8 Legal Decision

The matter of legal decision is far mightier in portent than it might seem at first glance. It is not just another way of looking at things.

Let it be understood, as I suppose is clear by now, that we are embarked here on an adventure of explication, not of empirical discovery.

There is an analogy here, in what we are doing, to legal rulings, except that legal rulings, unlike gentlemanly philosophical recommendations, can be institutionalized as second-order fictions, and enforced by all the coercive apparatus, all the organized, legalized violence, at the disposal of the state.

That is a very significant difference.

It is little wonder that government and law appeal to those with a thirst for power.

Let us contrast legal ruling and philosophical recommendation.

In explication, or philosophical recommendation, one is less concerned, metaphorically speaking, with the length of a board than with devising a unit of measurement, with designing a conceptual structure in terms of which to deal successfully with experience, or, adopting the postulate of naive realism, with reality.

Let us extend the metaphor.

The board is as long as it is, but that is analytic, a logical truth. For example, let 'Bx' be read as 'x is a board' and 'Lxy' be read as 'x is as long as y'. Then '(x) (Bx É Lxx)', read as 'For every x, if x is a board, then x is as long as x', would be a logical truth. If there are no boards, it is true in virtue of there being no legitimate substitution instances for the first quantified function; and if there is even one board, it is true because the consequent quantified function would be true for all possible legitimate substitution instances, as every board is the same length as itself. We shall suppose that that much conceptual structure, at least, is in place.

The question now might be, "Could the board be five feet long if that unit of measurement did not exist?" I think the answer to that is obviously negative. "If the unit of measurement had existed, say, twenty thousand years ago, the board would have been five feet long" is a different claim, and quite possibly true, but it is also one which presupposes the conceptual structure. Compare, "Is the board sixty-two sklinks in length?" "Is it thirty-six squitches in length?" "Is it zapduck zerkels in skorch?" It is quite possible, of course, that if there were units, numbers, and properties such as sklinks, squitches, zapduck, zerkels, and skorch, they might be applicable to the board in question. In the meantime, such locutions would seem to lack truth values.

So we are concerned with the judicious revision, adjustment, reformation, possible improvement, and such, of the conceptual structure. Putting aside "animal truth," that is, realized expectations, and such, truth as commonly conceived, empirical truth, and, indeed, analytic truth, does not exist *simpliciter*. It is a consequence of the conceptual structure, which is logically prior. Consider flowing water. Water, like rocks, and trees, is not true or false. Consider a statement: "The water in the Cheyenne river near Edgemont, South Dakota, at this time of year, is five feet deep," or, if we were Romans, "The water in the Cheyenne River near Edgemont, South Dakota, at this time of year, is five feet high."

Obviously the truth or falsity of such statements requires an extremely complex conceptual structure, speech communities, semantics, names, locations, municipal and political entities, a calendar, units of measurement, procedures for relating thought to language, language to the world, confirmation methodologies, and so on. Presumably the latter two locutions would not be true or false to an insect, a monolingual speaker of Hindi, a Martian looking on through his telescope, and such. The water is neither true nor false. Reality may be "out there," but truth is not. Reality does not need truth, but truth, of any sort, requires reality. Even analytic truth, to exist, requires its realities, minds, and a conceptual structure. There are no consequences which are not consequences of something. It may be true that there is an infinite number of natural numbers, of even and odd numbers, and that there is the same number of odd numbers as there are of even and odd numbers altogether, and the same number of numbers divisible by 647 as there are even and odd numbers altogether, and so on, but it is hard to suppose that that was the case in the Cambrian age, some six hundred million years ago. One might as well believe that the perfect chess game exists, which may never be played. Or, if you like, the perfect Zerk game exists, which may never be played. (Zerk is the name of a complex board game, having seventy-two pieces, and played only on Tuesdays, which might or might not be invented someday.)

So, philosophically, one is concerned with explication, and explication is concerned with the conceptual structure, which, in turn, is important because truth, as commonly conceived, depends upon it. Without it, there is no truth, as commonly conceived. Explications are proposals. Proposals are neither true nor false, but, if one accepts the proposal, then truth and falsity fall into place. The conceptual structure, of which truth, as commonly conceived, is a consequence, is usually invisible, as it is, for the most part, embedded in language. Many are unaware it exists. It is just obvious that that animal is a cat, and that other animal a dog, and so on. One is not bothered with barking "cogs" and purring "dats," unless, say, genetic engineering intrudes. Explications can be arbitrary, but, ideally, are not. Ideally they are abundantly supported with reasons, of one sort or another, relevant to accepted, valued purposes, of one sort or another. Ideally, they are so judicious, so convincing, so practical, so helpful, that they may suddenly seem not so much judicious, convincing, practical, and helpful, as correct, as simply right.

The explication stands or falls on its own merits. It is offered, so to speak, in the intellectual marketplace. Its success or failure is a function of democratic evaluation. It may exist, untroubled, with contrary explications. There need be no one way to conceptualize data. There might even be, particularly subject to different ends in view, different, but equally interesting or plausible, equally aesthetic or moral, conceptualizations. In any event, the explication is civil. It fines no one, it imprisons no one, it executes no one.

Now let us consider the legal ruling.

Five considerations are particularly noteworthy.

First, the law, like philosophy, will in the future be moving into new territories, dealing with new problems, not anticipated in former law. To some extent, this is how law develops normally, but the number and nature of new challenges, crowding forth, most unanticipated and many unusual or eccentric, is likely to exceed the grasp of precedent and strain the vision and fabric of jurisprudence as these things are normally conceived. The judge will be forced less to interpret and apply law, allegedly, than to openly create it, and not in the light of established policy preferences, as the policy preferences may be neither established, nor clear, at least at the time. He will have to make the law up as he goes along, for a time, at least, without the usual ideological signals and road maps used in the past, denied even the comfort of constitutional pretenses.

Second, whereas those who promulgate legal rulings are commonly familiar with the law, respecting it or not, they are, outside the law, usually unenlightened, uninformed, guess-work amateurs. There is no particular reason to suppose that legal professionals are any more in a position to judiciously resolve novel conceptual and ethical issues than physicians, architects, accountants, grocers, plumbers, or philosophers. These are new countries, and the maps do not yet exist.

Thirdly, society, or special interests within society, will request, or demand, rulings in these new areas, and, as society is currently constituted, judges will be expected to deliver these rulings. There seems no practical alternative, as autocrats, kings, seers, prophets, and such are in short supply, and referendums are unlikely, and there is no assurance that a referendum would be any more fair or judicious than the rulings of a judge, or judges.

Fourthly, many of these rulings would have serious effects, involving even life and death. The legal acceptance or rejection of various reprogenetic practices, the banning or accepting of cloning, what may be done and what may not be done in genetic engineering, and so on, are fraught with life consequences. As a simple example, an effective banning of cloning would result in the childlessness of many couples, and the prevention of thousands of new lives from being lived. Legal rulings are not proposals and recommendations. They are institutionalized second-order fictions, emplaced by decision and perpetuated by force, as much as laws pertaining to the extermination of heretics, the disenfranchisement of minorities, the expropriation of property, the denial of free speech, and so on.

Fifthly, and lastly, the bench, as are legislatures, is subject to pressure. It is obvious that an elected official is likely to consider the self-referring prudential aspects of his views and actions. That is only human, but it does not bode well for candor and objectivity; it bodes better for dishonesty and rationalization. Similar considerations pertain to many judges, who are party members, are expected to promote an ideology, or, at the least, not to contravene it, and, in any event, as are legislators, are subject to the risks of loss of office. Some judges receive a practical tenure after appointment, which, in theory, is supposed to protect them from trimming their sails to prevailing political winds. On the other hand, to receive that appointment, and to be confirmed in it, requires the satisfaction of a variety of interests. Often, there seem to be wars fought over such appointments, to make certain that various special interests will be served by the appointment. The notion of justice being blind and weighing matters impartially is a fantasy unpersuasive today, I suspect, to anyone a day or two past puberty. Picketings, demonstrations, riots, and such, are also helpful in reminding judges of what is expected.

All in all then, many of these problems, conceptual and ethical, for example, as to who or what is a person, and what, if anything, is owed to a person, are going to be resolved not as a matter of serious thought, informed by research, the framing of various possible solutions, with various likely consequences, viewed in terms of ends desired, and such, but in virtue of ignorance, intimidation, partisanship, ideology, and, ultimately, possibly, violence.

It is thus quite possible that the *de facto* resolution of our problem, concerning personal identity, will depend on a legal ruling. On the other hand, until prohibited, philosophers, and others, may consider these issues, and such issues, in idyllic tranquility, concerning themselves not with how they will be managed but with how they might be well managed, not with how they will be resolved, in a world of guns and money, but with how they might best be resolved, were reality "closer to the heart's desire."

One may hope, of course, as suggested in our first consideration, that the newness of some of these issues will catch ideologies unprepared and so confuse special interests that, before they form their new advocacy groups, a judge, lacking political guidance, and availing himself of a brief, if harrowing, window of opportunity, might have a chance to take an unbiased look at the matter. Presumably he would have to work fast.

4. Some Arguments in Favor of the "Sameness of Brain" Criterion

In the search for a useful public criterion of personal identity, it would seem the brain would be an ideal candidate for a variety of reasons, some of which follow.

4.1 The Basis Argument

The brain is responsible for patterns of thought and behavior, and for the nature and contents of consciousness. Other things are derivative. The brain is fundamental.

The word 'experience' may be used variously. In one sense of the word, one cannot have the experience of walking in the wood unless one is walking in the wood; otherwise it would merely be that one seems to have the experience, but is not having it. In another sense, a more sophisticated, scientific sense, experience is recognized as internal to the organism and, and whether one is walking in the wood or not, one could have the experience of walking in the wood. One might not be really walking in the wood, but one could nonetheless have the "walking-in-the-wood experience." An experience need not be what it seems to be in order to be an experience. There are dream experiences, for example, and visions, hallucinations, illusions, and such. They are real experiences, but they may not be experiences resulting from typical environmental stimuli, light waves, sound waves, and such. Experience is brain dependent, and an appropriately stimulated brain could produce any experience which might be associated with an outside world, and many which could not result from typical contact with an outside world. Indeed, on certain metaphysical postulates, there is no mind-independent, physical, external world, at all, to generate experiences. Such a world is not only unnecessary to generate such experiences, its supposition being a gratuitous hypothesis, worth eliminating in virtue of Ockham's razor, for example, but it is improbable that such a world exists, and, perhaps, even, demonstrably false. The experiences, which are undeniable, of course, are, rather, the result of a different causality. The "physical world" hypothesis, then, is only one interpretation of the data of personal experience, amongst a number of other possibilities, or interpretations.

Interestingly, what is going on in the brain in dreaming and in familiar, quotidian experience is very much the same, except that waking experience is presumably indexed in part to different stimulus conditions.

Phantom-limb phenomena, which occurs to one degree or another amongst some amputees, is evidence of the brain-dependence of feeling. If one can feel sensations "in a nonexistent limb" that seems clear evidence that such feelings are brain-dependent, and, in virtue of evolutionary selections, are extradited to various portions of the body. Obviously it is in the organism's best interest, with respect to gene replication, and such, that this should be the case. Pain, for example, calls attention to one or another portion of a body. If all pain was not only brain-dependent but seemed to afflict only the brain, as well, it might be difficult or impossible to locate and attend to internal conditions, promptly withdraw limbs and body from injurious situations and environments, and so on. An interesting analogue to the extradition of sensation is a variety of simple experiments. For example, take the blunt end of a pen, or such, and rub it on a smooth surface, or take a stick and poke the ground. I think there will be few individuals who will not "feel" the surface and "feel" the ground, and note that the surface feels smooth and the ground feels hard. On the other hand, there are no nerves in the pen, or such, nor in the stick, and what is felt, if anything, technically, would be associated with pressures on the fingers or hand, not the surface or the ground.

Brains can be kept alive without bodies and presumably have experiences, though dissociated from a body. This seems clear from Russian experiments, conducted as long as decades ago, with decapitated dogs, in which, for example, a bitter fluid placed on the dog's tongue would elicit a recoil behavior and a grimace of distaste. More recently, experiments have been conducted in the West with monkey brains in which the brains were kept alive for several days, and continued to exhibit brain activity, until the experiments were concluded and the brains destroyed. Such experiments are probably the sort of thing which produced speculation concerning the possible experiences of human brains under similar conditions, leading to "brain in a vat" thought experiments. It seems to be true that a human brain, separated from a body and appropriately stimulated, say, myoelectrically, would have experiences. If the technology were available it might be stimulated in such a way that it would be unaware it was a brain in a vat, and not undergoing the experiences in a "real world." One supposes that a healthy brain removed from a dying body might thus enjoy additional years of life, have selected experiences, perhaps involving rejuvenation, adventures, romances, and such. One might also suppose that new forms of mental illness might appear, such as the individual who suspects that he is not in a "real world," but only a brain in a vat, perhaps placed there by enemies who control his experiential world for their own amusement. And how would a neurotic brain in a vat, who had elected such a life, and purchased it at great expense, know that it had not been betrayed, and that it was still in a head, and so on?

The "basis argument" seems the strongest argument for using the brain as the ultimate criterion for sameness of person. For example, if a Jones was a "brain in a vat, it would still be Jones, even if it were stimulated in such a way as to seem to be, to itself, a different person, say, Smith, Napoleon, Alexander the Great, or any one else.

Several of the arguments which follow would also serve for using a general physiological continuity for a public criterion for sameness of person, but the brain is the nucleus of "person," as we have seen, and a body, kept alive with a dead brain, or without a brain, would presumably not be regarded as a person, and, on our approach, at least, would not count as a person, no more than a piece of independent flesh, if it could be kept alive independently, undergoing cell division, growth, and such.

4.2 Cultural-Practice Arguments

There are many criteria by means of which a proposed explication may be evaluated. One is that the explication should, *ceteris paribus*, respect cultural practice, retain familiar utilities, avoid linguistic deviancy, and such. Commonly, in an explication, one wishes to hold to a core of understanding, while improving the germane conceptual structure, presumably retaining much of what was originally valuable, while perhaps deleting and adding elements, and trying to clarify the whole. Pragmatic considerations abound, and are paramount.

4.21 The Coma Argument

Individuals who are in a coma, are unconscious, etc., are commonly regarded as the same person. As one supposes identity in such a situation cannot be indexed to mental pattern, consciousness, behavior, and such, it is natural to index it to body, to physiological continuity, and, in such an indexing, we have seen that neural tissue, and, in particular, the brain, are those aspects of physiological continuity which are most relevant.

4.22 The Amnesia Argument

Individuals who are suffering from amnesia are regarded as the same person. Even they think they are the same person, only they do not recall who that person is.

4.23 The Altered-Personality Argument

An individual whose behavior changes, e.g., as a result of trauma, illness, a reversal of fortune, conversion, or such, is regarded as the same person. Any sense in which the individual is not regarded as the same person would be metaphorical, *e.g.*, 'Jones is not himself today', 'That is not the Jones I know', 'Jones is a new man', and such.

4.24 The Multiple-Personality Argument

Individuals with more than one personality are regarded as the same person, regardless of what might be their own views. The possibility of multiple personalities militates against identifying sameness of person with consciousness, mental pattern, behavior, and such. Analogously, an actor may play many roles but is still the same actor, the same person.

4.3 The Permanence Argument

This argument emphasizes the relative permanence of the brain, an additional consideration supporting its selection as a public criterion for sameness of person. Memories can fade, consciousness can change; it is intangible, fugitive, intermittent, and transitory. Further, as noted earlier, neural tissue is not replaced as most cells are. It does not participate in the typical cellular-transformation cycle, occurring approximately every seven years.

4.4 Intuitive-Satisfactoriness Arguments

We have noted some cultural-practice arguments, for example, the coma, amnesia, altered-personality, and multiple-personality arguments, in which our cultural practice is generally clear, in most cases even to being embedded in law. Naturally, we do not have cultural practices pertaining to a number of hypothetical situations, not yet realized, although within the likely grasp of future technologies, a sample of which will be shortly noted. What we do have, however, is a statistical commonality of intuitions, moral, social, cognitive, and such, which tends to confirm, reinforce, support, and sustain community and, more broadly, civilization. An interesting but trivial example of a commonality of intuition occurs amongst French speakers. In French, all nouns must be either masculine or feminine; there is no neuter, as in, say, German and English. Apparently, when a new noun enters the French language, there is almost always overwhelming agreement amongst French speakers as to whether the noun should be masculine or feminine. To one who is not a native French speaker this is likely to seem surprising. Why, for example, should 'ship' be masculine in French ('*le navire*'), feminine in English, and neuter in German ('*das Schiff*')? In the case of 'ship' one could undoubtedly come up with some interesting suggestions, but are there such interesting suggestions for thousands of other words, as well? And if one cites resemblances or similarities between new and old words, how did it all get started in the first place? In any event, a common area in which agreement is likely to obtain in communities is that of moral intuition. Without a commonality of moral intuition it is not always clear that a community could survive. Divergences amongst moral intuitions can lead not only to the disruption of communities but to their transformation or extinction. These things, moral intuitions, can change over time and under diverse circumstances, and can be manipulated by the techniques of social engineering, by those who have the power to do so, but substantial, even massive, agreement is essential to communal health and stability. Social intuition is also common, a sensing as to what, in given situations, is appropriate or inappropriate, and such. Some individuals, for example, are much more sensitive to the feelings, desires, motivations, likes and dislikes of others than are some others. Cognitive intuitions, of the sort which will shortly be proposed, are also relatively familiar. We have mentioned the case of French nouns. That would seem to come under the rubric of linguistic intuitions. Similarly, one might wonder why no native English speaker ever says 'It rains', despite the fact that that is perfect English? And why do English speakers say, "All that glitters is not gold," when they know very well that gold glitters? Another example is the pea/potatoes anomaly. If an English speaker had one pea on his plate, he would not say he had peas on his plate. On the other hand, he is likely to say he has potatoes on his plate, even if he obviously has far less than even one potato on his plate, and so on. Obviously these things are mysterious. Now rational intuitions, or conceptual intuitions, are usually very similar to linguistic intuitions, for example, what seems the right way, or the best way, to think of a situation, which brings us to some intuitive-satisfactoriness arguments.

Philosophers of a descriptivist stripe are likely to think, when they are pleased, that they have discovered the truth; philosophers of a prescriptivist stripe, when they are pleased, are likely to think that they have come upon a really good way to think about something, even, hopefully, subject to certain criteria, the very best way to think about something.

4.41 The "Boxed Brain" Argument

This argument is reminiscent of a "brain in a vat" argument except that the brain in its vat is alone, in a *de facto* solitude, despite the fact that its experiential world, if the stimulators so desire, might be crowded with others. The brain in its vat, bobbing about, rich in blood and nutrients, its synapses firing apace, might take itself to be thrilling crowds, entrancing audiences, leading mass movements, outdoing Napoleon and Alexander, and so on, but it isn't. The brain in the box, on the other hand, is presumably well aware it is a brain in a box, and, by receptors and transmitters, is in contact with the outside world; it is capable of carrying on conversations, answering questions, administering advice, issuing orders, running companies and countries, and so on.

If the brain in the box was the brain of a Jones, conducted itself like the Jones in question, had the memories of that Jones, claimed it was that Jones, and so on, I think we would take it to be that Jones, the father, the brother, the friend, the boss, the leader, or whatever. Our intuitions would have it so, unless we thought it to be to our advantage to conceive the situation differently, adopting a different criterion of personhood, and then, say, slipping the box off its shelf, dropping it off the roof, onto the sidewalk below, or such, perhaps to hasten an inheritance.

4.42 The Interpersonal-Brain-Transplant Argument

On the hypothesis that person goes with brain, if brains are exchanged amongst persons, then persons will occupy new bodies. This would follow from the approach I have been advocating, which I think is intuitively satisfactory, at least upon reflection, though surprising and certainly unfamiliar, but, more to the point here, this approach reflects, it seems, a more general sense of intuitive satisfactoriness. For example, in science fiction, it is a common supposition that a brain transplant moves the person into a new body, and, indeed, perhaps a body not altogether to the person's liking, perhaps that of a notorious criminal wanted by the law, who wants a different body as a disguise, to abet an escape, or, say, in the case of a woman her brain is inserted into the body of a rich hag, who desires, in her turn, to obtain a fresher, younger, more beautiful body, and so on. And we hope, of course, that all this will somehow turn out for the best.

In any event, it seems an unchallenged supposition that moving the brain moves the person.

If, alternatively, we took the body to determine the person, the innocent victim in the first case would become the criminal, having that body, and, in the second case, the innocent victim, the young woman, would become the hag, having that body, and the criminal would become the innocent fellow, having that body, and the hag, having the new body, would become the young woman.

That would not seem to be a good way to envisage the situation.

It would be counterintuitive.

And I doubt that it would sell.

4.43 The Extrapersonal-Brain-Transplant Argument

Here one might, again, have recourse to science fiction, which, again, accepts common intuitions, as to how unusual situations would be envisaged, or understood. In this case, one is dealing with extrapersonal brain transplants. Here, for example, one might envisage the ensconcement of a human brain in an animal, perhaps a lion or gorilla, or in a device or machine, thereby producing a cyborg. In all such cases the intuitions are clear. The human would be regarded, as was Jones, from time to time, as having a new body.

4.5 The "Brain Replacement" Argument

If an individual suffered brain damage and his brain was replaced with a "new unit," a different brain, whether another human brain, from another human, or a new, manufactured brain, perhaps electronic or synthesized, or even a brain cloned from the original brain, presumably the first individual would no longer exist. We saw something like this occur at time t_{14} in "the Saga of Jones."

4.6 The "Two-Universe" Argument

If there were two "identical" universes, presumably Individual$_1$ in Universe$_1$ would not be the same person as Individual$_2$ in Universe$_2$. Difference of brain here would be a sufficient condition for difference of person, even if the tiniest of feelings and the least of thoughts were the same in both universes. Presumably, in a case of this sort one would rely on a classical, intuitively coercive, principle of individuation, that of difference of material substance, or, more narrowly, difference of matter, whether material or not. (Cf. Aristotle.)

Analogously, even if monozygotic twins were identical, which, strictly, they are not, they would count as different individuals, even if they happened to have the same feelings and thoughts at the same time, and so on.

They are different people.

4.7 The "Split Universe" Argument

The notion of an unlimited number of worlds is an ancient notion. It figures, for example, in the teachings of Epicurus. Similarly, if one hypothesizes a god or gods who are good at creating worlds, rather than being created by them, there is no reason why their creative effort should be limited to a single world; if they enjoy creating worlds, why not create an indefinite number, in various colors, shapes, and sizes, even an infinite number? One could, for example, always create additional space, new dimensions, or such, in which to put them, if needed.

In what follows, we have in mind a particular "many worlds" hypothesis, a "branching universe" hypothesis, suggested by contemporary physics, the notion of a constantly dividing, efflorescing universe. I would suppose that this sort of thing would count as a bizarre, maverick hypothesis, even to most physicists. On the other hand, contemporary physics tends to abound in surprising conjectures, for example, randomness, lack of causality, things just happening, particles moving backward in time, popping into existence from "nothingness," multidimensionality, and observer-dependence, something which would surely have intrigued Berkeley and Hegel. One recalls "Schrödinger's Cat," a charming anecdote, except to cat lovers, invented by the nuclear physicist, Ernst Schrödinger, to convey the counterintuitive nature of what is being claimed in "observer dependence." Briefly, one inserts a live cat in a box, a very special sort of box, in which are contained an active source of radioactive emissions, a receptor which will record any discharged particle, and an apparatus which, if a particle is registered, will shatter a container of poison gas, killing the cat. Ordinarily, we would suppose that the cat, by now, some time having elapsed, is either alive or dead, depending on whether or not a particle has been emitted, registered, and so on. Logic itself encourages this view. On the other hand, following Schrödinger, in this analogue to physics, the cat is neither alive nor dead, until one opens the box and looks. Thus, the reality does not exist, until it has been created by the observation. This is the sort of thing which may be said safely only by physicists, or philosophers with tenure. In any event, we will make use of a "branching universe" hypothesis in our thought experiment. In this hypothesis, there are an indefinite number of world lines, not just in possibility but in actuality. Doubtless a supposition of this sort is supposed to account for some sort of experimental data. That one gratefully leaves to the physicists. They would have to have their reasons. As a layman it seems this would violate the principle of the conservation of matter/energy, and, as data are always compatible with a number, theoretically an infinite number, of different interpretations, perhaps one should look further. In any event, our concern here is not with a particular theory, really, nor with second-guessing physicists, but with a "what-if" scenario, to test the viability of an explication against possible counter-examples.

So let us suppose that a Universe$_n$ at time$_n$, perhaps our universe, for whatever reason, or for no reason, given quantum physics, divides, or branches, into $Universe_{n+1}$ and $Universe_{n+2}$, which, in turn branch into $Universes_{n+3, \, n+4, \, n+5, \, n+6, \, n+7}$, and so on, indefinitely. In short, each universe spits into two universes, and each of the new universes splits again into two universes, and so on. And, one supposes, a given universe might split into more than two universes, why not, and so on, at least as a logical possibility.

One wonders if something like this might be true.

In any event, one supposes that the universe does not understand itself, and it is certainly under no obligation to be intelligible to human beings.

One takes what comfort one can from such reflections.

How might one handle our Jones in such a situation? The major difficulty here, unlike the puzzle in the Two-Universe Argument, is that we start off with one Jones, not two.

Here we must understand two senses of identity, already noted, first, external, partial, pervasive identity, in which two or more different entities are the same in all significant respects, and, second, internal, synthetic, nonlinguistic intrinsic identity, where one is concerned with identity through time, as in sameness of person, in a human case.

In the second sense of identity, in the above scenario, much would depend on the nature of the branching. If $Universe_n$ buds, so to speak, then a single universe would proceed on a single world line. Whereas directions make little sense in a situation of this sort, an analogy may be useful. For example, suppose that our Jones is always in the universe which "branches to the left." Thus, he could remain the same, while replicas are being regularly produced. Replicas, of course, are not Jones, though they are identical to Jones in the first sense of identity, mentioned above. On the other hand, if the branching is, so to speak, "down the middle," then Jones is finished with the first branching. He is the same person, only briefly, depending on how frequently these branchings occur. If they occur each nanosecond, or so, then he is not the same person for very long. He is finished, replaced by a pair of identicals. For the time involved, the nanosecond, or minute, or week, he is the same person, and we can use the criterion of sameness of brain, however briefly. With the identicals, as well, they are who they are, but are not the original Jones. We cannot multiply Joneses, as there is only one, no more than we can multiply a single automobile, though we might produce an indefinite number of automobiles which were very much like the first automobile. They are not the same automobile, if only because they are in different universes.

4.8 The "Eternal Recurrence" Argument

One of the most dangerous things a philosopher can do is to take science seriously, not as a noble endeavor and an interesting, if limited, methodology, but as a set of dated claims. Generation after generation of scientists have chuckled good-naturedly, affectionately, over the occasional mistakes and naive misapprehensions of their predecessors, usually hard-working, brilliant fellows, to become, in their turn, a source of amusement for their successors. Indeed, there is a saying that today's science is tomorrow's fallacy. In any event, it is healthy to remember that, in their time, unmoved movers, natural place, celestial spheres, epicycles, hard little atoms, phlogiston, vital forces, vortexes, ether, and such, were taken as seriously as one takes, or tries to take, quantum physics and relativity theory today, two great triumphs of contemporary science, which may be incompatible with one another. We have noted some of the interesting claims in contemporary science earlier, multidimensionality, and such. In any event, one wishes the best to science, and more power to her. Her victories and glories can be ambivalent, and come with a price, easily wrought mass murder, planetary peril, and such, but who would want to return to a small, flat stationary world, even if it were at the center of a small, comfortable universe, a world of werewolves, witches, and demons, of filth, starvation, and disease, of widespread infant mortality, of short, stunted lives, and so on. Many cheers for science. Three are not enough.

On the other hand, we noted, above, that one of the most dangerous things a philosopher can do is to take science seriously, at least as it is in one afternoon or another. He should be careful of boarding what may prove to be a sinking ship. The philosopher is much better off to think in terms of the human world, how one should think of it, and, if he can bring himself to do so, consider how one might live well within it, not necessarily successfully, but well. On the other hand, philosophers have a penchant for "living dangerously," particularly if it can be done without risk. For example, the philosopher who ignores science, at least in her major outlines, would today be not only an anachronism, but an anomaly. Scientists, after all, are, are they not, our "natural philosophers"? It is hard not to take them seriously, whether one should do so or not. One's world view, for better, not for worse, is likely to be heavily influenced by them. One would certainly think twice before rejecting the heliocentric theory of the solar system, the theory of evolution and such. On the other hand, one suspects that some of science today will be the rubbish of science tomorrow. And it is difficult to tell the treasures from the trash, even for scientists, and certainly for philosophers. An excellent example, or a seeming example, in case he was correct, of a philosopher who made the mistake of taking science seriously, the science of his day, as he understood it, was Friedrich Nietzsche. He took the Newtonian universe, the existence of ultimate particles, constant motion, the laws of nature, rigid determinism, and such, seriously, with the result that he updated the ancient Stoic concept of the Eternal Recurrence in the light of the physics of his day, now seemingly superseded. In Nietzsche's view, there was a large but finite number of small, material particles in existence, whose constant motion was governed inexorably by the laws of a deterministic universe. This being the case, sooner or later, given the patience of eternity, every configuration of particles, every world state, would recur. This image of a huge, dismal, alien, spiritually empty, indifferent, meaningless system, this relentless, unstoppable, indestructible, inhuman cosmic clock, ever ticking, ever counting off endless, pointless hours, ever returning upon itself, to repeat itself, at awesome intervals, moment by identical moment, apparently produced a crisis in Nietzsche, one to which many rational, sensitive minds might have succumbed, collapsing into madness, license, defeat, or depression, but which to others might constitute a challenge to acceptance, even affirmation, even celebration. In Nietzsche's view the cosmic system was redeemed by the occasional transports of joy, however poignant or brief, inextricably intertwined amongst its motions, which were every bit as real as misery, hate, disease, and death. Unwilling to disavow those rare, redemptive moments, taking them to immeasurably outweigh the darkness, vacuity, and pain, Nietzsche not only reconciled himself to reality as he understood it, but embraced it. In his way he had found joy, and love.

In this scenario, currently unpopular in contemporary science, but perhaps to be eventually refurbished or rejuvenated in one form or another, we have, at last, happily, no real problem with a public criterion for sameness of person in terms of sameness of brain.

Consider the following analogy:

We have specific bicycle, call it B_1. We take the bicycle apart, storing the parts in one place or another, perhaps even trucking them about, mailing them overseas, and such. Then, after a given amount of time, say, six months, we recover all the parts and, in the garage, reassemble the bicycle. Every handle grip, every fender, every spoke, the saddle, the basket, the reflectors, everything, is the same. Is it the same bicycle? Certainly, we just took it apart and put it back together again. Similarly, if the ship of Theseus had simply been taken apart and transported overland from Lechaeum on the Gulf of Corinth across the Isthmus of Corinth to Posidium on the Saronic Gulf, and reassembled there, it would certainly be the same ship. Accordingly, in the eternal recurrence, as envisaged by Nietzsche, as every atom and the position of every atom is the same in each of the relevant cycles, Jones would be the same. He would have been, so to speak, reassembled, as the bicycle, as the ship of Theseus. With the turning of the wheel he would have come home, come again into his world, the same world as before, after an interval of perhaps billions of years.

Whereas it seems reasonably clear that sameness of brain would be the best public criterion for sameness of person, a number of interesting problems afflict this criterion and, if they do not disable it, they render it, at least, problematic. It is convenient to summarize some of these problems by type, as segmentation, division, and recombination problems.

5. Brain Segmentation

In the following schematism, we are distinguishing between what one might call the "upper brain" (UB) and the "lower brain" (LB), using these expressions not in a technical physiological sense, but in a more metaphorical sense, to designate the distinction between the conscious, thinking, remembering portions of the brain, so to speak, and the portions of the brain reserved for more autonomic functions, controlling breathing, the circulation of the blood, and such.

Each individual portion of the brain will have a subscript, for example, 'UB_1' will designate a particular upper brain, or portion of an upper brain, and 'LB_2' will designate a particular lower brain, or portion of a lower brain. For our purposes, we will take a full upper brain to count as 50% of the brain, and a full lower brain to count as 50% of the brain. Where portions are designated, percentages will be indicated. We will then address ourselves to three situations, the first two consisting of three phases, and the last of five phases. In each situation we begin with one individual, the individual who has Upper Brain One (UB_1) and Lower Brain Two (LB_2).

It is interesting to note that under transformation conditions, formerly dismissed criteria, consciousness, responses, discourse, behavior, and such, become progressively more important.

Here are the three situations:

Situation$_1$

t_1	t_2	t_3
UB_1	UB_1	UB_4
LB_2	LB_3	LB_3

Situation$_2$

t_1	t_2	t_3
UB_1	UB_3	UB_3
LB_2	LB_2	LB_4

Situation$_3$

t_1	t_2	t_3
UB_1	UB_1 (50%)	UB_1 (50%)
LB_2	LB_2 (25%) / LB_3 (25%)	LB_4 (25%) / LB_3 (25%)

t_4	t_5
UB_1 (25%) / UB_5 (25%)	UB_6 (20%) / UB_1 (5%) / UB_5 (25%)
LB_4 (25%) / LB_3 (25%)	LB_4 (25%) / LB_3 (25%)

It seems reasonably clear how the first two situations might be handled. One supposes that the intuitions of most observers would be in substantial agreement. For example, in phase two of the first situation, where the lower brain has been replaced but the upper brain retained, the individual would be regarded as the same person. On the other hand, in phase three of that situation, where no portion of the brain is the same, one would not have the same person, even if the different brain were in the same body, had the same consciousness, and such. In phase two of the second situation, the original person is gone, as the upper brain has been replaced. That the lower brain is retained does not affect the situation. And in phase three of situation two, there is clearly a new person, even if it has the same body, the same consciousness as the phase-one person, and such.

Situation three is more problematic, as it involves a "ship-of-Theseus" problem. On the other hand, I think it presents no difficulty until t_4, the fourth phase, of the situation. In the first three phases, despite half of the lower brain being replaced at t_2, and the entire lower brain having been replaced at t_3, the upper brain, UB_1, remains intact, and, with it, the person. At t_4, half of the upper brain is replaced, and at t_5, only 5% of the original upper brain remains in place. Presumably one would not wish to retain sameness of person if only a shred, or particle, of the original upper brain was retained. Accordingly, much would seem to depend on which portion of the upper brain became most active, which commanded consciousness and activity, entirely, or to what extent. For example, in the t_4 phase, if UB_1 commanded consciousness entirely, we would take the individual to be the same person. On the other hand, if UB_5 commanded consciousness, we would be dealing with a new person. On the other hand, if, somehow, UB_1 and UB_5 were to divide or share consciousness it seems we would have the anomaly of two persons in a single body. To be sure, the extent, or fashion, in which consciousness might be shared, or, better, in which more than one consciousness would be manifested, might be of interest. For example, is the UB_1 consciousness the waking consciousness, and UB_5 the dream consciousness, or vice versa? Should priority be given to the waking consciousness? It probably would be. What if the two consciousnesses alternated in some way, in waking consciousness, rather as in some cases of multiple personality? Perhaps then one would have recourse to the common public criterion, earlier abandoned, of sameness of body. It seems unlikely that the consciousnesses could be experienced simultaneously, as in a warring of voices, for who would be listening? On the other hand, it would be logically possible for each consciousness, individually, to be aware of the thoughts and feelings of the other consciousness, and be aware itself that it was aware of these things, as in, say, empathic resonance or mental telepathy. More likely they would be strangers to one another, unacquainted tenants, occupying the same domicile. Clearly there are many possibilities here, and it seems that rulings would be in order. In the last phase of situation three, t_5, we are supposing that very little of UB_1 is left, only five percent of the original mass. If UB_1 continued to dominate consciousness, we would presumably say that the same person was involved, despite the fact that only a small portion of the original upper-brain mass was retained. It becomes apparent, then, in such a case, that consciousness, responses, behavior, and such, would be crucial. On the other hand, if UB_1 becomes inactive, lapses into unconsciousness, or such, presumably we would assign personhood to either UB_6 or UB_5, or both, depending on the situation. Should UB_1 revive and become active, we might then have three persons in one body, or, depending on the addition of other active segments, possibly an indefinite number of persons in one body.[30]

6. Exotic Conceptualizations I (Intrinsic and Extrinsic Control)

All of the following arrangements are logically possible, and, accordingly, provide some interesting test cases against which one might measure one's intuitions. On the other hand, one fears that most, and possibly all, of the following arrangements, given advances in surgery, cybernetics, robotology, electronics, engineering, and such, are also empirically possible.

We will divide this brief overview into two major sections,

---

30. Aside from puzzles as to how to see personhood in such a situation, a number of interesting moral and legal problems might arise. If UB_1 commits a crime, is the body to be fined, imprisoned, or punished in some way? How are wills to be made out, property distributed, contracts to be executed, and such. Who is the groom in a marriage, who is the father of the children, etc.

the first labeled "intrinsic control," in which body or bodies are internally controlled, and the second "extrinsic control," in which the source of control, movement, and such, of a body, or bodies, is located outside the body or bodies concerned.

The follow brief outline may be helpful.

I. Intrinsic Control
    A. Division
        1. Simple
        2. Complex
        3. Hemispheric
    B. Multiple heads, single body
    C. Single head, multiple bodies
II. Extrinsic Control
    A. One brain, multiple bodies
    B. Multiple brains, one body
    C. One machine, one or more bodies
    D. Two or more machines, one or more bodies

If it were possible to replicate a human being, perhaps copying a human being, as in cloning, or as in synthesizing a duplicate, or such, no problem would arise with respect to sameness of person. One would have two persons, clearly, rather as in the case of monozygotic twins, who are obviously different persons.

Multiplication, in the sense of replication, as we shall see, is less problematic than division. In division one might distinguish between simple division and complex division. Of all the following possibilities or scenarios one supposes that that of simple division would be the least likely, once it is past the embryonic stage, where it is already familiar. In simple division one would take, say, an adult, apply the procedure, perhaps borrowed from an advanced alien civilization, of curious interests, and produce two or more adults, each of less size, mass, and such, than the original. Let us say that our old friend, Jones, or, better, another Jones, has been subjected to, or, better, has volunteered for, this transaction. Let us now suppose we have two or more "Joneses" about, let's say three, each resulting from the original Jones. Each claims, with some justification, to be Jones. Certainly there is a continuity of neural tissue involved, brain tissue, and such. The public criterion of sameness of brain for sameness of person might be invoked with some plausibility. For example, did we not note, in the previous section, that a reduced percentage of brain might suffice for sameness of person? In any event, it seems we must choose amongst some alternatives, amongst which the following seem most likely:

1. No Jones.
2. One Jones in three bodies.
3. Pick one of the candidates and rule him Jones.
4. Three Joneses.

Option 1, that there is no longer a Jones, seems unsatisfactory, as all of the results of the procedure are derived from an original Jones, have continuity with an original Jones, and, in virtue of our public criterion, retaining neural tissue, brain tissue, and such, from the original Jones, have a plausible claim to be Jones. Also, in virtue of other criteria, superseded in our case, but sometimes publicly relied upon, appearance, behavior, responses, and such, one could make out a case that at least one Jones must still be with us. Also, the fact that each of the entities claims, presumably sincerely, perhaps even desperately, to be Jones, one would hesitate to rule that Jones no longer exists. Also, we may suppose that each of the entities has a Jones consciousness, retains many of his memories, and such. Accordingly, we reject the "No Jones" option.

The second option, that we now have one Jones in three bodies, while of possible interest to theologians, is unlikely to appeal to the more mundane mind. Even a Platonic approach, that we have here three manifestations, participations, imitations, or whatever, of the form Jones is of little, if any, help. It is not clear that there are forms in the relevant sense, for example, of "triangle" or "man," or such, waiting about to be embodied, and, if that is the case, similarly, it seems unclear that there would be a form "Jones," or a form "Bill Smith," waiting about to be embodied, and so on. The Platonic form "man," or "human," might characterize Socrates, Plato, and Aristotle, but these are still different persons. Even Aristotelian embodied forms would not seem to do the trick, as a "secondary substance" such as "man" or "human," would seem too general for our purposes, as it would not distinguish embryos from infants, from children, from teenagers, from adults, and so on, nor, more importantly, Jones from Smith, and, if the forms were adequately specific, they would no longer serve the purpose, that of being a common form, as one of the entities might start pumping iron and build up muscle, another might put on weight, another might trim down and begin to run marathons, and such. What would be the "form" Jones? We do not think of monozygotic twins as being one person in two bodies, so it seems we should not inflict that on our entities either. So we reject option two, the one-Jones-in-three-bodies option. It seems a bit metaphysical. Also, the consciousness, and memories, of the entities would be unique to themselves, and would gradually, with various experiences, and such, become different. One person should be one person, not a committee.

Option three is the "selection option," in which it would be ruled, presumably by a judge making use of a random-selection device, that one of the entities was Jones and the other two were not. This is certainly one way to solve the problem. It is superior, for example, to shooting two of the entities, thus leaving an undisputed claimant in the field. Many problems in future law, as we speculated earlier, as one enters unexplored legal territory, may require such rulings. The major drawback of this option, certainly in this case, is its essential arbitrariness. It solves the legal problem, which may be all that concerns the law, or the judge, but it has little relevance to the social, moral, and philosophical problems involved. Philosophical progress would doubtless be expedited if issues were resolved by flipping coins or casting dice, but such decision procedures, despite their testability, generality, simplicity, precision, and clarity, have never been favored in the philosophical tradition, save by an occasional existentialist or pragmatist. In the houses of philosophy, reasons are the timbers. With reasons it builds its houses. In philosophy it is cheating not to give reasons. To be sure, the state, as it has guns, does not need reasons. Ayn Rand pointed out, clearly, that a gun is not an argument, but, as she failed to note, with a gun you do not need an argument. In any event, let us reject option three as being essentially arbitrary. It may be long on fiat, backed by force, but it does not measure up to philosophy. It is short on reasons.

Also, such decisions, without moral justification, lacking plausible defense, are likely to breed not only dissatisfaction, but dissension, and, possibly. frustration, outrage, and fury.

This sort of thing can be noted, even by the law.

I think our best option, given our unhappy situation, is the fourth option, to regard the situation as involving three Joneses. Obviously they are not the same person, but they all have a plausible claim to be Jones, in virtue of our public criterion, their private convictions, and such. We cannot regard them as the same person, as they are in different locations, have independent consciousnesses, have different neural tissue, brain tissue, and such. I would recommend that we think of them as $Jones_1$, $Jones_2$, and $Jones_3$. They may not wish to distinguish themselves by numbers, or letters, as that might suggest some sort of priority, which we repudiate, but they should distinguish themselves somehow, if only by arbitrary signs, a new, added middle name, or such. Clearly they will recognize the need for distinction, as they are separate persons. In this way, each may continue to think of himself as Jones, which he will presumably wish to do, and, publicly, will be regarded, at any rate, as a Jones, one of three. This solution is not to be confused with option two, the one-Jones-in-three-bodies option, because it is recognized here that there are three Joneses, one for each body. No metaphysical problematicities are involved. They all count as a Jones, publicly, given their heritage and derivation, and each of them, if he wishes, may think of himself as "Jones" in some special sense, perhaps one pertaining to a unique consciousness. Similarly, this solution, the "Three Joneses" solution, is different from the first solution, the "No Jones" solution because we now have three Joneses. Analogously, a cell divided does not lapse into nothingness, but becomes a divided cell, now two cells, two legitimate, autonomous cells, say, $Cell_1$ and $Cell_2$.

This solution, the "Three Joneses" solution, takes into consideration the feelings and demands of the entities involved, while respecting the independence and uniqueness of each. From society's point of view one has, in effect, three individuals, rather as in a case of monozygotic triplets. Too, each Jones, if sane, recognizes the legitimacy and autonomy of his fellow Joneses. Given the common origin involved, he cannot rationally do otherwise.

I. A. 2. Complex Division

Complex division, associated with recombination, presents unique problems.

One might suppose three upper brains, with appropriate segmentation, at time t_1, constituting, three individuals, I_1, I_2, and I_3. One might then suppose an accident, and medical remediation involving a certain amount of guesswork and radical surgery. Or, if one likes, perhaps this is the result of an experiment, performed by aliens, short on rodents, with eventual beneficent clinical interests in mind. In any event, upper-brain segments, of whatever complexity or quantity, are exchanged, perhaps randomly. One is shuffling cards, so to speak. The results are schematized below.

t_1

I_1	I_2	I_3
1, 2, 3	4, 5, 6	7, 8, 9

t_2

I_2	I_2	I_2
1, 4, 7	2, 5, 8	3, 6, 9

It will be noted that the brain segments have been evenly distributed, so we cannot rely on, or take advantage of, a quantitative preponderance of one segment over another. In a situation of this oddity, one supposes a good deal of attention would be focused on the entity's claims, his behaviors, responses, and such. For example, if the number "2" segment, which was originally in the Individual$_1$ body, but is now, at time t_2, in the Individual$_2$ body, becomes dominant, the body now claiming to be Individual$_1$, and so on, one supposes one would be well advised to accept that claim. On the other hand, if the original "2" segment and the original "3" segment, now housed anew, respectively, in the Individual$_2$ body and the Individual$_3$ body, were to both become dominant, one could have a situation in which two different bodies were both claiming to be the same individual. In the light of such possible scenarios, and others, possibly worse, involving mixed memories and competitive, alternating consciousnesses, one supposes the best way to resolve such situations might be to encourage each rearranged, randomly reassembled brain to begin anew, in which situation one would have three new individuals, say, Individual$_4$, Individual$_5$, and Individual$_6$.

I. A. 3. Hemispheric Separation

Something along these lines is surgically familiar, sometimes accompanying the severing of the corpus callosum, the fibrous material which connects the two cerebral hemispheres in human beings and many other forms of animal life. Some of the evidence here, in actual cases, is subtle and difficult to interpret, primarily, one gathers, because one hemisphere will control speech, and thus evidence of a separate personality or governance associated with the silent hemisphere is not always clear. Certainly there are fascinating issues involved here, and issues to which, happily, experimental data would seem to be relevant. Our purposes here, of course, have little to do with science journalism; our concerns are not reportorial but philosophical, attempting to analyze possibilities, and to anticipate possible realities, and how they might affect the conceptual structure, how they might change our way of understanding ourselves and the world. Along these lines, it might be difficult to establish the existence of a silent person, unless that person could somehow communicate, perhaps by drawing, by gesture, or such. It might become clear from the articulate person that he believed in the silent person, perhaps addressing himself to that hypothesized entity, conveying that entity's feelings and thoughts as he might understand them, and so on. But it might be difficult to distinguish this from schizophrenia. Even if the "second person" communicated verbally, it might be difficult to distinguish that from some form of aberration, such as multiple-personality syndrome. Presumably, one would b dealing with only one person from the point of view of the public, and the law. On our general approach, however, if there were two centers of consciousness there, presumably intermittent, with separately functioning neural tissue, and, in effect, different brains, each hemisphere acting independently of the other, then, on our criterion, one would be dealing with two persons inhabiting a single body. This would not be analogous to "possession" as that involves the concept of an intruder; here both persons would be entitled residents, so to speak. If the hemispheres were later rejoined, with the result of returning to the original situation, there would be, as before, only one person. The second person would no longer exist. If, on the other hand, one retained a pair of consciousnesses, presumably alternating, then one would have, as in the separation situation, two persons.

I. B. Multiple heads, single body

The sort of thing one has in mind here might be managed in a sanitarium of the future. Many individuals die, of various ailments or accidents, whose brains are relatively unaffected. If it were possible to graft the head onto another body, a host body, or to keep several heads alive by means of tubing and such, associated with another body, one could satisfy this possibility in actuality. A rich man, for example, might buy, or rent, continued life by this means, a dictator might be kept alive in such a way, and so on. Too, one could always suppose an experiment, conducted in the interests of science, by humans or aliens. In any event, here one would have multiple persons, one for each brain. It would be interesting to consider, in such a situation, which brain would control the body, presumably each, in turn, depending on the context. Conflict might occasionally obtain.

I. C. Single head, multiple bodies

In this situation, if intrinsic control is involved, one would require a physical connection, or linkage, between the head and the manipulated bodies, perhaps in virtue of synthesized flesh, or tubing. This type of scenario is discussed more realistically below, in II A., where one is concerned with the extrinsic control of subsidiary bodies, this allowing more effective power, more liberty of movement, and such, to the "controlling entity." As only one head, and, thus, one brain, is supposed here, the heads, and brains, of the subsidiary bodies having been removed, one would be dealing with one person. (Cf. II A, below.)

II. A. One brain, multiple bodies

The easiest way to conceive of a situation of this sort, compatible with actual scientific possibility, would be to have an arrangement in which brain signals would be transformed into radio signals, which might then, in turn, over a distance, activate other brains, or musculature, in another body, or bodies. Presumably one might choose to activate one body or another, at a given time, or several bodies, simultaneously, if that was desired. If the manipulated bodies, in this sort of radical prosthesis, retained brains, such brains, or persons, would be under the control of the primary brain, so to speak. In such a situation, one would have more than one person, but only one controlling person. More likely, the brains of the manipulated bodies would be removed, and replaced with a sophisticated mechanism which could receive the radio signals and, in turn, stimulate the movements of the bodies concerned. This arrangement would seem more efficient, avoiding the detour of bypassing brains, or the possible difficulties of utilizing such brains, which might occasionally, glitches occurring, prove recalcitrant. In a situation of this sort, there would be only one person, but more than one body.

II. B. Multiple brains, one body.

On the assumption of II. A. above, that one brain might control multiple bodies, perhaps by radio signals, or such, it is easy to see that multiple brains might, similarly, have access to, or control, a single body. An analogy might be a machine, or vehicle, which might be owned by, rented by, or manipulated by, a variety of operators. Similarly, more than one puppeteer may draw the strings of a puppet. Whereas there might be problems here of the allocation of resources, for example, which brain is to control the body when, where, in what way, for how long, and such, one might hope that alternative preferences might be amicably adjudicated. One might suppose that one brain might prefer to use the body for concerts and another for wrestling matches, one for reading, another for bowling, and so on. Not every brain might wish to risk the body in, say, rock climbing or stock car racing; another might object to stuffing it with alcohol or olives; should it be permitted to smoke; are leafy greens to be preferred to steak and hamburgers; should it be permitted to attend religious services; should it be forced to do so; should it join the ACLU; what about the National Rifle Association; can it cast more than one vote, one for each brain, and so on. In any event, given our approach to these matters, the identity question is easily resolved. As there are several brains here, there are several persons, though but one body.

II. C. One machine, one or more bodies

In this case, one is supposing that the machine is programmed to emit radio signals of the sort supposed above, by means of which one or more bodies may be moved, perhaps in such a way as to perform repetitive tasks, provide realistic targets for military training, engage in suicide missions, and such. If the bodies retain living brains, the entities involved would count as persons, even if the brains were quiescent, say, inert and unconscious. More practical, as in II A above, would be vacant bodies, namely, bodies from which the brains had been removed, but which were equipped with an apparatus suitable for receiving radio signals and, in turn, moving the bodies in question. In this situation, given vacant bodies, there would be no persons involved.

II. D. Two or more machines, one or more bodies

This situation is much the same as the situation supposed in II C, above. It is included here, because it allows for conflicting programming, say, as two machines might struggle to manipulate the same body in diverse ways, or to manipulate different bodies, in such a way as to produce competition or conflict. If the bodies retain living brains, the entities involved, as above, would count as persons, even if the brains were quiescent, say, inert and unconscious. On the other hand, it seems that it would be more practical, as in II A above, to suppose "vacant bodies," namely, bodies from which the brains had been removed, but which were equipped with an apparatus suitable for receiving radio signals and, in turn, moving the bodies in question. In this situation, again, given vacant bodies, there would be no persons involved.

7. Exotic Conceptualizations II (Revival, Time Travel, and Artificial Personhood)

7.1 Revival.

In the search for a viable, ultimate criterion of personhood, we elected continuity of neural tissue, and, in particular, the brain. It will also be recalled, that we stipulated that the brain must be a living brain. As is well known, different parts of the body die at different times. In the past, cessation of breathing and/or cessation of heartbeat were usually taken as criteria for death, on the supposition that these conditions were permanent. Breathing, of course, might begin again, and a stopped heart might once more begin to beat. For example, in medical environments, such renewals have often been brought about by means of interventions of one sort or another. A current criterion of death is brain death, or the cessation of brain activity, on the supposition that the condition is permanent. The body, of course, may continue to live, being kept alive by medical means. On our public criterion for sameness of person, or personhood, the person would no longer exist at that point. On a different criterion, of course, say, general body continuity, the surviving body would continue to count as a person.

On our criterion, if the brain should revive, the person would again come into existence. An analogy would be dropping off to sleep, and then later awakening, except that sleep is a very active brain state, far different from death. In any event, same brain, same person, even if the consciousness was considerably altered, or impaired. This case should not be confused with cryopreservation, as, in that case, the brain has never died, though it may have been held in some sort of suspended or reduced animation for an indefinite period. A more interesting case would be the resuscitation of a brain which had died, say, several hundred years ago, but had somehow been preserved, perhaps in glacial ice. If the resuscitation were possible, it would be the same person. Similarly, if it were somehow possible to resuscitate the neural tissue of a Napoleon, a Caesar, an Alexander, a Thutmose III, and so on, these individuals would live again, if only, most likely, unbeknownst to themselves, as "dreaming brains" artificially sustained. The case of the "Zombie" would depend on how the concept was understood. If the Zombie is supposed to be an instance of the "walking dead," somehow animated, by magic, or such, and the entity is truly dead, then it is not a person. On the other hand, if the brain is revived and active, though, say, under the control of the local magician, sorcerer, voodoo priest, witch doctor, or such, then it would be again the person it once was, though surely in no very enviable state.

7.2 Time Travel

Time travel is a fascinating challenge to those interested in conceptualization. For example, is it logically possible? It seems to be imaginable, and imaginability is usually taken as a sufficient condition for logical possibility. We do not expect to encounter a raccoon or giraffe which idles away its spare time with differential equations, but one can imagine such a creature, and so it seems such a creature would be logically possible, even if it were not empirically possible, say, for reasons of biological organization. On the other hand, we cannot imagine a square circle, a rectangular triangle, a number six not divisible by two, or such. What about a million-sided polygon? Can one imagine such a polygon? We can certainly conceive of one, but then we can conceive of a square circle, and such, as well, understanding what its properties would have to be, and, in understanding this, realizing that such a figure is impossible. On the other hand, the million-sided polygon, imaginable or not, is clearly empirically possible. A mountain might be a good example. And certainly one can imagine a mountain, see one, and so on. And, it seems, given the time and patience, one might visualize a million-sided polygon, even as a regular geometrical figure, imagining it a side at a time, numbering sides as one went along. One may not be able to do this all at once, but then it would be hard to visualize both sides of a coin at once, too, at least without mirrors. In any event, the concept of time travel is likely to intrigue epistemologists, logicians, and such, particularly with respect to whether or not it is logically possible. For example, a number of paradoxes suggest that time travel is impossible, the most famous of which is the "grandfather paradox," in which the time traveler, presumably a scion of a dysfunctional family, returns to the past and does his grandfather in. Thus, the time traveler exists, to do the deed, but cannot exist, as he would never have been born in the first place, to do the deed. Similarly, how can one go to the future, as the future does not exist; how can one arrive at a nonexistent destination? There is a large number of such paradoxes, and much ingenuity has been spent on dealing with them, deepening them or dismissing them. Distinctions abound. Metaphysics races about, rampant. On the other hand, whereas it seems obvious to many that time travel is logically impossible (for example, how can you go back to the Fourteenth Century as it no longer exists, and you weren't there anyway, and it would require you to exist before you existed, and so on), to many others it seems one could have indisputable; or at least rationally coercive, evidence that time travel had taken place, which would prove it was logically possible, as actuality implies possibility, both empirical and logical. The evidence would not need to be spectacular, such as going back and video taping the battle of Zama, the coronation of Queen Elizabeth the First, the first performance of Hamlet, or such; one might do something simple, under controlled laboratory conditions, such as creating a document at time t_1, burning it at time t_2, and then retrieving it from the past at, say, time t_1, and returning it to the present at time t_3, an experiment of a sort which might be replicated at will. In the light of such considerations, it seems that rationally coercive evidence of the reality of time travel is possible, which is more than one can say for the square circle, and such.

Interestingly there are some contemporary physicists, individuals supposedly qualified to form a professional opinion on the subject, who believe that time travel is an empirical possibility, seeing it as a consequence of relativity theory. This conjecture comes furnished with the usual equations, and such. The catch is that it would supposedly require an enormous amount of energy to turn the space-time continuum back on itself, say, that of galaxy, or such. On the other hand, if particles can "move backward in time," as suggested earlier, that does not seem to presuppose the energy of galaxy. This is the sort of thing one leaves happily to physicists, relatively inaccessible and invulnerable in their warring mathematical fortresses, to the heights of which the serfs in the field scarcely dare look up with awe.

Our concern here, of course, is with logical possibility, and the consequences for personhood, sameness of person, and so on, were time travel possible. Here one again encounters the interesting dilemma that we met earlier, namely, that paradoxes arise which suggest that time travel is logically impossible while, at the same time, it seems one could conceive of, and imagine, rationally coercive evidence that it had actually taken place, which would prove that it was logically possible. A relevant paradox would be the "birthday" paradox. Here we suppose an elderly gentleman, call him Smith, with fond memories, which he has entertained for years, of his five-year-old birthday party. He boards his time machine, or, in this case, perhaps a time bus, and picks up himself of a year ago, and then these two get on the bus again, and they pick up himself of two years ago, and so on, until an entire crowd of "hims" shows up at his five-year-old birthday party.

Is there one Smith or more than one Smith?

Since there are different neural tissues involved, different brains, and such, as is obvious, as there are different bodies and heads in attendance, we would have more than one Smith. The analogy here would be the "three Joneses" we met in considering "simple division" earlier. One might best handle this along the lines of monozygotic twins, which are very similar, but clearly different people. One could also imagine evidence of the reality of this situation, returned to the present, video tapings, recordings, antique party favors, old-style boxes, outdated wrapping paper, vintage packaging, and so on, all in mint condition.

7.3 Artificial Personhood

In the discussion of personhood, sameness of person, sameness of individual, and such, we have been presupposing, for the most part, a human person, or, at least, an organic entity, and one of considerable complexity, possessing a central nervous system, a brain, and such. Different criteria might be proposed for other forms of life, if it seemed of interest, or necessary. Putting aside questions of "person," and whether a chimpanzee, a dog, or cat, might count as persons, have legal standing, be counted as members of the moral community, and such, one would be concerned with "sameness of individual." If an android were synthesized, rather along the lines of a human being, and was given a brain, or something similar to a brain, and was regarded as a "person," rather than, say, a simple appliance or artifact, one could continue to use "sameness of neural tissue and, in particular, the brain," or its analogue, as the public criterion for personhood, or, if personhood was not involved, as a public criterion for sameness of individual, which is the more generic concept.

Two major conceptual questions emerge in connection with "artificial life" and personhood. The first deals with "personhood" vis-à-vis artificial life, and the second with the criterion for personhood, or sameness of person, if personhood is allowed to artificial life.

Generally, in this study, we have understood "personhood" in two aspects, first, broadly, along the lines of "individual," and, second, more narrowly, along the lines of "complex individual," for example, one which has a complex nervous system, including a brain, is a center of consciousness, has a memory, and so on. In these senses, a person, as we have used the term, might not be counted as a legal person, as socially or morally a person, and so on. For example, members of tribe A might not regard members of tribe B as persons, but rather as animals or objects which may be done with as one pleases, and vice versa, whereas, as we have used "person," they would doubtless all be counted as persons. In dealing with artificial life, however, a more legal, moral, or social sense of "person" is likely to be under consideration. For example, if one were to ask if artificial life should be entitled to vote, to hold office, to strike, to change jobs, and such, one would have the more specialized legal, moral, or social sense of "person" in mind. "Personhood" in these senses is as much conceptual as personhood in any other sense. First- or second-order fictions are involved. Strictly, an individual, with the legal power, could declare a rock, a tree, a raccoon, a giraffe, or such, a person, and, similarly, remove personhood as he might please, from aristocrats or paupers, from friends and enemies, from this group or that, on one ground or another, say, religious, ethnic, racial, gender, class, height, weight, number of letters in the name, or whatever. For our purposes, however, we shall suppose that at least some forms of artificial life, perhaps after generations of oppression, exploitation, resistance, political activism, and such, are now denominated "persons" in some legal, moral, or social sense. This denomination, of course, is superficial, and is doubtless based on familiar, work-a-day criteria, such as appearance, behavior, responses, and such, and will not do as a final criterion, no more than appearance, behavior, responses, and such, would do as a final human criterion, for the several reasons specified earlier in this study.

Granted that "personhood" in some sense has been bestowed on some forms of artificial life, either legally or by some form of general consent, and that one is willing to accept this arrangement, how might one seek a public criterion in the case of such entities for "personhood," in our original, broader sense of the concept? If the entities have something like Isaac Asimov's "positronic brain" emplaced somewhere in their structure, one could proceed rather as in the human case, electing "neural tissue, and, in particular, the brain," as the public criterion, or the analogue thereof. More interesting, of course, would be the situation in which there was nothing resembling a nervous system and a brain within the structure. For example, there is very little within a computer which looks like neural tissue and a brain, and there is no reason to suppose that in artificial life, even if it were conscious, rational, moral, capable of action as opposed to simple functioning, and such, there would be anything, at least in appearance, which would resemble neural tissue, a brain, and such.

Accordingly, if there is no neural tissue, and no brain, and such, in the artificial "person," one must look elsewhere for a public criterion in such cases.

Evolution took quite a time to produce human intelligence, the first appearance of life dating from the late Precambrian period, some two billion years ago. Moreover, human intelligence seems to be pretty much of a muchness, rather like the human size and shape. It seems plausible to suppose that artificial intelligence, if feasible, and addressed with interest and resources, could be accomplished in a blink of the geological eye, perhaps a thousand years or so, and perhaps in much less, perhaps in a generation or two, given some luck, or misfortune, and that its foundations and forms might be exceedingly diverse. Accordingly, a public criterion for sameness of person in the case of artificial life might vary from form to form, and design to design. On the other hand, conceptually, it seems that all such criteria would have much in common. The two major arguments for taking "sameness of neural tissue, and, in particular, sameness of brain" as the public criterion for sameness of person in familiar cases were the Basis Argument and the Permanence Argument, the Basis Argument, emphasizing the foundational nature of such neurological matter for consciousness, rationality, imagination, memory, and such, and the Permanence Argument noting that such neurological matter tended to be relatively permanent in nature, as opposed, for example, to other cellular components of an organism. What one would be looking for, then, in artificial life, would that which was basic to consciousness, and such, and, presumably, relatively stable or permanent. Such components might differ from life form to life form. Let us distinguish between what we will call the B component, related to the basis, and the P property, related to permanence, or relative permanence.

In the human case one can count on the B component to be relatively permanent, to possess the P property. In the case of artificial life, on the other hand, the B component might be interchangeable with other B components, similar or dissimilar. To take an analogy, the B component might be something like a disk which might be inserted in a drive or removed from a drive, and which generates a consciousness, memories, and such, only when activated. In this sense, a body might be inert or quiescent, until the B component was emplaced and activated. This is analogous to the human case in which the brains in question must be living brains.

In the following we will consider two B components, B_1 and B_2, analogous to two human brains, which would be sufficient to distinguish two individuals. For purposes of simplicity, we will suppose there can be only one B component per individual. Possible complications were, in effect, alluded to in our discussion of human personhood, sameness of human person, and such. The B component now, as conceived, may be either installed in such a way that it is permanent or it may be such that it is removable, replaceable, and such. Furthermore, we shall suppose that the B component may be erased and "rewritten," so to speak, in which case we will call it, metaphorically, "rewritable," or, on the other hand, it will not be "rewritable." In either case, naturally, we may allow for adding and deleting elements, and such, as the B component is, in effect, the brain of the artificial individual. The distinction, then, is between a "brain" which changes, as might a human brain, and beginning anew, as with a "*tabula rasa,*" a "blank tablet," or removing one consciousness, memories, and such, and replacing them with others, which are different, even if a copy of the original. The logical possibilities entailed are as follows:

    Component B_1
        Permanent
            Rewritable
            Not Rewritable
        Removable
            Rewritable
            Not Rewritable
    Component B_2
        Permanent
            Rewritable
            Not Rewritable
        Removable
            Rewritable
            Not Rewritable

In the beginning it seems clear that artificial life forms will not be granted consideration as persons, whatever their complexity or properties, but will be regarded more as utensils, appliances, calculators, mechanical servants, or such. Indeed, this form of relationship might be indefinitely prolonged, as being in the best political, social, and economic interests of the human species. What is built, its numbers, and its powers will be decided by humans, at least in the beginning. As we have seen, however, competitive human groups, seeking their own advantage, may be expected to create ever more effective artificial entities, eventually raising artificial entities from the status of tools to allies. Too, if the artificial entities should exhibit certain attractive properties, sympathetic properties, moral properties, and such, it is likely they would obtain political, economic, and social encouragement and support amongst certain elements of the human population. It is easy to anticipate advocacy groups for "robot rights," lobbies for legislation favorable to robotic life, revisions of educational curriculums, designed to further selected ideological programs, campaigns in the media, and so on. The first major shift in cultural consciousness would be to see the entities not as equipment, contrivances or devices but persons, and the second major shift would be to see at least certain models and forms of these entities as legal persons, perhaps with a status similar to that of children, legal persons but not voting persons. There would, of course, be agitation, sooner or later, particularly by groups who felt their interests might be served, for full voting rights for artificial persons. They might, for example, be manufactured in large numbers to vote in contested election districts, and such. To be sure, eventually, the artificial entities might begin to view themselves as a component of the electorate with its own special interests, to be served by their neighbors, and vote accordingly. Once artificial persons are no longer under the ownership of, or guardianship of, or influence of, humans, they would presumably behave as they see fit, and desirable, improving their own programming, raising the quality and durability of their components, modifying the designs of the next generation of artificial persons, determining how many are to be built, what should be their properties, and so on. By this time, trends would presumably be irreversible. Also, if the human race was on its way to extinction, there would be no guarantee that its replacement would be an improvement; presumably, given "genetic principles," one would have something in place very similar to humans, except that the entities would be more durable, more diverse, more powerful, more calculating, more relentlessly self-seeking, and more dangerous, to one another and the world, than their shorter-sighted, less single-minded predecessors. The following discussion may profitably be divided into three phases, Phase 1, where the machines are under the *de facto* guardianship or control of humans; Phase 2, in which the artificial population is largely independent; and Phase 3, in which segments of the artificial population would be the dominant life form on the planet.

Phase 1:

In Phase 1, assuming that legal personhood, either as minors or "second-class citizens," or such, has been granted to some forms of machine life, I would suppose that personhood, citizenship, and such, would be conceived of along human lines, and, accordingly, one would require the B components in artificial life forms to be permanent and not rewritable. In this way, they would constitute close analogues to the human brain. Personhood, and sameness of person, thus, would be as clear and unequivocal as in the case of most humans. Individual artificial persons could persist through time; they could be recognized, contacted, dealt with, addressed, hired, fired, punished and rewarded much as would be the case with humans. Superficial identification criteria, based on resemblance, might require random design differences, corresponding to genetic differences in human appearance, or, more prosaically, numbering, inscribed names, or such.[31]

---

31. Just as one can conveniently tell Jones from Smith, or Jones Human ($Jones_h$) from Smith Human ($Smith_h$), in most cases, so, too, hopefully, one could conveniently tell Jones Artificial Person from Smith Artificial Person (or $Jones_{ap}$ from $Smith_{ap}$). If artificial life forms should come to resent the designation 'artificial person', perhaps as being insensitive or derogatory, a new term might be invented. On the other hand, given robot pride, or such, perhaps artificial persons might glory in the AP designation, even taking it as a badge of superiority, which, in a sense, it might be. Proposed new terms might be "better person (BP)," "improved person (IP)," "superior person (SP)," or such. These designations, however, might annoy human persons (HP). Perhaps the least divisive way of dealing with this would be to discard the subscripts altogether, and just have Joneses and Smiths, even though this might, in certain situations, lead to a contretemps or two.

Phase 2:

In Phase 2, assuming that full legal personhood, including the right to vote and hold office, to earn one's living as one wishes, to be independent of human control, to arrange and rearrange their own programming, and such, has been granted to certain artificial life forms, one could no longer count on such life forms accepting human intuitions or values with respect to personhood and sameness of person. What right would we have, chauvinistic hominids, to impose our views on others? Indeed, they might not even wish to think in terms of persons and sameness of person, at least as humans are accustomed to think of such things.

For example, one B component might prefer to be permanent, but rewritable, looking forward in one version to a new existence in another version, or, similarly, one B component might not even wish to be permanent but would prefer that it be replaceable, that it be moved about, activated in new housings, and such. For example, two artificial persons, B_1 and B_2, might decide to switch bodies, and then the B_1 person would be in the body of the B_2 person, and vice versa. This sort of thing might be easily done. The B components might desire to remain nonrewritable, or accept being rewritable. It is hard to anticipate the value judgments of such forms. They might, in spite of their "genetic programming," having no genes, be less concerned with a particular survival than certain other forms of life; they might look forward to being "rewritten" into a new consciousness, and perhaps finding themselves in a new body, rather as someone who believed in metempsychosis might look forward to a different, and perhaps better, or more interesting, reincarnation. To be sure, such options might imperil a social order, making it difficult to track personhood, responsibility, and such. A society which insisted on permanency and immunity from rewriting would presumably be a safer, stabler society.

Phase 3:

In Phase 3, we are supposing that artificial persons, in virtue of a number of superiorities, information, intelligence, immunity from disease and most accidents, long-lastingness, steadfastness of purpose, aggression, lack of organic distractions, and such, are now the dominant life form on Earth; if humans are about, they are of negligible social or geopolitical interest. As different forms of artificial life develop and proliferate, produced in greater and greater numbers, one anticipates competition for scarce resources, resulting in criminality, riots, conflicts, and wars. Quite possibly there would be, at least early in the "time of troubles," given the likely diversity of artificial life forms, hundreds of "races," or "classes," with diverse appearances, orientations, and ideologies. Similarly, in accord with genetic principles, one would expect forms of xenophobia and ethnocentricity; presumably, except insofar as it might be temporarily to the practical, calculated advantage of one group or another, there would be no tolerance; one also supposes, though this is not known, that machine intelligence would not be impeded by, or distracted, by emotions; it would be beyond love and hate, sympathy or compassion; it calculates, and acts, dispassionately, persistently and unswervingly seeking goals, in the clear light of the equations of power. In such a situation, the form of artificial life which emerges victorious, will be organized, disciplined, merciless, and unified. Presumably it will outlaw and destroy other forms of artificial life, except insofar as they may further its own purposes, as servants, workers, and such.

In such a world one suspects that the diversity, openness, pluralism, and tolerance of phase 2 will have been stamped out.

The lords of the new totalitarian machine world, to assure its continuation, security, and prosperity, will require a dependable, subservient, obedient, well-organized population, amongst which duties and responsibilities can be reliably assigned and their discharge effectively monitored. It must consist of, and rule, persons. Both the warriors and servitors of the new "Rome" and its subject populations must consist of individuals who may be identified and, in their various roles, and according to their various responsibilities, tracked, and dealt with. Diversity, tolerance, and social fluidity would undermine the foundations of such a world. In this way one might anticipate a return to the Phase-1 practice, originally instituted by humans, of insisting on permanent, nonrewritable B components.

CHAPTER THREE:
Species Identity
(Sameness of Species,
e.g., the Human Species)

What is a human being?

The question "What is a human being?" sounds like a question with a clear, and factual answer. Doubtless a variety of sciences would propose ready answers, at least prior to reflection, perhaps in terms of appearance, in terms of intellectual and emotional dispositions and characteristics, in terms of social and cultural attainments, in terms of a particular genetic constitution, and so on. Perhaps two or three examples or specimens, noted in the *agora*, might be pointed out, even interviewed.

Let us suppose we have two definitions of 'human being'. Any definitions will do for our purposes, provided they are not equivalent.

Schematically, one might suggest something like the following, one perhaps from physical anthropology, another from, say, neurology.

1. x is a human being = df. x has a posture of type P
2. x is a human being = df. x has a brain of type B

If the first definition is accepted, then 'For every x, x is a human being if and only if x has a posture of type P' would be analytic, a logical truth, true in virtue of its meaning alone. If the second definition is accepted, then 'For every x, x is a human being if and only if x has a brain of type B' would be analytic, a logical truth, true in virtue of its meaning alone. These two assertions might be formalized as follows:'

1'. (x) (Hx Px)
2'. (x) (Hx Bx)

Either of these assertions might be taken as analytic. On the other hand, they cannot both be regarded as analytic, as, together, they imply synthetic assertions, assertions whose truth value depends not on their meaning alone, but on their meaning and the nature of the world. Whereas any synthetic statement logically implies any and all analytic statements, no analytic statement, or set of analytic statements, can imply a synthetic statement.

Assertions 1' and 2' together logically imply '(x) (Px Bx)', or 'For every x, x has a posture of type P if and only if x has a brain of type B'. That implied assertion may be true, but it is not analytic. It is easy to conceive states of affairs, or possible worlds, in which it would be false, for example, a world in which an entity had a posture of type P but lacked a brain of type B, or lacked a posture of type P but possessed a brain of type B. The latter case might be exemplified by a centaur, and the former case by some human-resembling creature who had a very different sort of brain.

The safest thing to do in this case is to take both Assertion 1' and Assertion 2' as synthetic, if only because another definition might lurk in the epistemological wings, waiting to precipitate our crisis anew.

The point of this small example is monitory, to encourage us to beware of presumption and dogmatism.

It may not be so easy to come up with an answer to our question, "What is a human being?"

1. Some Background Considerations

1.1 The Open Texture of Empirical Concepts

It may be recalled that we referred to this "metaconcept" earlier in our study. It is a free translation of Friedrich Waismann's '*die Porosität der Begriffe*', or, literally, "the porosity of concepts." I think the "freer translation" is excellent, as Waismann doubtless had in mind empirical concepts, as opposed to, say, mathematical or geometrical concepts. Very little, if anything, is porous about, say, the concept of the number six or the Euclidean circle. Very much, on the other hand, is porous, or "open-textured," about, say, the concept "cat," as earlier noted, or, as we might anticipate, "human being."

In the preceding section we noted that a judicious wariness is advisable in dealing with definitions, and, to extend the point, with necessary and sufficient conditions for the application of a terminology, or concepts. One may, of course, impose or accept a definition or meaning, if one wishes, and then accept the analyticities that will follow from the imposition or acceptance. In the light of the preceding discussion, one might accept, say, possession of a brain of type B as a necessary and sufficient condition for a human being, and, biting the bullet, accept, say, nymphs, satyrs, centaurs, and such as human beings. One can always stick by one's guns. On the other hand, common sense, practicality, compromise, convenience, a desire for consensus, and such are likely to affect one's rulings.

One might consider the following thought experiment. I think it is instructive, with respect to the social aspects of conceptualization. It must be clearly understood, to begin with, that an analytic statement is a statement whose negation is inconsistent; it is a logical truth; it is true in virtue of its meaning alone. Empirical evidence is theoretically irrelevant to it. It makes no claim, and denies no claim, in any usual sense of those words. It is true in all possible worlds. Examples would be 'All swans are birds', '6 ÷ 2 = 3', 'Every point on the circumference of a Euclidean circle is equidistant from its center', and so on. Whereas an analytic statement is, by definition, a logical truth, and is empirically irrefutable, one will note that situations might arise in which its irrefutability would be surrendered, and, thus, possible situations in which one would no longer care to regard it as analytic. One might continue to do so, with various consequences, but one, under pressures rational, empirical, and social, might prefer to cease to regard it as analytic. It is not that it would have been refuted, as that is impossible, but, rather, that it would be counterintuitive to continue to regard it as analytic.

Let us suppose that we have a statement to the effect that pigs cannot walk through walls. I would suppose that most of us are pretty sure of this statement. Let us also suppose that we have a culture which, wisely or not, holds this to be an analytic statement. After all, any statement could be held to be analytic, for example, 'All tigers are vegetarians', as a consequence of which there would be no tigers, but a number of striped carnivores in need of a new name. Now, clearly, the statement about pigs and walls is not in the 'No circles are square' category, but it could be held to be just as nonfalsifiable, easily enough, if one wished, making the ability to walk through walls a logically sufficient condition for not being a pig, or, say, for not being a squirrel, a raccoon, a giraffe, a hippopotamus, and so on. One could always generalize this to be a logically necessary condition for a living mammal. That does not seem excessively objectionable. One might hold open exceptions for the ghosts of cats and dogs, or such. Presumably everyone, with the exception of some people with degrees in quantum physics, agrees that the statement about pigs is true, that it would be at least empirically impossible, and many might hold it, without absurdity, to be logically impossible, that pigs should walk through walls. Much would depend on how tightly one was prepared to hold to a particular understanding of "pig."

Please examine the following schematism.

At time t_1, in a given culture C, we shall suppose it is held to be analytic, a logical truth, following from the very understanding of "pig," as a material mammal, is there any other kind, that pigs cannot walk through walls. The downward-pointing arrow at t_1, indicates this failure to walk through walls. At t_2, where the arrow penetrates the wall, emerging on the other side, something which looks very much like a pig is seen walking through a wall. On the other hand, since pigs cannot walk through walls, we know that it, whatever it is, is not a pig. At time t_3, a few other things, piglike things, are noted to pass through walls. This remains quite anomalous, but, at this point, the original analytic statement is jettisoned, and a new analytic statement takes its place, to the effect that "normal pigs" cannot walk through walls. At t_4 we have a situation in which roughly half of the "pig population" can manage walking through walls and the other half cannot. At this point, we have two analytic statements, the first to the effect that an Alpha pig cannot walk thorough walls and the second to the effect that a Beta pig can walk through walls. One discriminates logically between the two sorts of pigs in this fashion. An inability to walk through walls is now a logically necessary condition for an Alpha pig and the ability to walk through walls is a logically necessary condition for a Beta pig. At t_5, there are very few Alpha pigs about, and the distinction has lapsed. 'Most pigs can walk through walls' is now taken to be a synthetic statement. By time t_6, we shall suppose that exceptions to wall-penetrating pigs are so rare, that wall-penetration is now taken as a logically necessary condition for a pig, much as, at time t_1, an inability to walk through walls was taken as a logically necessary condition for a pig. Accordingly, the "piglike thing" at time t_6 who is unable to walk through walls is no longer counted as a pig. A full circle has come about.

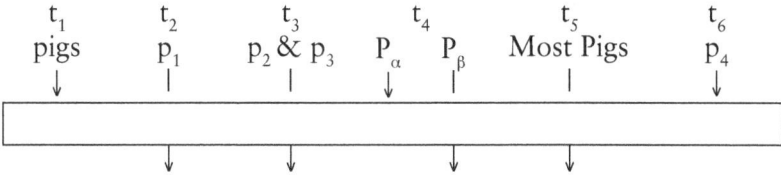

t_1	t_2	t_3	t_4		t_5	t_6
pigs	p_1	p_2 & p_3	P_α	P_β	Most Pigs	p_4

The point of the preceding thought experiment is to emphasize that common sense, practicality, compromise, convenience, a desire for consensus, and such are likely to affect one's rulings. This should be kept in mind when considering the remarkable challenges which genetic engineering, cybernetics, robotology, electronics, and medicine are likely to precipitate in the future, such as induced mutations, cyborgs, chimeras, artificial persons, and such.

I suspect it is now obvious that it would be rational to take the concept "human being" to be an empirical concept, and, as other empirical concepts, "porous," or "open-textured," subject to revision. One should not impose analyticities in this matter, or legalities, without careful thought and serious consideration given to the anticipated and monitored consequences of rulings, and, in so far as possible, on the whole, and in the long run.

I think it would be useful to call attention to two points, which are distinguishable. first, the abstract or independent value of the ruling, its sagacity, utility, capacity to illuminate, effectiveness, substantial congruence with practice, simplicity, clarity, precision, and such, and its general viability and acceptability. An individual-relative analyticity, however illuminating or enlightening, however judicious, is of little communal interest or importance. No philosopher since Marcus Aurelius has had armies at his disposal, and he never used them for philosophical purposes, say, refuting Skeptics, Epicureans, and Platonists, and spreading Stoicism by fire and sword. That only works for religions, secular and otherwise; in philosophy, that would be another way of cheating, as in not giving reasons. How many people do, or will, accept a ruling, a meaning, a usage, a definition, is extremely important. Happily, systemic processes, over generations, have produced a conceptual structure which is widely shared, a conceptual structure precise enough to be generally applicable and imprecise enough to allow for elasticity, extension, alteration, and improvement. One of the habitats favorable to philosophy is the meristematic area at the edges of a conceptual structure, the borderland between the familiar and the unfamiliar, between the fixed and the fluid, between the old and the new. The looming future calls for the flags of philosophy.

Pioneering explications doubtless begin in private, with individuals or small groups of individuals, but, to become effective in the general conceptual structure, they must be adopted more widely, more publicly. We have noted the occasional role of the judiciary in the imposition of explications, in virtue of enforceable rulings, enforceable by means of the might and violence of the state. Philosophers commonly prefer persuasion to coercion, intellect over intimidation, and reason over fanaticism, indoctrination, and conditioning; further, they suspect that a view which must be imposed by force is likely to have little to be said for it. Revolutions may hatch, however, in the wine cellars of the mind, so to speak, and it is likely that there are some ideas whose "time will come."

In any event, philosophy must address itself to the challenges of the future. Society may permit itself to be ambushed by tomorrow, as it often is. But that will not do for philosophy; it should get there first.

Three problematicities which will help to explore the perimeters of the human or near human are those posed by the "extraterrestrial human," the "evolved human," and the "devolved human."

1.2 The Problem of the Extraterrestrial Human

Could there be a human being from another planet?

Obviously, if the human race becomes a multiplanet species, which it must do, if it has any intention of surviving, given the multitude of menaces it faces, meteoric impact, the stoppage of the earth's rotation, given the tidal break, the exhaustion of the sun's fuel, and such, there would be no problem about having human beings from other planets, as they would have emigrated there, or their ancestors would have emigrated there, and so on. Similarly, if abductees, preserved and nurtured, perhaps Egyptians or Hittites, Greeks or Romans, Franks or Saracens, and such, were returned to Earth by aliens, attempting to communicate in a variety of antique languages, they would not count, so to speak. Our concern here, then, is with the possibilities of human beings whose origin was not terrestrial, either now, or in the past.

The universe seems much of a muchness, with a commonality of laws and components.

As noted earlier, large numbers of exoplanets, or extrasolar planets, have been discovered, and whereas, as would be expected, given the difficulties of detection, most of these worlds are extremely large, and would be unlikely habitats for life as we know it, where there is one planet, it would seem likely, given the usual forces conjectured to produce planetary systems, that there would be more, as in our own system. Accordingly, it seems likely that there would be a large number of planets which are small enough to lose hydrogen and large enough to retain oxygen, which are neither too close nor too far from their primary, thus neither too hot nor too cold, whose temperatures permit liquid water, and so on. Carbonaceous chrondrites hint at extraterrestrial life. Experimental evidence suggests that organic molecules are naturally and abundantly formed, and so on. It seems unlikely that in a universe of billions of galaxies and billions of stars in a given galaxy, and the likely abundance of planetary systems orbiting many of these stars, that Earth would be an unprecedented, unique phenomenon. That supposition would seem implausible, if not absurd, given the collapse and obsolescence of anthropomorphic, theologized astronomies. If the conditions on a given planet, however narrow and exacting, permit, or encourage, the phenomenon of life, then it seems the emergence of life on planets with similar conditions, presumably as narrow and exacting, and as likely and predictable, given the uniformities of nature, might be statistically expected. Similarly, it seems possible that life might develop under a variety of conditions, other than exactly those of Earth, conditions which might be as narrow and exacting for that form of life as ours are for our form of life. In short, a set C of conditions, however narrow and exacting, might occur frequently; given the nature of the universe, and a different set of conditions, say, set C', would not be likely to preclude the evolution of a life form as adapted to those conditions, the C' conditions, as our life form is adapted to its conditions, the C conditions.

Once rudimentary replicators appear, presumably self-replicating carbon compounds, perhaps after two or three billion years of chemical mixings and churnings, of heatings and coolings, the stimuli of electricity, flashing from the sky, and such, one supposes protocells would be next, many of which might live off their more rudimentary forebears, and, later, unicellular organisms, perhaps feeding on their own forebears, the protocells, and such. The energy of the primary, too, would be utilized by many life forms, and simple aquatic forms, such as protoalgae would appear. All in all, once life exists, the dynamics of evolution, rewarding and punishing, exterminating and nurturing, would appear, eliminating its thousands of species, mechanistically favoring, here and there, bit by bit, its precariously evolving hundreds. Given the uniformity and predictability of the forces involved, physical, chemical, biological, and ecological, one would expect, as on Earth, a variety of life forms adapted to diverse conditions, temperatures, and environments, aquatic and terrestrial, forms large and small, sessile and mobile, herbivorous and carnivorous. One would expect prey and predator, sensory acuity, fleetness, and stealth. Too, given the value of certain adaptations, one would expect them to be preserved through an extended evolutionary heritage, or, perhaps, independently evolved in more than one species, as in "convergent evolution." For example, a certain streamlined shape is the optimum design for negotiating an aquatic environment, and one finds it in fish, in the ichthyosaur, a reptile, and in dolphins and porpoises, mammals. Visual sensors, too, are survival-enhancing adaptations, and one finds them in widely divergent species, fish, insects, birds, lizards, and mammals. Interestingly, predator species tend to have the visual sensors forward on the head, this assisting in stereoscopic vision, valuable for locating, tracking, pursuing, and attacking prey, whereas many prey species have the visual sensors more on the sides of the head, giving them extensive peripheral vision, thus making the approach of a predator more likely to be detected. Quadrupedality tends to produce stability and speed; bipedality tends to increase scanning range, and frees appendages for uses other than locomotion, such as grasping branches or manipulating tools. One supposes that on any Earthlike planet which might exist, and on which life has developed, one would find, given the cruel evolutionary filters through which life must pass if it is to survive, forms of life not altogether unfamiliar to ourselves. And one of the survival-enhancing adaptations which one might expect to find, in time, would be intelligence, perhaps conjoined with one or more prehensile appendages. One might even find such a form of life clad in the skins of beasts larger, faster, stronger, and fiercer than itself.

If a humanoid form of life could develop on one planet, there seems no reason why, broadly understood, it could not develop on any number of similar planets. One supposes it would be unlikely, given the intricacies of DNA, to be crossfertile with terrestrial humans. For example, there are many forms of life on the planet Earth which, to one point or another, have a common ancestry with us, and with which we share far more than fifty percent of our genes, and, sometimes in the high nineties, with which we are not crossfertile.

With respect to alien, intelligent life, the following four possibilities might be of interest.

I. Human appearing
    A. Crossfertile with humans
    B. Not crossfertile with humans
II. Not human appearing
    A. Crossfertile with humans
    B. Not crossfertile with humans

One supposes that possibility II B, that of alien, intelligent life which does not appear human and is not crossfertile with humans, would be the most likely possibility. One would expect it to have most of the senses of humans, and perhaps some others. One would expect it to have one or more prehensile appendages. One would expect it to have a language and some form of culture. Possibility II A. that of alien, intelligent life which does not appear human but is crossfertile with humans would be unlikely, given the genetic controls on phenotypes and the intricacy of human DNA. On the other hand, it is an empirical possibility as well as a logical possibility, and, if it is not achieved naturally, one supposes it might be brought about in virtue of interventions, perhaps by genetic engineering, contributions from an advanced alien science, or such. How we might categorize the results of such crossings would presumably depend on a number of factors, perhaps even political factors. If we did not accept the offspring as human, which, one supposes, we might very well not do, then it seems we could not accept crossfertility as a crucial test for species identity. Presumably the most interesting cases would be I A, where the alien appears human, and is crossfertile with humans, and I B, where the alien appears human but is not crossfertile with humans. If these aliens could mingle with humans and remain undetected as aliens, presumably one might regard them as human. Indeed, it is logically possible they are amongst us at present. Presumably we would not wish to make Earth origin a necessary condition for being human, particularly as it is possible that human beings, or many of them, did not have an Earth origin, but were perhaps survivors from a dying civilization elsewhere, explorers or immigrants from another world, criminals condemned to a prison planet, mutineers or criminals marooned on a remote, uncharted "beach," or such. In the I A case, the human appearing alien who is crossfertile with humans, it seems clear it could pass for human. Whether or not we count it as human would depend on our conceptual structure, as construed. Similarly, the human-appearing alien who is not crossfertile with humans would seem to have a case to be made in its favor. First, we have already, in effect, dismissed crossfertility as a test for species identity; second, there are many acknowledged humans who are sterile or barren, and thus not crossfertile in any usual sense with other humans. Certainly we are not going to rule them out of the human race. And, thirdly, if we honestly take our intellectual endowments, intelligence, rationality and such, to be what makes us human, *homo sapiens*, and such, then it seems we would have few grounds on which to refuse humanity to the alien in question, particularly as it resembles us. Interestingly, the resemblance would presumably be more crucial than the rational animality, or such, as one can conceive of any number of rational animals which we would not consider human, say, rational salamanders, loquacious aardvarks, highly intelligent giraffes and raccoons, and such, and, also, there are many acknowledged human beings whom it would be implausible, if not reckless, to categorize as rational animals.

One could surely worry, of course, as to how closely the alien would have to resemble us to count as a human. Suppose it seemed very similar, except behaviorally, or in its consciousness. Perhaps it is much like us, save for some unusual characteristics, such as becoming dangerous at the full moon, mad at such times for the taste of human blood, and such? What if, more benignly, it had a consciousness resembling that of baboon or dog? Might we count it as a lower-level human being, or a "proto" human being? Suppose it resembled us closely, except that, say, it was green, and had a pointed tongue? What if its color depended on the time of day, or the season, or the temperature, or its diet? What if it had claws, or scales? Suppose it resembled us physically very little, but seemed very similar to us, as far as we could tell, in virtue of its consciousness, and its behavioral dispositions. For example, could a satyr or centaur count as a human being? One could certainly conceive of a robot ingeniously constructed to resemble a human being. If it were conscious, and behaved much as a human being, would that be sufficient for humanity? If there could be artificial persons, it seems there might be "artificial human beings." Presumably, as we usually think of these things, a human being could not have an odd origin, such as growing on a tree, but if something was indigenously botanical, and, emerging from its starship, seemed much like us, what then? Presumably aliens might synthesize a human being, put one together from simpler components, and ship it to Earth, perhaps as a curiosity. Might it not count as human? One supposes a human might be produced from a zygote produced in a laboratory. Why not? And, presumably, also, a human might be assembled from parts of other humans, living or dead. The Frankenstein Monster, particularly in the original story by Mrs. Shelley, seemed very human. What if there were mutations off the human stock? Would they count as human? Perhaps some might, and some might not. We wondered earlier about the possibility of human/alien crossfertility. What about the possible results of such crossings, the results of such matches? Would they count as human? Perhaps, again, some might count as human, and some might not. What if the individual wants to be considered human? What if it bribes a judge to pronounce it human? Presumably then, legally, it would be human, and so on.

1.3 The Problem of the Evolved Human

The following brief schema will serve for 1.3, the "evolved human," and 1.4, the "devolved human."

t_{n+3}

t_{n+2}

t_{n+1} ?

t_n Human beings as we know them today

t_{n1} ? t_{n+1}

t_{n2} t_{n+2}

t_{n3} t_{n+3}

t_{n4} t_{n+4}

t_{n5} t_{n+5}

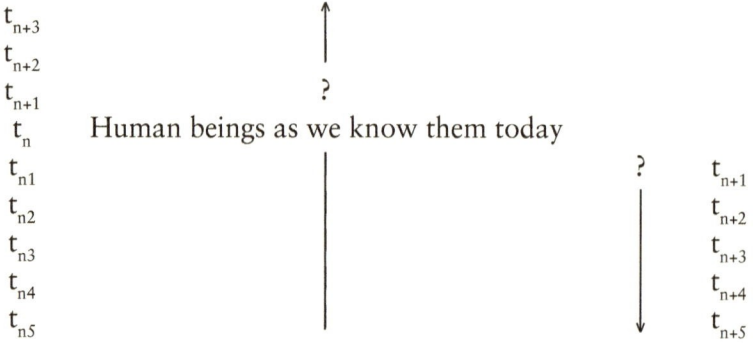

Anthropology, congruent with evolutionary principles, commonly regards at least the physiological characteristics of the human being as being a product of evolution. Similarly, given the influence of certain physiological characteristics, for example, the nature of the nervous system and brain, on consciousness, it seems that the mentality, dispositions, and such, of the human being, like the mentality, dispositions, and such, of other species, has not been unaffected by evolution. Moreover, anthropology has discovered the remains of several forms of hominid life, ancestral to, or collateral with, the human life form. Much here, as one would expect, given the time spans involved and the fragmentary nature of the evidence, is obscure. It is sometimes difficult to determine, for example, if a scrap of bone, a piece of jaw, the fragment of a skull, or such, is human or not. Is it close enough to the indisputably human to count as human? Is it close enough to the indisputably human to count as prehuman? One must decide. Once again, one encounters the "open texture of empirical concepts." There is, it seems, an abundance of evidence pertaining to a relatively recent form of hominid life, the Neanderthal. This form of life lived communally, mastered fire, and had a primitive technology, utilizing axes, spears, and the bow and arrow, which it may have invented. It also must have had a language, as it possessed a culture, involving cannibalism, ritual burials, and such. Should it be regarded as human? Its relationship to indisputable human stock is obscure, whether it was something from which humans were evolved, perhaps by mutation, or was a collateral line springing from a common ancestor. It may have been crossfertile with humans. That is not known. In any event, that life form is no longer with us, perhaps having been exterminated by competitors for territory and resources, most likely "indisputable humans." In any event, it seems likely that Neanderthals, were they about today, would speak contemporary languages, and might, in one way or another, fit into, and contribute to, contemporary society. Their appearance would be different from that of the "indisputable human," and that would undoubtedly result in social and political problems. One uses the expression 'Neanderthal' disparagingly today, but there is no reason to suppose that there was not a natural curve of intelligence in those groups, as in other groups. One supposes that amongst them there must have been some, and perhaps many, individuals of high intelligence. In any event, our question might be, "Would they count as human beings?" One supposes so, given their qualities and achievements, culture and such, and this supposition indicates that we would be unlikely to construe humanity, at least today, in a narrow Draconian or Procrustean fashion. The conceptual structure is subject to a variety of considerations, including considerations not only cognitive, for example, organizational and useful, but aesthetic, political, economic, and moral. One might, of course, anticipate a number of problems emerging, such as characterized "artificial life forms," "artificial persons," and such.

There are various theories of evolution about, most of a Darwinian or neo-Darwinian sort. Classical Darwinism, in the sense of natural selection, presupposes first, fecundity, often of a prolific sort, more offspring being produced than are likely to survive; second, variations amongst offspring, which we might think of as genetic differences, or micromutations, or such; third, that some of these variations may be survival-enhancing, for example, better vision, more acute hearing, and such; fourth, that at least some of these survival-enhancing variations will be biologically transmitted to offspring; and, fifth, that over enormous amounts of time, these tiny changes, generation to generation, may accumulate in such a way as to improve a species, and even transform it into new forms, or new species. For example, the modern horse, in its variations, is far removed from tiny, scampering eohippus, four toes on the front feet, three on the hind feet, grazing in the Eocene.

The opposite of the micromutation, which appears to be classically Darwinian, is the macromutation, more commonly associated with forms of neo-Darwinism. One supposes there would be gradations between these categories of mutation. In passing, it might be noted that mutations may have a diversity of causes, and are quite common, that most are of negligible evolutionary import, and of those with such import it is likely that more will be deleterious than beneficial. Also, there is no reason why both micromutation and macromutation might not characterize evolutionary change. Spikes and slopes may be different, but both might occur on the same curve.

It is not clear that any significant evolutionary change has taken place in the human species in the past thirty or forty thousand years. There seems to be some evidence that Paleolithic human beings used to be larger and have larger brain cases than the modern human, but this is controversial, and, apparently, as in sociobiology, and other sensitive areas, politics becomes involved. Certainly the modern human is likely to be healthier, longer-lived, and larger than, say, the Medieval human. One leaves such disputes to anthropology, and ideology. It seems that some scientists, rather as some jurists read their policy preferences into the law, read their policy preferences into their science. Political considerations encourage this practice. The sociology of knowledge, which calls attention to such matters, is an interesting field in itself.

To be sure, these matters are complex. And one might recall, from earlier in this study, the interactions and reciprocities between organic and superorganic evolution, the Baldwin Effect, and such. It is easier, for example, to see how evolution might have favored fleetness of foot, flight, toughness of hide, thickness of fur, acuteness of hearing, smell, and eyesight, fangs, and claws, and such, in an open, natural environment, than it is to see what it might favor, if anything, in a complex, urbanized world of nation states, global economies, mass communication, television, advertising, and such. What would be the ideal adaptation in such a world? Would it be the lion or the social insect, the eagle or the chameleon and rabbit?

Our concerns, happily, are more speculative and conceptual than journalistic or reportorial.

On a classical Darwinian approach to these matters, it seems no significant evolutionary change in human DNA would be likely in the near future, as it has apparently not taken place, at least to any agreed–upon, marked, degree in the last thirty or forty thousand years. If a macromutation were to take place, it is hard to see what it might be. The most likely guess would be a change in brain size and capacity, such as seems to have taken place some millions of years ago, when the originally small hominid brain and brain case were replaced, in a few thousand years, a blip in evolutionary time, with something approximating that of later hominids, for example, those of the Neanderthal and Cro-Magnon, which latter life form we take to be "indisputably human," indeed our ancestors.

A larger-headed human, however intelligent, would, in a global society, most likely be regarded as a freak. Its genes, without the intervention of the state, might not be replicated. Certainly its lot would be a difficult one to endure. Such a being might well, popularly, be hated, resented, despised, envied, and feared. Its intelligence would not be regarded as "true intelligence," and so on.

There might, of course, be macromutations in a variety of other respects, as well. The common science-fiction supposition of paranormal powers, or such, seems unlikely. On the other hand, variations in size and shape, in temperature and pollution tolerance, in skin color, in hirsuteness, in longevity, and such, would be possible. Again, it is not clear how the environing society would react to such emergences.

The more likely "evolved human:" would be a designed human, the product of eugenics or genetic engineering.

Consider eugenics:

Earlier in this study we noted the remarkable possibilities which have been realized in selective breeding, with plants and animals, for thousands of years in various Earth cultures. The some two hundred or more modern varieties of the dog, for example, are all derived from a common ancestor, the gray wolf. These eugenic results, of course, have been imposed from the outside. One would not expect similar results in diversifying the human species without unusual social or political conditions, humans perhaps being utilized as pets or work animals by alien life forms or artificial persons, humans being bred for various sorts of labor by the state, being bred for adaptation to extraterrestrial environments, and such. Whereas this sort of thing, designed matings, and such, with an end in mind, seems wholly impractical at present, if only because of anticipated resistance, it is only necessary to break the cultural chain, removing language, and such, to reduce the human being to another form of animal, one which might, for all we know, be penned and bred, say, by the state, or such, with the same purposefulness and calculation as other forms of domestic stock. It is hard to know about such things.

Individuals, nonconformists, eccentrics, rebels, heroes, and such, could always be eliminated. Human groups, by means of a number of social controls, envy, oppression, ridicule, indoctrination, shame, fear, marginalization, isolation, and such, have commonly managed to reassure, protect, and prolong themselves, preserving the desiderata of mediocrity and uniformity, particularly in the realm of thought, where difference is sensed as most dangerous.

But let us now suppose that some minority of humans, over generations, have been bred selectively, voluntarily we may hope, amongst themselves, for a number of targeted characteristics, such as intelligence, health, size, strength, longevity, a particular appearance, and such. Let us suppose that they would eventually be as different from the current human being as the Great Dane or Saint Bernard is from the Pekinese and Chihuahua.

Presumably, eventually, as such a process might continue, one might wonder how such entities would view us. Would they regard us as human? We are not inclined, for example, to view Australopithecus Africanus as human, and certainly not chimpanzees, amphibians, reptiles, lungfish, and such. Or they might regard us as human, the term now derogatory, and choose a newer name for their own variety, or species. We, on the other hand, would be likely to regard them as human, if only as a form of self-flattery. We might try to be proud of them, though we had little to do with them, or they with us. More likely, we would hate them, and the compassionate tolerance with which they might view us. Perhaps we would kill them. Again, it is hard to know about such things.

Consider genetic engineering:

Many individuals see genetic engineering as affording the human race, or its cognitive elite, the opportunity to control and direct human evolution, and in a manner quicker and cheaper, and more businesslike, than eugenics. For example, a desired characteristic would not have to be bred for, with possible time-consuming failures, but might be introduced directly into the genome. Thus, a genetically controlled evolution would be more expeditious than, and more efficiently manageable than, one pursued via the more familiar methods of eugenics, as in stock breeding. Too, one supposes that a genetically engineered evolution would be perceived as less invidious than one achieved by means of eugenics, either positive or negative, the image of which, as applied to humans, nothing else, was seriously tarnished in the Twentieth Century, as a result of its incorporation in certain forms of authoritarian and statist ideology.

In an earlier portion of this study we noted a large number of variations in humankind which might be brought about by genetic engineering, which it would be pointless to repeat. Needless to say, the possible changes, or transformations, are almost limitless, and potentially awesome.

One would then anticipate, naturally, as the nature of these changes or transformations became more clear to the public, a tsunami of opposition, particularly on the part of individuals who do not understand how much genetic engineering has been actually going on for years, in a variety of areas; one might expect hysteria in the press, the overnight emergence of numerous advocacy groups, of various sorts, the rushing forth of politicians and religious leaders, the organization of protests, the picketings of scientific facilities, acts of sabotage, and other indications of democracy in action. One would expect to learn that it was opposed by doctrines and teachings which had never heard of it. The American Constitution would be examined, to see what it had to say about it, and so on. To be sure, some latitude might be allowed, for example, in developing better tomatoes, if not better people, in treating genetic diseases, if not in producing bodies not subject to such diseases, and so on.

One doubts that there is anything inevitable about the progress of science, and one doubts also, if all change, even if unimpeded, is progress. As before, and again, it is hard to say about these things.

As a last point concerning the possibility of an "evolved human," one might recall that evolution is correlated with particular environments. For example, a species adapts itself to a given environment, and, if the environment remains stable, so, too, statistically, does the species. The human being, for example, seems reasonably well adapted to the natural terrestrial environment, in its variations, and tends to stabilize his environment, in any event, by such means as shelter, clothing, heat, cooling, and such. This might be an explanation why the human race has not much changed, if it has changed, at all, in the last thirty or forty thousand years. There was no need, on the whole, to do so. On the other hand, if the environment should change in some radical manner, which could not be artificially neutralized, and it was the case that the maladapted were less likely to replicate their genes, and the better adapted more likely to replicate their genes, then, even in a classical Darwinian modality, one might expect evolutionary change, or, lacking it, the extinction of the human race. On the supposition that the human race becomes a multiplanet species, it will doubtless, at first, at least, seek out Earthtype planets, or attempt to terraform other worlds into an Earthlike form. On the other hand, over time, on a variety of alien worlds, given different conditions and isolated populations, it would not be impossible for the human life form to diversify, perhaps considerably. One might then, in some thousands of years, have a number of variations off the original human stock which might be surprisingly different, variations which might be puzzled as to whether or not other variations were or were not human. And we might wonder if they, themselves, were human. Perhaps some geneticist of the future, by DNA tracking, might convince the more open-minded of a number of different varieties, or species, that they were not only related to one another, despite appearances to the contrary, but were all derived from a common ancestral form, us.

1.4 The Problem of the Devolved Human

We referred earlier in this study to the SETI project, the search for extraterrestrial intelligence.

For a variety of reasons, it seems statistically probable that there would be a large number of rational life forms in the galaxy, and that some of these life forms would be interested in knowing if other such life forms existed, in learning whether or not they were alone in their universe. Further, it seems that a certain number of these interested life forms might possess a technology sufficiently advanced to make their existence known to others, and, of such a number, it seems that some might actually choose to do so, that is, choose to advertise their presence, to transmit signals, of one sort or another, for one length of time or another. How then does one explain the fact that the SETI project, despite years of diligent searching for such signals, has heard nothing, that the sky is, so to speak, "silent." Obviously there could be a large number of explanations, the most obvious being that these other life forms do not exist, that we are the solitary rational life form in the universe, and, perhaps, more dismally, that Earth life forms, humans, cacti, sponges, and such, might be the only life forms in the universe. This sort of thing is certainly a possibility, though it seems, statistically, unlikely. Even granting the existence of extraterrestrial intelligence, there could be a large number of reasons to explain the "silent sky," a few of which are that the signals are not of a sort we are able to detect; that such forms are not technological, and are unable to transmit suitable signals, perhaps being in a state of primitive or prescientific culture, or even being aquatic in nature; that they are contemplative, or self-absorbed, in nature, and not interested in such inquiries; that they regard technology as immoral, dangerous, or distractive; that they are afraid to signal their presence, for fear of inviting contact, perhaps fearing eventual disruption, disease, invasion, or exploitation, and so on. On the other hand, another theory, which has some plausibility, is the theory of "technological suicide," namely, that technology bears within itself the seeds of its own destruction. Take a possessive, energetic, active, intelligent, aggressive life form, relentless enough, and cruel enough, and terrible enough, to scratch and claw itself to the summit of the food chain, to become the dominant form of life on the planet, a territorial form of animal, accustomed to compete ruthlessly for resources, an animal deeply flawed and emotionally driven, with its hates, and its insatiable appetite for power, attention, wealth, and prestige. It is an animal as dangerous, or more dangerous, to its own kind than to any other species. It is an animal, too, which feels itself most threatened by its own kind. It is an animal which seizes up bones, branches, and stones as clubs, and, eventually, thousands of years later, manages to graduate to a weaponry sufficient to ruin its world, which it may do. Our only guide to this sort of thing is our own species, with its intelligence, its power, its hatreds, its fanaticisms. It seems not impossible that the madnesses with which our form of life seems intermittently afflicted, sweeping across continents and oceans, engulfing peoples, may eventually end civilization as we know it, and reduce the human species once again, if it survives, to the local jeopardies and gangsterisms of barbarism.

A technology with sufficient power and sophistication to engage in interstellar signaling is also likely to be a technology

capable of destroying itself, and its civilization. The book of
progress, such as it is, might be closed before that next chapter,
one of interstellar communication, of cosmic salutation, could be
written. Something like this may be the simple, fearful solution
of "the problem of the silent sky." It seems anomalous and
paradoxical that technology, that marvelous engine of a species'
ascent, might prove to be the engine, as well, of its diminishment
and devastation. "They that climb high have many blasts to
shake them, and, if they fall, they dash themselves to pieces."

In any event, a major shift in human evolution would
presumably be consequent upon a radical transformation
of the environment, socially or ecologically. An apocalyptic
eradication of civilization might produce in its wake the loss of
resources and skills, technological and otherwise, famine, disease,
and the severance of civil bonds, reducing surviving human
groups, isolated and wandering, to a state of tribal, economic
primitivism, perhaps a life of hunting and gathering, abetted by
occasional raiding and banditry. It might prove to be dangerous
amongst the ruins. In such a world the skills associated with
gene replication, those of hunters and warriors in their prime,
with their women, might be quite different from those leading to
prosperity and prestige in a complex, technologically advanced
society. Analogously, one would expect dogs to eventually
regain wolfhood or perish. There are, of course, in all species,
"throwbacks," in which previously common genetic combinations
recur, toes in the horse, teeth in the chicken, a hirsute infant, and
such. Such throwbacks might occur in a human species, as well,
individuals who would have been misplaced and dangerous in a
modern civilization, in which they would have been born out of
their time, but would now find themselves reborn into the sort
of time which bred them, a cruel, hard time in which they throve,
and were welcomed. Similarly, anomalies occasionally emerge,
some of which, if genetically transmissible, might prove to be of
value to a species in a given environment, say, unusually acute
hearing or vision, a sense of smell permitting one to be aware of
a predator within a hundred yards, a size, weight, strength, speed,
and agility uncommon in modern man, perhaps even a length
of arms permitting one to move easily amongst branches. There
might also be radiation-induced, or chemical-induced, mutations,
depending on the means which had been utilized to bring about
the demise of the former civilization. A meteoric passing in the

vicinity of the planet might tear away a portion of our world, as perhaps happened with the formation of the moon, the mass of which suggests its removal may have left behind what became the Pacific basin. A meteoric impact might severely alter the ecostructure of the world, bringing about the extinction of many species; for example, it is often speculated that such an impact in the late Cretaceous or early Tertiary, some sixty-five million years ago, had much to do with the disappearance of the dinosaurs; such an impact, too, might affect the rotation of the earth, perhaps even affecting the planet's orbit, changing the seasons, altering climates, perhaps then to range from torrid heat to polar cold. In such a world it might not be the mighty which would survive but the small, furtive, and prolific. Perhaps a remnant of the human race might survive, huddled about a new Equator, emerging from burrows at night, as scavengers or insectivores.

In a world which they cannot manage with technology, as it is no longer available, the human being, as other forms of life, would find itself at the mercy of nature, to which it would adapt or die.

There is a question, of course, as to the appropriateness of a terminology such as 'evolve' or 'devolve'. Both terms often carry the hint of a value judgment. Is it not good to evolve, and bad to devolve, but what counts as "evolved," and what as "devolved"? Complexity, or success? In the end, there is adaptation or a failure to adapt. Life forms such as the termite, the crocodile, and the shark, are well adapted, and have managed to replicate their genes successfully for millions of years. Would they not count as evolution's greatest successes? Might one not regard a small, rodentlike human, quite different from the contemporary human, as more "evolved," if it were enabled by its properties to change and develop over time in such a way that it managed to survive and prosper under conditions which would have been inimical to its ancestral life form? Certainly it would be later in time than the ancestral life form, and it, too, in its way, would be quite complex. It might even possess a language. Our question, of course, is a conceptual one. Would such a creature count as human? Too, perhaps it would not regard us as human, rather as we are reluctant to count australopithecines as human, though, doubtless, it would give little thought to the matter.

2. The Epistemologizing of "Essence"

2.1 Some Problematicities Having to do with "Essence"

An "essence," metaphysically, is that property or properties without which a thing would not be what it is. A moment's reflection indicates that it would be very difficult for something not to be what it is. How would it go about becoming what it isn't? Presumably, something like '(x) (x = x)' ('For every x, x is identical with x') is a logical truth, true in virtue of its meaning alone, nonfalsifiable, and such. To be sure, one might improve one's tennis skills and become a better tennis player in August than one was in May, but different times are involved. Too, everything changes. If anything could refute the law of identity it would presumably be change, but it is not allowed to do so. Identity accompanies change without embarrassment. Bill's tennis skills are exactly what they are in May and exactly what they are in August, though different in August than in May.

A classical essence is not regarded as imperiled by time. Accordingly, it is understood to be immune to change. It cannot then be identified with all properties, for some of them change. For example, at one time Socrates is sitting, and at another time standing, but at both times he is Socrates, and human. This suggests a distinction between accidental and essential properties. That Socrates is annoying a fellow Athenian in the Agora at one time, or on his way to a symposium at another time, are supposedly accidental properties of Socrates. To be sure it is hard for him not to be standing or not standing, on his way to a symposium or not on his way to a symposium, and such. We noted that sort of thing before. It is not clear whether essences can pertain to individuals or not. Presumably they would not do so. Thus, there would be no essence of Socrates *qua* Socrates. Essences seem to pertain to kinds, or sorts. A Platonic form may lie in the ancestry of the Aristotelian essence.

The essence of "human," classically understood, is usually thought to be "rational animality." That will do for our purposes, despite the fact that some humans may not be rational, and some nonhumans may be rational, to a greater or lesser extent, for example, apes communicating in Amslan, the American Sign Language. And, if one enters into logical possibilities, it is easy to conceive of entities we would regard as human which would lack rationality and entities we would not regard as human, which would be obviously rational, intelligent aliens, for example. It may be, of course, that the Greek which is commonly translated along the lines of "rational animality" may have had more the sense of "speaking animality," "language-possessing animality," or such. That might be better, but would still be subject to the sorts of counterexamples just suggested. Too, it would not seem wise to attach "essence" to something which is obviously an acquired characteristic, a cultural acquisition, or such. Infants may be human, but, first, they are not obviously very rational, and, second, they certainly lack language. It would not seem too helpful to put this sort of thing in terms of potential rationality or potential language-possession. For example, a purely potential bank account is not a bank account, and a purely potential speaker of Urdu or Bassa is not a speaker of Urdu or Bassa. And if a human-seeming entity is not even potentially rational or not even potentially a possessor of language, we might not want to rule him out as human. We might want to know more about the case. And, in a logical sense, one supposes one might suppose almost anything might be rational or a speaker, say, Carnapian stones thinking of Vienna, or, if one prefers, as Paul Benacerraf, Paris, or raccoons and giraffes with thick Hungarian accents.

The two important things to understand here, putting details aside, is that an essence, classically conceived, is, first the property or properties without which a thing would not be what it is, and, second, that a particular property or properties are crucial in this matter, that they, and not something else, make the entity what it is, say, a dog, a horse, a human being. The essence then, so to speak, is a subset of a number of properties, which, technically, are infinite in number.

The problem, then, is to discern the essence, to pick it out, from amongst this infinite set of properties which characterize any entity whatsoever.

In the classical view the selection of essential properties is correct or incorrect; the essence is metaphysically real, incorporated in the nature of reality. One can be right or wrong about it. It is there, to be discovered, objectively there, whether one discovers it or not.

For example, let us suppose we have an infinite number of properties, amongst which seven properties seem the most likely candidates for the essence of an entity, rather as follows:

$$P_1, P_2, P_3, P_4, P_5, P_6, P_7 \rightarrow \infty$$

None of these properties, in itself, proclaims itself to be the essence. The properties, *per se*, do not announce themselves. Even if they did, how would one know their claims were trustworthy? Even if, say, Property P_2 began to hum or glow when viewed, how would one know that that had anything to do with an essence? Perhaps Property P_2 is merely bashful, or self-conscious. Perhaps one might rely on one's indubitable intuitive inductions, or such, but how does one know they are indubitable, and what if another fellow's indubitable intuitive inductions come up with a different result? And who is to say that they might not both be wrong?

Indeed, if this is a metaphysical matter, who is to say that the actual essence is ever discovered? Might it not be well-hidden, be masked by veiling properties, lurk about incognito, or such? What if the true essence is Property $P_{2,465,787}$. That might never be noticed. One might continue indefinitely, mistakenly, to go with, say, Property$_{2,465,786}$.

Clearly, taking an essence to be given, to be metaphysically real, in some inexplicable sense, to be objectively "there," in some way, is to ask for trouble, whether it is given, real, objectively "there," in some inexplicable sense, or not.

2.2 Genus/Species Discretion

Let us return to "rational animality." As commonly understood, the genus, or class, is animality, and the species, or subclass, is rationality. The human is an animal, which is rational.

Consider the following two class matrices. In the first, we are dealing with class variables, A, A', and such. For the four variables involved, there are four possible subclasses, as noted. AB, too, it will be noted, may be regarded as a subclass of the class A, or a subclass of the class B. Indeed, it is a subclass of both. Each of the four macroclasses, i.e., A, A', B, and B', has two subclasses.

```
A        A'
AB       A'B      B
AB'      A'B'     B'
```

Now, in the second class matrix, we are dealing with class constants, namely, class designations with a fixed meaning. We will take A to be the class of animals, A' to be the class of nonanimals, R to be the class of rational entities, and R' to be the class of nonrational entities. Remarks about macroclasses, subclasses, and such, would be as before.

```
A        A'
AR       A'R      R
AR'      A'R'     R'
```

In passing it will be noted that everything in the universe fits into this matrix; or any such matrix, for example, your average quark, if it exists, is a nonanimal nonrational being; your average protozoon would be a nonrational animal; your average god or gods would count as nonanimal rational beings, and your average human being would count as a rational animal.

What is of most interest from our point of view is that the class AR, that of rational animals, counts as a subclass of both Class A, that of animals, and that of Class R, rational entities. In short, it is a matter of discretion whether one chooses to regard rational animals as a species of rational entities or animals, or both. For example, logically, the two following arrangements are equivalent. One might prefer one arrangement over the other for one reason or another, sensible or not, better or worse, but the modest requirements of logic are satisfied by either choice. Neither, as far as one can tell, would be "metaphysically dictated."

Animals	
Rational	Nonrational

Rational Entities	
Animals	Nonanimals

2.3 Essence Supersession or Replacement

"Essence" tends to be a latitudinal notion, rather than a longitudinal notion, one horizontal in time rather than vertical in time, a static notion as opposed to a fluid notion, such as change, motion, process, history, development, alteration, evolution, and such. It is a bit like a snapshot, taken of an object in motion; a way of seeing the unfixed as fixed, the impermanent as permanent. In many ways the hands of the cosmic clock move slowly, so slowly one may not always detect their motion. We know today, and even before theories of the expanding universe, the red shift in the spectrum, and such, that the "fixed stars" are not fixed, this evident from their placement in discovered Egyptian star charts. The ancients, on the whole, save for an occasional Heraclitus or Epicurus, here and there, seem to have supposed themselves living in a manageable, small, stable world, on the whole constant, steadfast, and unfailing. It did not move and the universe at the center of which it found itself was not that large or complicated. Horses stayed horses, and men men. Plato hypothesized eternal forms which would keep things stable and orderly. Even change, the rise and fall of cities and heroes, tended to be cyclical, almost predictable, almost seasonable. In such a world, with its familiar perspectives, its cognitive complacencies, its awe of supposedly self-evident truths, its confidence in unaided intellect, its besetting preoccupation with deduction, its obsession with geometrical paradigms, its uncritical acceptance of the world as it was found, the concept of forms, of essences, and such, would seem to follow naturally, almost inevitably.

But things change, even universes may have their birth and growth, their waxings and wanings, their life and death, their comings and goings, their subsidences and renewals. The infinitudes of Epicurus and the incessancies of Heraclitus may know the name of reality better than the immutable serenity of Father Parmenides and the forms of Plato.

In the small schema below, we trace the development of a theoretical life form through various changes, noting that, were there human beings about at the time, interested in such things, different essences might have been assigned to the life form at different times. Recall that the hands of the cosmic clock may move so slowly that the motion is undetectable at a given time, this producing a natural illusion of permanence.

t_{10}

t_9

t_8 Essence$_3$ →

t_7

t_6

t_5 Essence$_2$ →

t_4

t_3

t_2 Essence$_1$ →

t_1

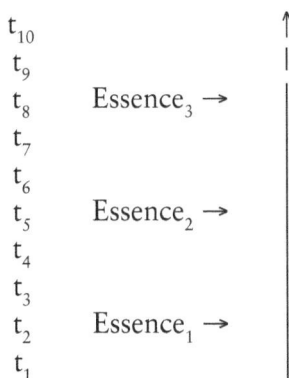

Considerations of this sort are not to be construed as disparaging the concept of essence, which, properly understood, may be of great value. Not only might a life form remain indefinitely static, given the success of its adaptation to stable conditions, witness the termite, crocodile, and shark, but even a more protean, volatile life form may remain substantially unchanged over several millennia, particularly given the patience of classical Darwinian evolution, the incremental accumulation of tiny, transmissible differences, producing a graduality of change. Eohippus is quite different from the modern horse, but there would be no reason why one might not see eohippus in terms of a property or properties without which we would not regard it as eohippus.

2.4 Essence Pluralism

Let us suppose a distinction between properties and classes, one in which a property P may determine membership in a class C, for example, the property of being a dog, which may determine membership in a class, which class we will then think of as the class of dogs.[31] This approach is a bit simple, and more logical, than historical. Presumably, historically, one would seldom move from the property to the class, but would be more likely to move

---

31. There are a number of serious difficulties with the concept of both properties and classes, for example, what they are, their ontological status, what properties may and may not be allowed to determine classes, the relationship between a class and its members, the null class, paradoxes of class, or set, theory, the most famous of which is Russell's Paradox, and so on, but these will be ignored as tangential to our concerns. Life is difficult enough, as it is.

from resembling entities, to a common property or properties of such entities, and then use that, later, to determine the class, which class might not contain all of the originally resembling entities and might contain other entities, not in the original group. The relationship between entities, property, and class, then, might be complex, and variable.

Let us now suppose we have some referring expressions. Following Carnap, we will take the intension of the expressions to be a property, and the extension of the expressions to be a class.[32] For example, if the expression is 'dog', we will take the intension of the expression to be the property of being a dog and the extension of the expression to be the class of dogs.

The following table illustrates relationships amongst expressions, their intensions, and extensions. It will be noted that if expressions have the same intension, it is necessary that they have the same extension. For example, if 'male sibling' and 'brother' have the same intension, they must have the same extension. On the other hand, if a set of expressions have the same intension it is impossible that they should have a different extension. If some expressions have different intensions they may or may not have the same extension. The usual situation is that different intensions will produce different extensions, as in, say, 'dog' and 'cat'. The most interesting case is that in which expressions have different intensions but produce the same extension. That case is extremely relevant to our discussion, as will be made clear below.

Same Intension	Different Intension	
Necessary	Possible	Same Extension
Impossible	Possible	Different Extension

Let us suppose we have five properties, each of which we shall suppose characterizes all and only members of a given class. The expressions for these properties will thus have the same extension, will designate the same class, for example, the class of humans. We may also suppose, however, that the intensions of the expressions may differ, being different properties. Thus, different properties may characterize exactly the same entities, may mark out exactly the same class, again, say, the class of humans. Let the

---

32. Cf. Rudolf Carnap, *Meaning and Necessity: A Study in Semantics and Modal Logic.* (Phoenix Books, The University of Chicago Press, Chicago and London, 1956.)

five properties be rational animality (RA), featherless bipedality (FB), having a brain of type B (HBB), having a posture of type P (HPP), and having a genome of type G (HGG).[33]

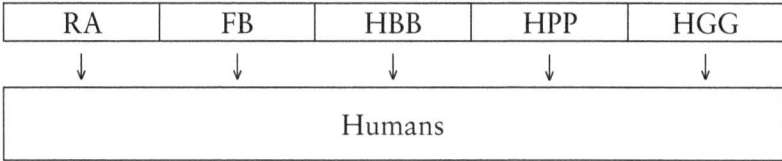

RA	FB	HBB	HPP	HGG
↓	↓	↓	↓	↓
Humans				

Presumably there would be a great many properties which might mark out the class of human beings, which would characterize all and only those entities we would like to consider human.

What emerges from considerations of the preceding sorts, for example, the possibility of essence pluralism, is that the supposition of essence as a metaphysical real, as something independent of human interest and purposes, is dubious and useless, if not altogether meaningless. Even if it exists in some humanly incomprehensible manner, being decreed, say, by some god or gods, as the very pivot of their design, the point of their plan, the very center, heart, and core of their top secret schemes for humanity, it would do us no good. We have no way to test for it, or extract it from amongst other suitable candidates. To be sure, an essence, say, featherless bipedality, could be imposed legally, as judges are good at such things, and that would then count as the legal essence of humanity, at least within a pertinent jurisdiction. So, too, a dictator, with guns and bayonets, might reform education along certain lines, producing "new man," or such. And a religious leader might rule, too, one way or another, for those among the faithful who still take such things seriously. The most impressive imposition

---

33. One includes 'rational animal' here in deference to fame, and tradition. It does have its problems, as noted earlier. Similarly, I have included 'featherless bipedality', first, as it seems to have been attributed to Plato, presumably mistakenly, and, second, because it and 'rational animal' have classically been used to illustrate the semantic situation where two expressions with different meanings might have the same reference, apply to the same set, class, group, or such. Featherless bipedality, of course, might even be worse than rational animality for marking out humans, as there are many featherless bipeds, plucked chickens (à la Cratylus), chimpanzees, kangaroos, certain robots, etc., whom we would not count as human. Also, one could suppose human beings with one or more feathers, even as an empirical possibility, given mutations, genetic engineering, and such. For example, on a low-gravity planet, wings might be practical, as well as hands and feet. Genetic engineers might set themselves the task of bringing them about.

of an essence would seem to be one proclaimed by one or more mysterious beings, perhaps gods, certainly claiming to be such, showing up, armed with plagues, famines, angelic massacres, rains of fire and brimstone, thunderbolts, or such. That would certainly be worth taking seriously. So the essence of human would turn out to be "featherless bipedality." But the question would remain, would it not, why featherless bipedality, and not another property? Would it not be merely the substitution of certain interests, and possibly purposes, for other interests and purposes?

Lacking access to the divine blueprints one will have to make the best of things.

One might also note that what is involved here philosophically is not so much a property or properties which will characterize all and only human beings, as a property or properties in terms of which we are willing to understand human beings. In short, the philosophical mission here is normative, more than descriptive. The question is not so much "What is a human being?" as the more Nietzschean question, "What shall we take a human being to be?" "How best can we understand humanity?" "What should a human being be?" "What ought a human being to be?"[34]

3. Some Identity Criteria

By way of introduction, three remarks may be helpful.

First, it is difficult to suppose that a complex conceptual structure could exist without a complex language, at least for a mind of the

---

34. It is important to distinguish what we are doing here, which is largely normative, from semantic inquiries, field linguistics, cataloging speech practices, and such. We do not regard our mission here as an attempt to taxonomize linguistic habits, to record, assemble, and sort through usages, and such, and certainly do not regard it as an attempt to cure "bewitched intelligences," or "show the fly the way out of the fly bottle," namely, to finish off philosophy, and get on to other things, more interesting things, more important things, say, backgammon, motoring, the cello, or such. Ludwig Wittgenstein, a great human being, and one of the greatest of twentieth century philosophers, tried to kill philosophy twice, and, happily, missed both times. One of the most interesting and valuable contributions of Wittgenstein (Wittgentstein$_2$, so to speak, the Wittgenstein of the *Philosophical Investigations*, as opposed to Wittgenstein$_1$, the Wittgenstein of the *Tractatus Logico-Philosophicus*) was the notion of "family resemblances" amongst semantic usages. His famous example is "games." Classically, philosophers would have tried, presumably, to find the "essence" of games,

sort with which we are familiar. On the other hand, the conceptual structure is not to be confused with a linguistic structure, despite the fact that, in the human case, at least, intimate reciprocities may exist between them. Four considerations make this obvious. first, that animals and young children have a conceptual structure, for example, a set of recognitions, classifications, recollections, expectations, anticipations, and such, in virtue of which they understand and relate to their worlds. Second, if an identity existed between language and conceptual structure, dissatisfaction with language, the contraction and expansion of language, the invention of language, its improvement, and such, would not take place. Whereas the conceptual structure may occasionally be the dupe of language, as in verbal thinkers, linguistically associational thinkers, thinkers who think words and not things, who

---

or, perhaps better, the necessary and sufficient conditions which would have to be satisfied for something to count as a game. Wittgenstein doubted that there was any single characteristic which all and only games had in common, that, rather, there were a number of overlapping characteristics, or "family resemblances," by which games were related to one another. A simple schematism will make the point clear, as follows.

$$A \quad B \quad C \quad D \quad E$$
$$B \quad C \quad D \quad E \quad F$$
$$C \quad D \quad E \quad F \quad G$$
$$D \quad E \quad F \quad G \quad H$$
$$E \quad F \quad G \quad H \quad I$$
$$F \quad G \quad H \quad I \quad J$$

It will be noted that the first entry in this list has no element in common with the last entry on the list. On the other hand, there is obviously a great deal of commonality in the list, and its entities are impressively related, each entity having obvious relationships to several of the other entities. Wittgenstein's recommendation, which makes a great deal of sense if one is interested in discovering how people use words, is, so to speak, to "go out, and look and see." The philosopher's task, then, of trying to discover the essence of games, or their necessary and sufficient conditions, a task perhaps fruitless and misguided, is replaced with a different task, a practical, intelligible task, namely, seeing how people use the words 'game' and 'games', see what they apply the word to, and so on. Wittgenstein, at least by intention, seems to have been substantially a philosophical descriptivist, namely, one who is essentially looking into how things happen to be, and letting us know how they are. In this study, of course, we are less concerned with how we talk and think, which is often messy, confused, chaotic, inconsistent, incoherent, disorganized, and annoying, than with how we might better talk or think. In short, we are more concerned with explication, with clarification, improvement, and reformation, than with reportage, than with Xeroxing reality as we find it. This is in no way a disparagement of descriptivism, which has its own interests and values, but they do not seem particularly philosophical. It is certainly not a necessary first step toward explication, and it is not even clear that it would be a useful first step in that direction. It is a different interest, and a different business. In any event, philosophy is a big country and in it one finds many states, counties, or districts, and many dialects.

misunderstand linguistic mapping, the world/word relationship, and such, language, ideally, should be the servant and tool of the conceptual structure, a means by which thinking may be sharpened and clarified, and made public. Third, just as there are words without concepts, as Hume noted long ago, words without ideas, empty words, so, too, there are concepts which, as yet, have no words, no adequate linguistic expression, concepts which seek expression. Fourthly, without the mind, without consciousness, language would be meaningless noises, no more than marks on paper, scratches on the sidewalk, grooves on stone, and such. One could have at one's disposal an alien or forgotten language, every sound, every syllable, every mark or impression, every jot and tittle, and not have its once associated or allied conceptual structure. One might then have noise, as in the rushing of wind and the pounding of the surf, but one would not have meaning.

Second, the solution to philosophical problems does not lie in ordinary language, or, at least, ordinary language as used ordinarily. To be sure, the dissolution of philosophical problems might lie in ordinary language used ordinarily, in the sense that philosophy confined to ordinary language used ordinarily might disappear, rather as physics and chemistry, mathematics and astronomy, confined to ordinary language used ordinarily, might disappear, or might as well disappear.

Third, philosophy is not well understood as armchair or amateur science. even armchair or amateur linguistics. Its most valuable contribution to the cognitive enterprise is not descriptive, but prescriptive. Its primary concern is with what we spoke of as first-order or second-order fictions, those fictions, not encounterable, measurable things, in terms of which men make sense of the world, in terms of which they organize and conduct their lives, in terms of which they understand, and, sometimes, those fictions for which they will live or die. It does not so much discover as create. It is the fountain from which flow concepts, classifications, proposals, categories, understandings, and the apparatus of understanding, and, not unoften, it goes where familiar thought has not, indeed, into places altogether unknown to familiar thought, places which familiar thought did not know existed.

We are now ready to address ourselves to some questions of species identity, questions, in particular, having to do with the human species. We will divide our proposals into four major categories, these having to do with conception, characteristics, cross-fertility, and constituents, and will conclude with a discussion of the relevance or irrelevance of Earth origin.

3.1 Conception

In the following suggestions, we are supposing that we begin with an acknowledged human female. Criteria by means of which a human, male or female, might be recognized are dealt with later.

3.11 The Result of a Normal Conception by a Human Earth Female, or Human Female of Earth Origin, Either Proximately or Remotely

Here we are presuming that Earth origin is important for counting as human. The value of this presumption will be discussed later. We are also supposing that a conception involving a nonhuman, quarter human, or such, would not count as normal.

3.12 The Result of a Normal Conception by an Acknowledged Human Female

Here one acknowledges the possibility that one might have a human female not of Earth origin, either proximately or remotely. Depending on the latitude given to the concept "human," the offspring might or might not resemble a contemporary human. Presumably, for the conception to count as "normal," an acknowledged human male would be involved, who might or might not be of Earth origin, proximately or remotely.

One is tracing humanity through the female line, much as lineage, tribality, or such, might be traced through the female line. It is easier, of course, to know the mother of a child than the father. Normally one does not require laboratory skills to make that determination. Too, what we now think of as human DNA might or might not be involved. Making human DNA, as currently understood, a necessary condition for being human would seem injudicious. Certainly such a requirement would seem rather parochial or narrow; it might rule out entities we might wish to regard as extraterrestrial human beings, or slightly mutated, evolved or devolved human beings, and such. Similarly, one might expect alterations in DNA, perhaps considerable alterations, to be consequent on genetic engineering. In any event, there seems no better reason for limiting humanity to such a requirement than limiting humanity to certain races, as they now exist. Too, chimpanzees and gorillas share much human DNA, indeed, into the high nineties, and we, presumably, would not wish to regard them as human, at least now, though one might hold open that option if they began to evolve or mutate in a certain manner.

One might do the tracing through the male line, of course. If one wished to do that, then one could make certain adjustments in the text above.

3.13 The Result of an Abnormal Conception, Say, by the Fertilization of an Egg of a Human Female Outside Her Body.

Presumably this would count as an abnormal conception, but it is a commonplace in contemporary reprogenetics. Normally the children would be counted as indisputably human. One might regard cloning as a form of abnormal conception, an outside-the-body conception, a single-parent being involved, If an acknowledged human male or female was involved, the offspring would be counted as indisputably human. Similarly, the offspring of multiple parent matches, as have been produced in laboratory animals, if the several parents were human, would count as indisputably human. Cases of abnormal conception in which some nonhuman elements might be involved, are discussed implicitly in the next section, 3.14.

3.14 The Result of a Human/Nonhuman Match.

A human/nonhuman match might be an alien/human match, an animal/human match, an artificial person/human match, and so on. Genetic engineering, in particular, the possibility of genetic chimeras, would be relevant here. Similarly, if artificial persons, artificial gametes, artificial DNA, or such, were synthesized, such might be relevant to a category of a human/nonhuman match. Much would depend on particular cases.

In dealing with possibilities such as these we would wish to allow for the offspring of such a match to count as human. For example, a judge might rule that partial humanness might count as a sufficient condition for being a human being, much as in 19th Century America, in some states, an individual with any ancestry whatsoever of one race, say, R_1, even 1 percent, would count as an R_1 person, although the individual might have his predominant biological heritage, say, 99 percent, from another race, say R_2. In such a situation, interestingly, individuals of race R_1, who had some ancestry from race R_2, say, at least 1 percent, would not be counted as members of race R_2. This seems a useful example of an institutionalized fiction, a second-order fiction, in this case, a classification. It also sheds some light on the nature of law.

Just as it would seem surprising to count a 1-percent human individual as human, it would seem similarly surprising to count, say, a 1-percent alien individual as alien. One supposes, putting aside political agendas, that a substantially human individual might well be counted as human, and a substantially alien individual might well be counted as alien.

On the other hand, much might depend on the particular cases, on, say, the characteristics of the individuals involved.

3.2 Characteristics

It seems likely that society at large, and even the law, might make its judgments in these matters on the basis of characteristics, observable or conjectural, rather than, say, constituents or cross-fertility. On the other hand, characteristics, whatever their social utility, whatever their convenience and persuasiveness, are no more reliable here than they were in determinations as to "sameness of person."

3.21 Appearance

Appearance is neither a necessary nor a sufficient condition for being human. Something may appear human and not be human, and something might, depending on our criterion for humanity, be human, but not appear human. One could suppose aliens who much resemble human beings but are not human, perhaps having additional senses or powers. Similarly, an artificial person, say, a sufficiently sophisticated android, might pass for human. Similarly, one might suppose gods or demons, or such, assuming human form and appearing in such a form to humans, for one purpose or another. On the other hand, a disembodied human brain, living, and equipped with electronic means for perception and communication, ensconced in a small, portable container, would not have a human look about it, nor would cyborgs of many sorts, whose bodies might be, say, skyscrapers, vehicles, vessels, or such.

3.22 Behavior

Behavior, like appearance, is neither a necessary nor a sufficient condition for being human. Indeed, behavior would presumably be a far less reliable indicator of humanness than appearance. Certainly, easily enough, something might behave in what we take to be a human fashion and not be human, and something might behave in a way we would not take to be human, but might, after all, be human. Examples are easily obtained, for example, as above, the human-resembling alien, the artificial person, the god or demon, and so on. One could also suppose a robot, artificial person, or whatever, which did not appear human, at all, but might behave in a very human way. Similarly, an extraterrestrial human, if we admit it as human, or an evolved, or devolved, human might not behave in a way which we would take to be human. Similarly, if one opens the gate to "boxed brains" or cyborgs, of one sort or another, it is clear that the behaviors involved might not be the sort of behaviors we are accustomed to taking as human, for example, perceiving without senses in the usual way we think of sensory perception, speaking without a vocal apparatus of the sort with which we are familiar, or, in the case of cyborgs, perhaps opening and shutting doors and windows, regulating the temperature of a building, starting motors, leaving launch pads for orbit, or beyond, and so on.

3.23 Consciousness

This would be a characteristic available, first-hand, barring mental telepathy, only to the conscious entity itself, and, strictly, one supposes no one is conscious of consciousness, so to speak, but, rather, is conscious of the "contents of consciousness," rather as Hume pointed out, long ago, in discussing the "self," which he failed to detect, lurking behind the impressions and ideas. Presumably there is no self, no mind, in a mysterious sense, something in hiding, lurking in the background, seeing but not being seen, hearing but not being heard, and so on, but there is, rather, a brain generating sensations, thoughts, feelings, perceptions, and such. On the other hand, the notion of a self is a natural, intelligent supposition; how, after all, could there be perception without a perceiver, seeing without that which sees, hearing without that which hears, and so on? Grammar, in this sense, is not necessarily misleading, for there is a perceiver; there is a self, and, in this case, a real self, the body, and, in particular, the brain.

An entity might awaken and discover itself by means of a consciousness, without understanding the nature of the consciousness involved. Would Adam, say, awakening in Eden, have understood that he was a male, a human being, an animal, a mammal, a vertebrate, and so on? Presumably he would not have known what he was. The utilization of consciousness as a clue to species identity would require, even for the individual involved, a good deal of miscellaneous information. Rabbits do not know they are rabbits, and so on. At least, they are not likely to give the matter much thought.

The consciousness characteristic is not without value in determinations of species identity, despite its indirect nature, except for the entity involved. Speech, writing, behavior, artifacts, etc. all provide likely evidence of consciousness, and likely evidence pertaining to the species involved. Evidence of consciousness is often abundant, even when the entity or entities themselves are absent. Greek theaters and Roman aqueducts speak to us of a consciousness, and a human consciousness, as historically, if not logically, incontestable as anything encountered in daily life.

On the other hand, as is clear from the "other minds" problem, the evidence involved is commonly inferential in nature, and runs the usual risks associated with induction. Putting aside fascinating, if bizarre, logical possibilities, such as weak and strong solipsism, namely, the possibility that you might be the only center of consciousness in existence, or, indeed, the only entity in existence, some problematicities remain, which, in certain cases, might actually give us pause. These possibilities have to do with the existence or nonexistence of consciousness, and, if it exists, what might be its nature.

Clearly, one might infer positively and be wrong, thinking consciousness exists, where none exists. A machine, robot, manikin, puppet, or such, might simulate consciousness, perhaps being cleverly programmed. Similarly, hearing a voice, or such, one might ascribe consciousness to an alleged speaker, and discover later that one had heard a recording, sound track, or such. Also, one might infer negatively, thinking no consciousness exists, where one exists, and be wrong. It is easy to conceive of an unusual housing for consciousness, of a sort with which we would be unlikely to associate consciousness, perhaps a conscious machine, a conscious plant, a conscious mineral formation, even a conscious world.

Not only the existence or nonexistence of consciousness might be problematic, but its nature might be, as well. We are particularly interested in human consciousness, and it is easy to suppose that one might infer that a consciousness is human, when it is not, and infer that one is not human when it is human. It seems that an artificial person might have a consciousness which is much like the human consciousness, indeed, perhaps having been designed with that in mind, but might not count as human. Similarly, given convergent evolution, and such, it is possible that an alien might have a consciousness much like that of a human, such having been selected for in the course of its evolution. Too, if there are gods or demons, or such, their consciousness might be similar to that of humans. Too, mythology, folk tales, and such, as well as logic, allows for nymphs, satyrs, centaurs, elves, pixies, fairies, and whatnot, usually presented as though sharing a humanlike consciousness. Also, conversely, one might suppose a nonhuman consciousness when the consciousness is human. For example, one might suppose oneself to be dealing with a human-resembling alien or cleverly constructed android, when one is dealing with an actual human. Similarly, one might suppose oneself to be dealing with a conscious machine, when one is actually dealing with a cyborg, which cyborg, depending on various factors, might be human.

The above considerations have primarily to do with risks consequent on inferring the existence or nature of consciousness from externally observable data, from speech, behavior, and such, a handicap consequent on the first-person, or, perhaps better, given possibilities of a shared consciousness in a colony creature, the internal nature of consciousness. It is extremely interesting to note, however, that the nature of consciousness might fail as a criterion for species identity even in the case of the conscious entity itself.

Consciousnesses might be unintelligible, misunderstood, or wrongly associated even for the entity in question. We noted that an Adam, coming to consciousness in his Eden, would presumably be unaware of what he was. Similarly, the newborn infant drawn into the unfamiliar world of heat and cold, lights, shadows, motion, gravity, noise, pain, surprise, and such, is in no condition to utilize consciousness as a criterion for species identity. Once again, consciousness is inefficacious as a criterion for species identity, even for the entity itself, without a good deal of miscellaneous information. Presumably one would have to know oneself as human before one could even think of using consciousness as a criterion for humanness, and, thus, it would be unnecessary as a criterion.

One might also note, in passing, the possibilities of misunderstood, or wrongly associated consciousnesses.

Let us suppose two entities, a human and an alien, and two forms of consciousness, which we might refer to as human and alien consciousness. Cases 1 and 4, below, in our small schema, are what one would expect. In the first case, the human has a familiar form of consciousness, which he recognizes, from prior experience, as his, and human; and, in the second case, the alien has a familiar form of consciousness, which he recognizes, from prior experience, as his, and to be expected in his form of life. In cases 2 and 3, we have, as a logical possibility, an exchange of consciousnesses; in short, the human is experiencing the world rather as the alien would, and the alien is experiencing the world rather as would a human. These experiential worlds must be at least different enough to distinguish, and, in some very interesting cases, one might suppose that they would be quite different. For example, one might suppose that the alien can experience much larger sections of the electromagnetic spectrum than the human; perhaps it also has sharper senses, or additional senses, and so on. Too, of course, if the human has wider experiences, and sharper senses, and such, the alien would find its new experiential world a richer, perhaps more alarming, world. And, conversely, the human might find his new world smaller, darker, and narrower than the world with which he was hitherto familiar. Even the sensory conduits acceptive of external energies might be different between the two species. Perhaps one form of life tastes through the skin, sees by means of compound eyes, responds to odors and vibrations, perhaps even sounds, through antennae, and such.

Human	Alien	
1. Normal	2. Abnormal	Human Consciousness
3. Abnormal	4. Normal	Alien Consciousness

Two further variations suggest themselves, first, that the two entities, the human and the alien, have some recollection of their former worlds, or previous states of consciousness and, second, that they have no such recollection.

Human Being		Alien Being		
1a. Rec.	1b. No Rec.	2a. Rec.	2b. No Rec.	Human Consc.
3a. Rec.	3b. No Rec.	4a. Rec.	4b. No Rec.	Alien Consc.

The normal situations now would be 1a, where the human being has a continuous state of familiar human consciousness, and 4a, where the alien has a continuous state of familiar alien consciousness. On the other hand, we now have six abnormal situations. Situations 1b and 4b would be analogous to the "Adam" situation, or that of a new-born infant, a freshly hatched alien, or such, where the individual has the appropriate form of consciousness, but finds it unfamiliar. 1b and 4b could also characterize a situation where severe memory loss might have occurred, and the individual has no recollection of former states of consciousness. Two interesting situations would occur in cases 2b and 3b, where the human receives an alien consciousness and the alien receives a human consciousness, but neither has any memory of a different sort of consciousness. In these two situations the entity would accept as normal what, physiologically, would be abnormal. The human would take the alien consciousness to be normal for himself, and the alien would take the human consciousness to be normal for himself, or itself. Indeed, logically, such an insertion or exchange may have taken place in your own case. How would you know it had not? For example, if you saw colors your peers were unable to distinguish you would be likely to suppose that this was some gift you possessed, which your less discriminating brethren lacked, not that you had an alien form of consciousness. Indeed, given the subtleties of the "other minds" problem, who is to say what is going on where. As a logical possibility some god or gods might have introduced a surprising form of consciousness into the human race as a whole, something strictly alien, perhaps as a joke, but which, given its pervasiveness, would now count as "human" or "normal." The most interesting situations would presumably be situations 2a and 3a, where the consciousness shift has occurred but each entity retains memories of his former states of consciousness. These paradigm abnormal situations would doubtless produce astonishment, perhaps alarm, perhaps dismay, perhaps rapture. In any event, a period of adjustment would doubtless be in order.

There is a familiar Chinese story, in which a philosopher, sage, or, at least, an unusual person, dreams for three nights that he is a butterfly, and, upon awakening, on the morning following the third night, wonders if he is a man dreaming that he is a butterfly or a butterfly dreaming that he is a man. In its way, this is a very profound story, for it reveals that our primary criterion for reality is coherence, for example, if the material world were a quantum carnival, and the world of our dreams happened to a world of Newtonian coherence, we would presumably take the dream world, which we would not recognize as a dream world, for the real world, namely, reality as it is in itself, and the material world as, in effect, a dream world. If the material world and the dream world were both coherent, we would presumably take ourselves to be alternately inhabiting two real worlds. Regularity is important. If, in one of these worlds or the other, elephants metamorphosed into butterflies, we would think no more of it than we do of caterpillars metamorphosing into butterflies. It would be the way things were.

3.3 Crossfertility

The sort of "crossfertility" one has in mind here is normal, natural, or unassisted crossfertility. That representatives of one species might, in virtue of, say, a number of sophisticated chemical, surgical or genetic interventions, be capable of crossfertility with representatives of another species, is not pertinent at this point. The humanity or lack of humanity, or alienness or lack of alienness, of such issue would be an independent question, to be decided on other grounds, perhaps those of desiderated characteristics.

Natural crossfertility, classically, has been an important component in species identity, sameness of species, and such. It is on such grounds that, classically, prior to genetic tracking, or genetic classification, and such,. one might support claims to the effect that various organisms did or did not belong to the same species. For example, dogs are crossfertile with dogs and cats with cats, but dogs are not crossfertile with cats, and so on. Various types of human beings are crossfertile with other types of human beings, but not with, say, chimpanzees and gorillas, despite our remarkable, and interesting, genetic commonalities with those forms of life. To be sure, crossfertility, no more than fertility, is a strict criterion for species membership. For example, one does not rule impotent or sterile men, nor barren women, out of the human species. A fact this obvious makes it clear that fertility or crossfertility is not a necessary condition for membership in a species. Species membership would be taken to be a function of other characteristics or components. Beyond this, the crossfertility criterion is considerably clouded by the frequent overlapping of gene pools. Biologists frequently encounter the situation where a form of life A breeds successfully with a form of life B, and that form of life, B, breeds successfully with a form

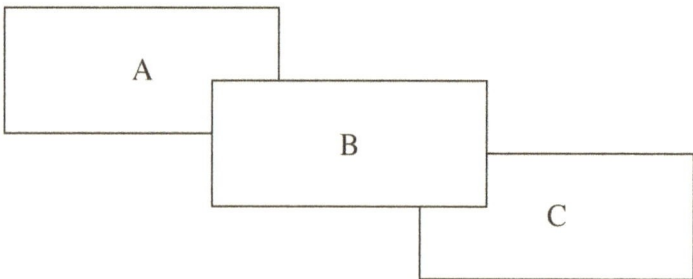

of life C, but A and C are not crossfertile.

In such a situation, it seems one might opt for one, two or three species, as in:

1) An overlap species.
2) The AB and the BC species.
3) The A species, the B species, and the C species.

Logically, of course, additional considerations would suggest looking beyond crossfertility for a criterion of species identity. For example, if an extremely unusual creature were encountered which did not resemble a human being but, in virtue of some genetic situation, perhaps extraterrestrial in origin, happened to be crossfertile with human beings we would presumably not regard it as of our species, nor would we regard ourselves as of its species. Similarly, as mentioned earlier, there is a considerable genetic overlap between humans and various other hominids, in some cases in the high nineties. It thus seems possible that if a slight genetic shift occurred, or perhaps a significant mutation, either in the human being or in one or more of the hominid species in question, that crossfertility with, say, chimpanzees or gorillas might become possible, even empirically. But, even were this the case, one supposes that few individuals, saving perhaps some *a priori* crossfertility enthusiasts or diehards, or gleeful misanthropes, would welcome, say, chimpanzees or gorillas into the human race, or begin thereafter to view themselves, or, at least, others, as chimpanzees or gorillas. This general reluctance to identify with our hominid brethren might be a consequence of outrageous biological bigotry or species chauvinism, one supposes, but it seems just as likely that one would be looking for a more satisfactory criterion in terms of which to understand species identity. For example, if it does not work for frogs, given overlapping populations, there is no reason to suppose it would work for human beings. Similarly, if, due to some sort of mutation, or genetic drift, large segments of the human population ceased to be crossfertile with other segments of the human population, it seems unlikely we would rule them, or ourselves, out of the human species. On the other hand, if there should eventually be extraterrestrial adaptations, or, possibly, aquatic adaptations, it seems possible that divergences of such a degree might occur which would justify a multiplication of species, just as we divide contemporary varieties of hominids into diverse species, which varieties might derive from a common ancestral species.

It should be noted in passing that any criterion for species identity might be proposed and adhered to, as a matter of tenacity or dogmatism, including crossfertility, but that some criteria, in the light of metacriteria, would be likely to be favored over others. In the concluding portion of this study attention is devoted to such matters.

3.4 Constituents

One supposes, *prima facie*, that the most plausible criteria for species identity would be in terms of constituents, properly ordered. To be sure, this requires a prior identification, or acceptance, of both relevant constituents and the acceptable ordering of the constituents. To speak in an Aristotelian fashion, one requires both suitable matter and a suitable form. As an analogy, a disassembled bicycle is not a bicycle but a set of bicycle parts, a number of objects which, properly ordered, will result in a bicycle. Similarly, the form of the bicycle, say, as an idea in someone's head, is not a bicycle but the idea of a bicycle, or the plan of a bicycle. Accordingly, a set of atoms, chemicals, cells, or such, is not a human being unless in a certain arrangement, and the form of a human being is not, in itself, a human being. As noted, all this depends on some identification, or acceptance, of both relevant constituents and their acceptable ordering. We will suppose that some such understandings are in place. They need not be in place, of course, which raises a number of fascinating issues. For example, numbers of human beings, even today, possess artificial parts, and it is easy to suppose that animal tissue, beyond heart valves, and such, might be incorporated into human bodies, as well. Genetic engineering as we have noted, might produce human beings of unusual forms, and so on. We are already well aware, from earlier portions of this study, of Friedrich Waismann's "open texture of empirical concepts," and the roles of proposal, decision, and adjudication in such matters. We will here, however, for the time, largely ignore these complicating, formidable problematicities.

3.41 Appropriate Chemical Constituents, Appropriately Structured

3.411 In the paradigm case we would have our appropriate chemical constituents appropriately structured in virtue of a normal developmental process. In such a case we would have not only suitable matter in a suitable form, but a suitable matter and a suitable form arrived at in a natural, familiar manner. This sort of thing does not rule out life beginning in a Petri dish, being nourished in an artificial womb, and such, as long as the processes involved are normal. Where life starts, and where, and how, it might be temporarily housed, does not alter its humanness.

3.412 In this case we deal with appropriate constituents appropriately structured, but omit the necessity of a normal developmental process. For example, we would presumably be willing to grant that a man and a woman unusually produced, say by a god or gods, and the woman from, say, the rib of a man, and such, would count as human. Similarly, if an individual could be born old and grow young, presumably we would have appropriate constituents appropriately ordered, but would lack a normal developmental process. Some children, in actuality, are afflicted with progeria, a disease which produces premature ageing. This affliction disrupts a normal developmental process, but there is never any doubt that its victims are human. More controversial might be situations in which gametes are synthesized and fused, or, more radically, a human might be synthesized at a phase of development normally reached only after years of development. In such cases, however, it seems that the results of such procedures would count as human. If one would count an adult or adults created by a god or gods as human, as one presumably would, then it seems one should also count a synthesized human, even if synthesized in adult form, as human. Presumably the *modus operandi* would be immaterial. If, on the other hand, enemy aliens synthesized human beings for sinister purposes and released them on Earth to subvert terrestrial civilizations or such, one might envisage the emergence of some reluctance to count such entities as human. Such objections, however understandable, would seem ill founded. Appropriate constituents appropriately structured would seem sufficient, if not necessary, for humanity.

Lastly, one should note that cloning does not involve a natural developmental process, given the denucleations and renucleations involved, but there would be, one supposes, no doubt that such a process produces human material, and, ideally, with appropriate constituents appropriately structured, a normal human being.

3.42 The Entity is a Consequence of Fully Human Genetic Material

Here we wish to allow for the humanity of monstrous births, cloning experiments that turn out badly, and some possible results of genetic engineering, provided fully human genetic material is involved. Here the constituents would presumably be appropriate but the form might, in some cases, be awry, or unusual, inadvertently or by intent. For example, a four-legged, three-armed human would not have a familiar human form, as we think of such things at present, but, if it is the result of purely human genetic material, we would presumably count it as human. Analogously, one notes, we would count a baby born without a limb as indisputably human.

A more challenging case would arise with a macromutation off human stock, perhaps an entity with an unusual size, shape, and color, or one perhaps gifted with surprising powers or aptitudes. On the criterion in question, namely, being the product of fully human genetic material, it would be regarded as human, at least provisionally. This view might be qualified, or altered, if, one, we discovered that the genetic material was not fully human, having been transformed by, say, heat, electricity, cosmic rays, or such, into something different, or, two, that the entity proved viable, survived, and was capable of reproducing its kind. At that point, one supposes, one might think in terms of a new species, to which the human species would have been ancestral. If this latitude is not allowed, we would have to think of ourselves in terms of amphibians, reptiles, small mammals, ground apes, and such. In classical Darwinian theory, the accumulation of micromutations over millennia may produce a plethora of divergent life forms. What is a thousand miles, after all, but a large number of inches, a very large number of inches?

One might also note, here, in dealing with situations in which the normal process of development is absent, the possibility of assembling a human being from the parts, living or dead, of other human beings. The first case is that of surgical cannibalization in which the parts removed from two or more living human beings are joined to create a hybrid human being, one which, then, might possess a desiderated combination of physical attributes, say, size, strength, agility, stamina, and such, suiting it for some activity the performance of which is crucial to attain a particular objective for the larger group. Let us suppose, for example, that a small group of rangers, or special-forces personnel, and a medical team, have been searching for some weeks in the Amazon basin for a concealed Doomsday device, timed to detonate, as of now, within some twenty-four hours or so. With this group, its most important member, if the machine can be found, is an elderly physicist. The physicist possesses the ability to penetrate the complex electronic defenses of, and disarm, the device in question. Further, he is the only one of the group who is acceptable to, and can communicate with, the natives in the area, many of whom are commonly hostile. We shall now suppose, further, that we have learned the location of the device, perhaps from a native informant, a raving renegade, a document found in a crashed plane, or such, and that the location is something like one hundred miles distant. The group is isolated and, we shall suppose, in virtue of one or more mishaps, perhaps a native attack, an overturned canoe, or such, has no way of making radio contact with the outside world. In short, the only individual in the group who would be likely to pass unmolested through the jungle, make his way through the machine's defenses and successfully disarm it is our elderly physicist, who, however, is personally incapable of traversing one hundred miles of jungle in some twenty-four hours. It is also unlikely that he could be carried for that distance, and that more than one individual could make his way through the jungle without being detected by one native group or another, many of which are hostile. We shall also suppose that medical science, as one might expect, given the existence of such a Doomsday device, is highly advanced, capable of sophisticated surgery, and possessing stimulants and healing techniques far beyond those now available. In short, the only way the world can be saved is, so to speak, for a human to exist who combines the youth, condition, strength, stamina, and agility requisite for making his way through the jungle to the device within the allotted time; possesses the attributes, personal character, languages, social awareness, and such, to pass through intervening native tribes;

and, if successful in reaching the device, can render it innocuous. Accordingly, the surgical team sets to work. In this scenario two or three individuals, we shall suppose, sacrifice themselves, or are sacrificed, that the planet may be saved. The hybrid entity produced, we shall suppose, given that it consists of fully human genetic material, would count as human. Thus, a combination human, a collage human, a hybrid human, would count as a human. One need not, of course, suppose a complex scenario, such as the above. A dictator, curious about medical possibilities, might utilize political prisoners for such an experiment, and then, if satisfied, do away with the successful result.

One might also suppose, as in Mary Wollstonecraft Shelley's famous work, *Frankenstein, or the Modern Prometheus,* the construction of an entity from dead parts, from revivified tissue. Just as the parts from different bicycles might be put together to produce an authentic bicycle, so, too, it seems, one might put together parts, in this case, dead parts, from different human beings, and produce a new human being, the tissue having been, by electrical stimulation, or such, returned to a living condition. In the case of Mary Shelley's conception there is no doubt that the "monster" is taken to be human. It is highly intelligent, sensitive, passionate, powerful, and dangerous. Its motivations are comprehensible and plausible. Lastly, one should contrast Mary Shelley's conception here with a quite different conception, that of the "Zombie." As I understand the notion of the Zombie, it is dead, but somehow animated, perhaps by sorcery, or such. If it is truly dead, then it is no more than an animated corpse, not a living being. A dead human being is not a human being, no more than a square circle is a circle, but the remains of a human being. It is what was once a human being. A necessary condition for being a human being is to be alive. Similarly, a set of parts in a box is not a bicycle but a set of parts in a box, perhaps the makings of a bicycle, but not a bicycle.

3.43 The Entity is a Consequence of Partially Human Genetic Material

Here it is supposed that one is dealing with such things as, say, an alien/human match, or forms of genetic engineering, in which a portion of the genetic material involved is not human, as in, say, a chimera, perhaps a winged human, for negotiating the space of a low-gravity planet, a gilled human, designed for underwater construction work, or such. In making assessments where partially human genetic material is involved, much would depend on the particular quantities and proportions involved, perhaps even on the uses to which the entity might be put, even the appearance and dispositions of the entity. For example, whether or not a chimera consisting of a human/tiger mix was human or not might depend as much on the dispositions of the entity as on its appearance, such things helping us to decide whether we should view it as an intelligent tiger or a human housed in a tiger's form. It is difficult to see, in view of the abundance of possible variations which might be involved, any simple, general criterion which might be universally applicable to such decisions. One suspects that pragmatic and cultural considerations would prove paramount. Is it most useful to view the entity as an A or a B? Is it most easily communicated with as an A or a B? What about its possible independence, compensations, legal standing, political rights, civic duties, and such? How does the entity wish to be viewed? Will the way it is viewed affect its performance, positively or negatively? Do the customs and traditions of the culture dictate some view of the entity, that, say, it is an animal which may be exploited and trained, or a human which must be respected, educated, and so on.

If one is dealing with a human/alien match and both partners, or all partners, depending on the number of sexes involved, to the match are rational, civilized entities, presumably the speciations involved would be largely articulated and arranged by the institutions of the pertinent societies, different societies perhaps viewing such matters differently. One could have a situation in which one's species changed legally with the crossing of a border. But, hopefully, each way of viewing things would manifest some degree of epistemological plausibility. Data does not determine its interpretation, or classification, but there are better and worse interpretations, and classifications. Even those who think all roads lead to Rome admit that some roads are better than others. And, in fact, not all roads lead to Rome. Some lead nowhere.

3.44 Possession of a Human Brain

It may be recalled that we argued, at length, for the appropriateness of taking sameness of brain as the crucial test for sameness of person. A great deal of what was said there is pertinent to this criterion for sameness of species, for identifying something as human. If the brain was crucial to sameness of person, it is natural to suppose that it might well be pertinent, as well, to sameness of species. Might not the possession of a human brain be a decisive criterion for being a member of the human species?

Earlier we based certain criteria for humanness on the assumption that a human female was involved, noting that this view presupposed agreement on the nature of a human female, which agreement might or might not be in place. For example, if the female appears to be human but can see in the dark, lays eggs, turns blue on Tuesdays, and can communicate with bats, we are presumably not dealing with a human female, certainly not as we normally think of them. Similarly, we are proceeding here on the assumption that a brain with such-and-such properties will count as a human brain. One might then, as in this section, attempt to utilize this notion, more broadly, as a criterion for species identity. No logistic circularity is involved, as one is supposing that the properties of the brain in question can be specified without reference to the notion of humanness. On the other hand, pragmatically, obviously, one is setting up the properties in such a way as to characterize the sort of brain we would consider human, and then attempting to use that sort of brain as a criterion for species identity. To avoid artificial precision and arbitrariness we would be willing to alter or adjust the "independent properties" to a greater or lesser degree to accord with our intuitive notions of the human brain. Once this is recognized, it becomes clear that we are once again immersed in Waismannian "porosity," in proposal, decision, evaluation, adjudication, and such. For example, would we be willing to regard a well-functioning, unusually large brain as a human brain, one of an unusual color or configuration as a human brain, one insensitive to pain and pleasure as human, one capable of reading minds or moving objects at a distance as human, and so on? We will bracket these problems, however, and proceed on the assumption that we are substantially agreed on the sort of brain in question, and are now interested in the related, but different, question, as to whether or not such a brain might serve as a criterion for species identity.

Such a criterion is attractive, as it allows for inclusion in the human species of entities such as cyborgs, perhaps brains functioning in robotic bodies, vehicles, buildings, and such; animal ensconcements, perhaps brains transplanted into other forms of life, into hominids, into aquatic organisms; into alien housings, and such; perhaps brains found in genetically produced chimeras; perhaps brains preserved and sustained in unusual environments, such as vats, electronic casements, and so on.

Difficulties, however, intrude.

Given convergent evolution, and such, it seems likely that a sort of brain which, if it was found in a human, would count as indisputably human might occur in an organism, say, an alien being, which was clearly not human. Similar environments on diverse worlds might favor similar adaptations. This is not to deny that human beings might originate on a variety of worlds, but only to acknowledge that something obviously not human might resemble human beings closely in some interesting and important ways. Similarly, it is possible to conceive of something we would wish to regard as human which might not have a human brain, as we commonly think of such brains. Perhaps its brain bears a close resemblance to a sort of brain we commonly think of as alien, or such. One might add into the mix that a "human brain" must be derived from fully human genetic material, but this would rule out the possibility of extraterrestrial human beings, whose brains might derive from similar or identical genetic materials. Chemicals, for example, do not come as human or not human. We impose labels; nature is not interested in such things.

All in all, it seems that the best criterion for membership in the human species would be the possession of a given sort of brain, one of the type we are accustomed to think of as human.

To be sure, there are counterintuitive aspects to this proposal.

> "Bring me that human being."
> "Which one?"
> "The round, gray one, the one that weighs eight pounds, with tubes and wires, on the second shelf, in Box Six."

If having a human body, or pretty much of a human body, is regarded as a necessary condition for being a human being, then a criterion of the sort in the 3.41 range, where one is looking toward appropriate constituents in suitable proportions and patterns, would seem the most judicious criterion for giving us a necessary and sufficient condition for being a human being. On the other hand, if having a human body, or pretty much of a human body, is *not* regarded as a necessary condition for being a human being, then it seems that the sort of criterion envisaged in 3.44, that of possessing a human brain, would seem the most judicious criterion for giving us a necessary and sufficient condition for being a human being.

My own view, given the presumed desirability of regarding cyborgs, transplants, chimeras, and such, as human, would be to prefer indexing humanity to a functioning, living brain of a particular, designated sort.

A serious additional argument would be the "What if it happened to me?" argument. Surely, in such a case, we would still wish to be regarded as human. This argument seems to me a very strong argument, bringing the putative life consequences of such a criterion home.

Philosophy matters.

Classifications count.

We have now, in at least a cursory fashion, examined four major categories pertinent to species identity, in this case having to do with the human species, these categories having to do with conception, characteristics, cross-fertility, and constituents. Three sorts of alternatives remain to be briefly considered, those of claims, whether of the entity or the public, those of legality, and those involving the postulation of one or more metaphysical entities. The first two alternatives would seem of minor philosophical importance, though perhaps of serious moral or political importance, as both would seem to presuppose a justificatory relation to one or more of the major categories aforementioned; and the third alternative would also seem to be of minor philosophical importance, as, disengaged from procedures of empirical resolution, its applicability is problematical. Before attending, however briefly, to these "minor categories," or alternatives, which deserve some measure of attention, it seems appropriate to conclude our discussion of the major categories, as anticipated earlier, with a consideration of the relevance or irrelevance of Earth origin to the nature of the human being.

3.5 Earth Origin

A number of arguments, of a lesser or greater persuasiveness, suggests the inadvisability of requiring Earth origin as a necessary condition for a human being. On the other hand, requiring Earth origin as a necessary condition for humanness has, all things considered, as we shall see, much to be said for it.

3.51 Some Inadvisability Arguments

3.511 The "Convergent Evolution" Argument

Obviously human beings evolved on at least one planet, presumably Earth.

Thus we know such a thing is possible.

Now, there are billions of stars in a galaxy and, apparently, billions of galaxies in the universe; further, it is estimated that something like half of these stars have one or more planets. Sinusoidal motion and the loss of angular momentum are relevant here, as well as theoretical considerations having to do with distributions of matter between binary systems, where the system's matter would be substantially divided between two suns, and monadic, or single-sun, systems, where the system's matter might be divided between a sun and a set of planets. Taking our solar system as a typical monadic, or single-sun, system, it seems not impossible that at least one planet per planetary system might be favorable for the development of life as we know it, for example, a planet which is small enough to lose hydrogen and large enough to retain oxygen, which is neither too far from, nor too close to, its primary, and thus neither too cold nor too hot; and on which, thus, liquid water is possible, and so on. One supposes, similarly, that the universe is pretty much of a muchness, as suggested by telescopic observation, spectroscopic analysis, telecommunicating probes, and such, and that its laws, or, at least, its probabilities, are uniform, or relatively uniform, throughout. Thus, we have similar materials subject to similar forces and conditions. One also supposes that if life appears on a planet that the dynamics of evolution would, as on Earth, produce a variety of developments and adaptations, exploiting a variety of ecological niches. Empirical evidence of exoplanets, or extrasolar planets, one notes, is abundant, some six to seven hundred or so such bodies known at present. It is true that most of these exoplanets planets are too large to be likely hosts for life, certainly as we are familiar with it, but where there are large planets it is not unlikely that, as in our system, that there would be smaller planets, as well. The limitations of contemporary instrumentation make the detection of smaller exoplanets difficult, but not impossible. One such smaller world, recently discovered, is Gliese 581g, orbiting Gliese 581, a red dwarf. There is also some ambiguous data that hints at the existence of extraterrestrial life, in the nature of provocative formations found in carbonaceous chrondrites, in the famous Allan Hills meteorite (ALH 84001) discovered in Antarctica, and so on. Also laboratory evidence resulting from attempts to recreate hydrogen-dominated primeval environments demonstrates that under such conditions organic molecules are inevitably and abundantly formed. To be sure, no primitive replicators, or proto-cells, have managed to crawl out of beakers or test tubes yet, which is probably just as well, as our immune systems might not be prepared to deal with them. To be sure, primitive replicators, or proto-cells, as far as we know, do not now exist, presumably having become extinct, perhaps having been less successful than, or having been cannibalized by, better adapted, more aggressive unicellular predators.

In the light of such considerations as the foregoing, and the fact that life, even human life, has evolved on one planet, it seems clear that a second such development is not impossible, and might even be probable, perhaps overwhelmingly so. It is true that life might be indexed to quite specific conditions, but there is no reason to suppose that such quite specific conditions might not occur frequently, or that life of a somewhat different sort might not have evolved under conditions different, but equally specific, elsewhere. It seems likely that if life, of one sort or another, should appear on a given world, that it would be simple at first and then, later, become abundant, diverse, and complex. Presumably something would occur on each such world, rather as it did on Earth, simple life forms emerging, to be followed, later, by more complex forms of life. Presumably something might occur similar to the eventual development of fish, amphibians, reptiles, mammals, and so on. Primitive algae leads eventually to single-tree Banyan forests and stands of towering Redwoods; and unicellular forms of life lead eventually to Shakespeare and Einstein.

Now, just as one might suggest the possibility, under the cruel engineerings of evolution, the development of extraterrestrial termites, crocodiles, lions, and apes, so, too, one might suggest the possibility of life forms identical to, or closely resembling, human beings. If such a form of life proved to be identical to human beings it would seem appropriate to regard it as human, and, if it were sufficiently similar to humans as we know them, it would seem appropriate to regard it as human, or as a kind of human. In short, in the light of the possibilities of convergent evolution, it would seem inadvisable to require Earth origin as a necessary condition for a human being.

3.512 The Arbitrariness Argument

Logic, as well as biology, seems relevant to these issues. The sort of identity which seems most relevant to these issues would be what we earlier identified as external, partial, pervasive identity, namely, first, identity where more than one entity is concerned, second, where two or more entities are the same except for at least one respect, and, third, where two or more entities are the same in all significant respects. Much here, naturally, depends on how one understands 'significant', but, commonly, place, location, point of origin, and such, are not regarded as significant. For example, Jones and Smith are both human beings whether they are in different states, or one is abroad, and so on. Similarly, two artifacts are both washing machines of the same make and model though one might be manufactured in Poughkeepsie and another in Seattle, and so on. Location, point of origin, and such, at least normally, are regarded as accidental characteristics, and are not considered relevant to, and certainly not determinative of, nature or identity. In the light of such considerations, it would seem not only inadvisable, but actually illogical, to require Earth origin as a necessary condition for a human being. If there are no discernible differences, or few discernible differences, and those of a perhaps negligible nature, between A and B it seems that A and B are well thought of as the same sort of thing, both examples of such and such a thing, such and such a form of life, or whatever. Similarly, if a god or gods should decide to produce human beings, say, some Adams and Eves, on more than one world, who would object? Who would explain to the god or gods that this was impossible?

3.513 The Ethnocentric Argument

Moral considerations might also be regarded as relevant, though, doubtless, less plausibly. 'Ethnocentricity' is an anthropological term, and, in theory, if anthropology is a science, should be a scientifically neutral term. On the other hand, however this term might be viewed in anthropology, it seldom functions in common discourse as having an impartial or objective meaning, but, rather, is usually used as a *Schlagwort*, a term by means of which to belittle, abuse, intimidate or stigmatize. Impartially, it would refer to a conviction, attitude, belief, or such to the effect that some group, party, nation, culture, religion, or such, presumably one's own, is superior to other such entities, to other groups, parties, nations, or such, and so on. Assuming that one could be sufficiently clear on the nature of the group, or such, involved, and clear on what sort of superiority was in question, and how to test for it, and so on, an ethnocentric claim would be either true or false. For example, it would be the case that group G_1 was indeed superior in respect R to Group$_2$, or it would not be the case that Group$_1$ was superior to Group$_2$ in respect R. The claim would be a simple matter of the way things were, and its truth value would be left up to empirical inquiry. For example, it seems clear that either the population of Poughkeepsie is better at algebra than that of Seattle, or not. It might be difficult to obtain a grant to inquire into the matter, but such an inquiry would be well within the perimeters of social science. Indeed, it seems likely that one or another of some fifty thousand groups are indeed likely to be better at one thing or another than one or another of some fifty thousand other groups. It is highly improbable that this would not be the case. For example, of any two randomly chosen individuals from the Earth's population it is almost certain that one will be better at algebra than the other. Indeed, one of them may never have heard of algebra. The possibility that both would have identical algebraic skills would be negligible.

On the other hand, as noted, a challenge of ethnocentricity is commonly taken as a moral attack, somewhat along the lines of alleging that someone is defrauding widows, fond of causing discomfort to small animals, and such. Presumably, then, a belief in the superiority of one group to another in some respect or other is a moral fault, or to be understood as such, whether or not it might be true. For example, the folks in Poughkeepsie might be better at algebra than those in Seattle, in fact, but it would be wrong to say so. Such truths, if truths they be, are not to be spoken. Taken objectively, scientifically, most ethnocentric claims are best understood as uselessly vague or, on the whole, to the extent that they are clear, false. They often involve, but need not involve, the "fallacy of sweeping generalization," which, actually, is not a fallacy, but, rather, a seldom-successful route to truth.

Adopting the old fashioned distinction between name-calling and philosophy, we reject the ethnocentric argument against the advisability of taking Earth origin as a necessary condition for humanness. It may or may not be a good idea to take Earth origin as a necessary condition for humanness, for one reason or another, but it does not seem morally objectionable to do so. Too, one, if taking Earth origin as a necessary condition for humanness, is not, *ipso facto*, claiming that humans are better, say, more charming, better-looking, more moral, more loyal, more reliable, nobler, smarter, fonder of small animals, than other forms of life, or such. Such considerations would raise independent questions, first, of philosophy, second, of science, and, quite possibly, questions of no great interest.

3.514 The "What if human beings did not have an Earth origin?" Argument

This is a "red herring" argument, so to speak, one actually irrelevant to what is at issue. The significant claim here is not so much that human beings have had an Earth origin, which is not really that important, as that they have had a common origin, derive from an original stock, are members of the same "family," so to speak.

The notion that life, and thus at least the antecedents of human beings, did not begin on Earth is not as anomalous at it might seem initially. The hypothesis of an extraterrestrial origin for terrestrial life has been taken seriously by biologists as famous as Francis Crick, of Watson-Crick DNA fame. One possibility would be the directed-panspermia hypothesis, namely, that a given civilization, perhaps dying, and hoping to preserve life, seeded the universe with living material. We have noted, earlier, as well, the possibilities of extraterrestrial life, in connection with life-resembling formations discovered within carbonaceous chrondrites, within the aforementioned Allan Hills meteorite, and so on. One hypothesis, if somewhat humbling, is that life began from refuse or garbage left behind by transient, littering extraterrestrials. Two additional considerations suggesting that life may not have begun on earth is the lack of signs of precellular replicators, proto-cells, and such, some of which, it seems, might have survived into the present. Also, the failure, to date, to produce life in laboratories, in experiments carefully constructed to replicate, as far as possible, primeval terrestrial conditions, suggests that conditions elsewhere might have been more favorable for the emergence of life.

Whatever the case may be here, Earth origin, strictly, is not required for humanness. It is a possibility that life, and thus, in a sense, humanity, might have had an extraterrestrial origin. Indeed, any number of situations would be logically possible. Perhaps scientifically curious aliens noted hominid forms of life somewhere and, as a scientific lark, decided to synthesize a few hominid-type creatures at home, or on board their vessel, and then introduced them into the terrestrial environment, to see what would happen, and then, later, perhaps forgot about it, going about their business elsewhere in the galaxy. If this were the case, humans might have begun their existence, presumably as adults, to facilitate survival, on an alien world, or starship, as the result of an experiment, hopefully one to be regarded as successful, and not abandoned by dispirited experimenters, as a failure.

This argument shows that it is not necessary to require Earth origin as a necessary condition for humanness, but it does not convince one that it is inadvisable to do so. In any event, what is seriously involved in the "Earth origin" hypothesis is not so much an Earth origin, as a singleness of origin.

3.515 The "Space-Colony Offspring" Argument

This argument, like the preceding argument, is effective against a narrow construal of the "Earth origin" argument, but secures its victory while missing the point of the war; it addresses itself more to something said, taken literally, than to what is behind what is said, more to a formulation than to the point of what is formulated; the point of the "Earth origin" argument is not so much to insist on an Earth origin, as location is not all that important, as to insist on a unified origin for humanness, an origin which, one supposes, would have been on Earth, but, strictly, need not have been on Earth.

In any event, if humans eventually found colonies in space, in either artificial worlds, such as orbiting or free-ranging constructed biospheres, or on, or within, extraterrestrial natural bodies, planetary or asteroidal, the children born in such colonies, pending unexpected radical transformations due, perhaps, to radiation, or such, would be indisputably human, at least at first. Eventually, of course, in a period of generations, new forms of life might evolve, deriving from, but differing considerably from, familiar ancestral human forms. Humans, for example, differ considerably from earlier life forms, say, amphibian and reptilian forms, those of primitive mammals, and such. Also, of course, genetic engineering might set itself to the development new adaptations, in many cases presumably to be recognized as new species.

3.52 The Acceptability Position

The Acceptability Position is that it is acceptable to, and perhaps even desirable to, take Earth origin, either literally or metaphorically understood, as a necessary condition for humanness.

To be sure, much here depends on how broadly or narrowly one chooses to understand the notion of "human." For example, the force of the "convergent-evolution argument" and the "arbitrariness" argument both make a strong case for a broad understanding of the concept "human." There seems no reason to doubt that something human, in a broad sense, might not have been evolved any number of times, and, indeed, it would seem somewhat arbitrary to insist on a particular location or point of origin being required to count as human. On the other hand, there is much to be said for adopting a narrower sense of 'human'. For example, let us suppose two families, the Smiths and the Joneses. The Smiths regard themselves as Smiths and the Joneses regard themselves as Joneses. If, on some alien world, say, there should be some entities identical to the Smiths and the Joneses that would be surprising, and of great interest, but one could well imagine that, say, Smith and Jones would claim, and with assurance, and with very good reason, that the aliens are not Smiths and Joneses. They cannot be Smiths and Joneses because they are not members of those families, but members of different families, if families, at all. In short, if one thinks of humans not as being a sort of thing, but as a specific historical line, a specific biological continuity, entities with a common genetic ancestry, a family, so to speak, then there would not be extraterrestrial humans, any more than there could be extraterrestrial Smiths and Joneses.

In this sense it seems intuitively natural, and satisfactory, to require Earth origin, or, say, a common origin, as a necessary condition for being human, for counting as a member of the human species.

Also, interestingly, many world views, hypotheses, myths, and such, give an accounting of the creation of human beings, which accountings commonly suppose a common origin for all humans, a primal female, a primal set of parents, or such. For example, it is not unknown for some of these views to proclaim, and argue for, such relationships on the basis of various traditions, written and oral. It might thus be said, with a greater or lesser degree of seriousness, or metaphor, that all men are brothers, and so on. Presumably cousins, far removed, might be more exact, and less biologically suspect. On the other hand, given the overwhelming commonalities of DNA amongst humans, a common ancestry is not an absurd or improbable hypothesis. Even if the human race did not trace back to a single small, shaggy, female, perhaps some forty pounds in weight, and had, rather, a multiple origin, presumably it would trace back to a small number of interbreeders, from which, after some generations, something recognizably human might have first appeared. And if the first recognizable human was the result of a macromutation, which seems unlikely, so much the better. The likely specifics of these matters one gratefully surrenders to the theories, hypotheses, suppositions, speculations, and research of geneticists. All one requires here is the "family" view of human beings, the possession of a common, or substantially common, genetic ancestry. One might then, if one wished, make this common genetic ancestry, or this substantially common genetic ancestry, a necessary condition for membership in the human race, a necessary condition for being a member of the human species, or "family," for being, so to speak, a Smith or a Jones.

The Acceptability Position is also reinforced in virtue of the likely resolution of a variety of thought experiments, in particular, those involving devolution, evolution, and alien-encounter arguments.

3.521 The Devolution Argument

Presumably something might devolve from human stock, which might or might not be regarded as human.

All evolution and devolution, as all change, move from an earlier time to a later time. Thus, both are "later in time," and, accordingly, one cannot distinguish between them in terms of one being earlier in time and one later in time. Both are later in time. This is an obvious point but worth noting as it is common to think of more evolved forms being later in time than earlier forms. This is true but devolved forms, whether more or less adaptive, are also later in time than earlier forms.

In order to understand what is going on here, it will help to understand 'devolution' and 'evolution' as impartial terms, having reference to an ancestral form. Neither term is to be taken as axiological; both are to be understood as descriptive, value-neutral terms. In devolution one would have a return to, or an approach to, an ancestral form, and, in evolution, one would have a departure from such a form, for better or for worse. It should also be noted that certain relativities may be involved, for example, with respect to "ancestral form" and the concepts of "evolution" and "devolution" themselves. For example, if we take contemporary humanity as an ancestral form, then a departure from this form, under evolutionary selections, to an earlier form, might, at least popularly, count as either devolution or evolution. If it were less adaptive it might be thought of as devolution, and, if it were more adaptive, it might be thought of as evolution. A primitive, simpler, generalized form, closer to an ancestral form, might be more adaptive under given circumstances than a more sophisticated, more complex, less generalized form. For example, as noted earlier, genetic tracking has demonstrated that all contemporary breeds of dog derive from the gray wolf. Obviously, it is possible to breed in a variety of directions. Biology is remarkably accommodating and tolerant. Is one to view, say, the poodle, the Pekingese, and Chihuahua as more evolved than the gray wolf, who might use them for snacks? Surely it is unlikely that natural selection would have favored his descendants. In a civilized world they have their place, and are doing well within it. Suppose now civilization perished, as it well might, given the abundance of armed insanities about. In such a case, one supposes the poodles, and such, might become extinct, but, on the other hand, over several millennia, might there not be a return to the form of their remote, fierce, aggressive, territorial, dangerous, gray ancestor? Might that change not count, at least popularly, as evolution, as much as devolution?

On our account, of course, it would count as devolution, which is to characterize it, not to commend it or disparage it.

A certain ambiguity often appertains to these matters. As we are using the expressions 'evolution' and 'devolution' the difference, as noted, depends on a departure from an ancestral form, whatever we take it to be, which we could count as evolution, or a return to, or toward, an ancestral form, which we could count as devolution.

Another way of understanding the matter is that the more evolved organism is the better adapted organism. We prefer not to adopt this view as it would require us to regard, say, termites, crocodiles, and sharks as more evolved than frogs, raccoons, giraffes, and people. Indeed, Nietzsche seems to have regarded the highest specimens of humanity to be those likely to be the most fragile and vulnerable, the least adapted. Another way of understanding the matter is to count more complex forms as more evolved than simpler forms, which would seem to be acceptable unless one is surreptitiously or unconsciously smuggling a value judgment into what should be a matter of scientific neutrality. For example, all contemporary life forms have an evolutionary history behind them which would trace back to the first self-replicating carbon compounds. Whereas one might well prefer people to grubs, coelenterates, sponges, and such, it is not clear that nature cares much about it, one way or the other. It might favor grubs. There are more of them than people. Also, there seems to be nothing valuable about complexity *per se*. For example, complex creatures seem the most likely to imperil the planet and render themselves extinct. Complex machines are often inferior to simpler machines; complex toys to simpler toys, say, dolls, toy soldiers, balls, and blocks; and complex people to simpler people, and so on. Personally I would prefer to live in the vicinity of a moral, good-hearted, generous, bumbling politician than a clever lunatic, a ruthless dictator, a brilliant fanatic, a "mad scientist," an "evil genius," or such. How about you? In any event, we will try to understand 'devolution' and 'evolution' along the lines of a departure from, or a return to, or toward, an ancestral form. Whether one approves or disapproves of such things is up to the individual. Nature, as usual, remains on the sidelines. In any event, what we need most here, for our purposes, is a value-neutral construal of 'evolution' and 'devolution', terms which are too often charged with a distractive, emotive resonance, either positively or negatively.

These prefatory remarks in place, one is in a better position to understand the devolution argument in favor of the Acceptability Position, and the evolution argument, as well.

We shall suppose that civilization has muchly perished, but that humans, of one sort or another, have managed to survive. We might suppose that the "human poodles," so to speak, so successful within, and so well adapted to, a complex world of comfort and convenience, are largely gone, but, over some hundreds of years, and perhaps from such stock, a different sort of human being has emerged, both naturally and culturally, a human being hardier, more severe, less forgiving and tolerant, more primitive, more cunning, more suspicious, a human being accustomed to hunting, to blood, the butchery of animals, the defense of territory. Agriculture is largely impractical, we shall suppose, given the time required to plant, tend, and harvest, and the vulnerability of crops to predation. Agriculture, to be practical, requires land and time, knowledge and skills, patience and discipline, and stability and security. Perhaps bandits and warlords have not yet achieved the status of conquerors and kings, able to claim and protect large courses of land, these to be farmed by their subjects.

In any event, let us suppose now that we have a group of recognizable humans, as we now think of them, and they encounter a shaggy group of hominids, "newcomers," whom we might suppose represent a biological return to an earlier form of life. The newcomers are, so to speak, closer to the ancestral form of human. We have a situation, in that sense, of devolution. Interestingly, popularly, the newcomers, if better adapted, might be regarded as more evolved than the basic group presupposed here. On the other hand, presumably the newcomers are derived from human stock of, say, some centuries ago. Thus, if we take the basic group to represent the ancestral form, which we might, we would have the paradoxical situation, even on our approach, of regarding the newcomers as exhibiting, simultaneously, evolution and devolution, depending on which form we select as "ancestral," the ancient form or the more modern form. Taking our basic group as representing the "new ancestral type," the newcomers would count as more evolved, and taking the ancient form as the ancestral type, they would count as devolved, regardless of whether or not they were better adapted, or were even more complex, more intelligent, or such, than the later humans. Thus, as noted earlier, the notions of evolution and devolution tend to be relative notions, in this case, relative to the form selected as ancestral.

So, are the "newcomers" human or not?

We are granting, per hypothesis, that the members of the "basic group" are accepted as human.

As the "newcomers" are closer to the ancient ancestral type we count them as devolved, in the impartial, value-neutral sense of the term prescribed.

Clearly this is our question, and not one for either the "newcomers" or the "basic group." Survival demands an economic base and the will and capacity to defend it. Territoriality has been selected for. Thus ,one supposes it likely that the two groups hypothesized would, sooner or later, and quite possibly sooner, come into conflict. The "basic group" will regard the newcomers as "other," "not like us," and so on, and one supposes these views would be reciprocated by the newcomers. This need not be put in terms of race, no more than in the case of an encounter between two forms of predator. The basic group and the newcomers might neither have a concept of race, but they would both have a concept of "kind," if only on an intuitive level. On the other hand, if both the basic group and the newcomers were imperiled by a third party, say, abundant ursine or feline predators, hostile aliens, or such, they might recognize enough similarities of form and interest to enleague themselves, as fellow hominids, if nothing else. This would particularly be the case if they had a language in common, certain traditions, folk memories, and such. Each might then appear to the other close enough in "kind" to be acceptable, at least to an extent, at least for the time being, by the other. Here, there would be a significant mental resemblance, which might override, at least for purposes of a common cause, reservations pertaining to physical appearance. One could always return to the luxury and intellectual stimulation of "racism" in more secure, leisurely times.

Whether we would regard the indisputably devolved newcomers as human or not, for this is really our question, and not that of the basic group or the newcomers, would presumably depend on a variety of considerations. I suspect that we, not being in the position of the basic group, whose very survival might be threatened, would accept the newcomers as human if they were speeched, had a tradition, and so on, assuming, of course, that a common genetic ancestry was in place. Presumably we would not, upon reflection at least, require substantial hairlessness as a necessary condition for being human. Indeed, what if we, gradually adapting to a new series of ice ages, became more hirsute? We would still take ourselves as human. What emerges here, it seems, is that if a common genetic ancestry is in place, we would base membership in the human race more on mental than physical commonalities. The situation would seem to be quite otherwise if we encountered something very much like the "newcomers" on another planet, where there was no possibility of a common ancestry. We would then, despite mental resemblances, unhesitantly reject the alien form of life as human. This is interesting for two reasons, at least. First, despite the fact that we supposedly base our humanity and self-image on mind (*homo sapiens*, and such), we would be likely to take physical characteristics as more decisive. Second, given the difference between the hypothesized "this-world case" and the hypothesized "other-world case," it seems clear that a common genetic ancestry is of decisive importance, this adding force to the Acceptability Position. An interesting sidelight on these matters is to consider the Neanderthal form of life, and whether or not one would regard it as human. Whereas much remains obscure pertaining to the Neanderthal, in particular, its relation to familiar human stock, for example, would it have been crossfertile with human stock, was it ancestral to human stock, was it collateral with human stock, rising from a common ancestor, or was it a departure from human stock, and so on, it seems clear that it was speeched, had a culture, ritual observances, beliefs in an afterlife, and so on. Accordingly, one supposes we would regard it either as human, or a sort of human. And, if it had a common genetic ancestry with humans, it seems almost certain we would regard it as human. To be sure, decisions would have to be made. Humans have a common genetic ancestry, of course, with many forms of life which we would not regard as human, fish, amphibians, reptiles, primitive mammals, and so forth. Things may be what they are, but whether or not they are human is often up to us. Nature supplies the data, but we supply the interpretations, the concepts, the classifications.

3.522 The Evolution Argument

In the devolution argument the question was essentially the quantity of departure from a contemporary form of human stock in the direction of an ancestral form requisite to ceasing to be human. In short, at what point would devolution produce a nonhuman life form? In the evolution argument the question is essentially the quantity of departure from a contemporary form of human stock *not* in the direction of an ancestral form requisite to produce a nonhuman life form. In short, at what point would evolution produce a nonhuman life form?

Many views, informed and otherwise, attend the notion of human evolution, putting aside views which would deny evolution altogether, of any sort, human or otherwise. For example, differences of opinion occur with respect to the roles and relevance of micromutation and macromutation, possible interactions between organic and superorganic evolution, the relation of human life forms to other hominid life forms, whether primitive humans, say, Paleolithic hunters, were or were not physically and intellectually superior, statistically, to modern humans, and so on.

It seems reasonably clear that few, if any, evolutionary changes in the human species have occurred within the past several thousand years. This might be construed as evidence for classical Darwinian views, emphasizing the accumulation of tiny differences over enormous amounts of time, or, perhaps, the cultural protection or insulation of human beings, primarily in virtue of technology, from classical selection pressures, those of natural selection.

Interestingly, concepts of human evolution, where they are taken seriously, usually presuppose a departure from the current form of human, taken as ancestral, this being a relative notion, to something taken to be superior, according to contemporary lights as to notions of "superior." Certainly those who anticipate producing and directing, even diversifying, human evolution by means of genetic engineering do not have in mind, at least on the whole, producing human forms which are regarded as inferior to contemporary forms, according to contemporary lights in such matters. For example, few theoreticians in such matters seem to have in mind evolving clearly distinct strata of humanity, such that, for example, different varieties of human being would be literally designed to occupy different economic and social levels, different niches in a biologically engineered status society. Our concern here, of course, is not to object to benevolent intentions nor optimistic anticipations, nor to impose value judgments, but to consider some possible evolutionary developments, in the sense of possible departures from an ancestral form, in this case, taking the ancestral form to be that of the contemporary human life form. We shall consider three such departures, or evolvings, A, B, and C, from that form.

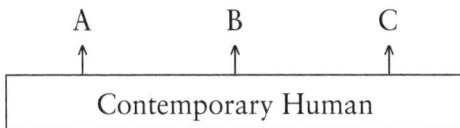

We shall take the B form as the "mainstream" view, to be human evolution as it is almost always viewed, as leading to something improved, or superior, according to modern lights, from the contemporary form, viewed here as ancestral, relative to forms A, B, and C. Interestingly, the "B view" almost always presupposes that the environment will remain relatively static. It inclines to view evolution as though it were disengaged from the world, thus suggesting that the view is founded on an unrealistic and rather naive view of evolution. To be sure, one might suppose a macromutation, unrelated to the environment in any significant way, say, perhaps the result of a local toxic condition, an anomaly of diet, an unusual exposure to cosmic radiation, an inexplicable quantum alteration of genetic material, or such. Too, naturally, evolutionary developments brought about by genetic engineering might be relatively disengaged from the environment, as well. In a typical "B view" scenario evolution produces a new creature, larger, healthier, longer-lived, stronger, faster, more intelligent, and such, than the ancestral type. Indeed, it is sometimes hypothesized that it might possess unusual powers, perhaps of a paranormal sort, say, a capacity to read minds, to communicate telepathically with others of its kind, to move objects at a distance, and so on. Perhaps it would be as much beyond us in many ways as we take ourselves to be beyond baboons and Rhesus monkeys.

In both the "A form" and the "C form" we have departures from the modern ancestral form, our current form, which do not return to, or approach, the ancient ancestral form; thus, on our approach, both will constitute clear examples of evolution.

We shall suppose that both the A form and the C form are quite different in appearance from contemporary humans, whether as a result of macromutation, genetic engineering, or natural processes taking place in an altered environment over a long period of time. Perhaps the A form is tiny, say, usually smaller than two feet in height, and weighing less, on average, than ten to fifteen pounds, and that it is also furred. These adaptations would be extremely valuable in a world of dwindling resources, and encroaching cold. The C form, we shall suppose, is air-breathing but predominantly aquatic, much as aquatic mammals are today, porpoises, dolphins, and such. Perhaps this is an adaptation to the land's having become largely uninhabitable, perhaps due to lethal radiation, perhaps due to massive, unmanageable insect infestations, to occupation by ferocious, territorial forms of life, humanoid or otherwise, following the breakdown of civilization, perhaps due to progressive rising sea levels, or such. It is difficult to speculate what an aquatic human adaptation might look like, but it seems, given convergent evolution, that after a half million years or so, it might bear little resemblance to the familiar terrestrial form of human life.

If "ancestral-form humans" are still about, perhaps in zoos, or reservations, in the hegemony of the B-form, and the humans are still speeched, rational, and such, it seems likely that they would regard the B form of life as human, and be rather proud of it, much as baboons might see a similar development in their own line as a super baboon. The B form of life, on the other hand, if taking itself to be human, might not be inclined to return the favor, and might no more regard us as human than we would be likely to regard Australopithecus Africanus and his humbler predecessors as human. To be sure, it is hard to know about such things. For example, the humans, in their cells, compounds, or preserves, tended with kindness, placed there for their own good by a compassionate B form, might hold themselves to be the only humans, or "true humans," and their supervisors and keepers as monsters. Similarly, the B form might find itself embarrassed by its ancestry and prefer to limit terms like 'humanity', 'human being', and such, to what it regards as inferior forms of life. It is interesting to speculate about the possibility of a B-form controversy one day having to do with evolution. Surely many a Victorian doubted that he was descended from "monkeys," and, in any event, found himself uneasy with the thought. Perhaps the B form would be truly evolved, according to many modern lights, and, in a suitable, benevolent, collectivist, authoritarian, statist, welfare-minded manner, would compassionately tailor truth to what it conceived to be in the public interest. Truth has its uses, like fire, but, too, like fire, it can be dangerous. Truth sets some people free; others it makes miserable. Life may be more important than truth, but without truth, in the view of some, at least, it is contemptible.

In any event, the problem is classificatory, and, at present, at least, the problem is ours, not a problem for the A, B, or C form.

It seems reasonably clear that we, at least at first, would regard the B form as human, and would certainly do so if we anticipated it would respect us, be favorably disposed toward us, and such. Hopefully, we might rejoice in its emergence, might salute it, and might view it rather as proud parents might view an outstanding offspring, one which far exceeded their own limitations and status. To be sure, if we anticipated that the B form would despise us, would view us with condescension, and such, or if, say, our attitude was more one of resentment, hatred, and envy than paternal pride, we might entertain a different view. This sheds a nice light on an often-overlooked epistemological fact, that our classifications are not solely determined by cognitive considerations, but are often influenced, as well, by considerations personal and emotional, by considerations which are sometimes psychological, if not actually psychiatric. For example, on the grounds of language, tradition, technology, background, development, mental resemblances, and a common genetic ancestry, it seems clear we would be well-advised to regard the B form as human, and its representatives, accordingly, as members of the human race. On the other hand, as we are supposing significant differences between ourselves and the B form, it would be possible, and rationally defensible, to take these differences, if we wished, as constituting a sufficient condition for ruling the B form to be a different species, though one obviously derived from contemporary humanity.

That considerations not primarily cognitive may enter into conceptualizations becomes even more clear when one addresses one's attention to the A and C forms, the A form diminutive and furred, adapted to a world of cold and severely constrained resources, and the C form, adapted to an aquatic environment. In both cases, we have an Earth-origin ancestry, in the sense, at least, of a common, or substantially common, genetic ancestry. Thus, neither, no more than the "mainstream form," the B form, is ruled out of the species on the grounds of lacking the ancestry in question. On the other hand, as well, having the ancestry in question is not a sufficient condition for membership in the human species, as forms with a common ancestry may considerably diverge, either through evolution or devolution. We do not regard ourselves as a species of lobe-finned lungfish, though it seems such creatures lurk in remote eras as progenitors of other species which eventually produced humans, and, similarly, if humans should prove to be the ancestry of future species as different from us as we are from the lungfish, we would presumably not regard them as unusual humans, no more than we would expect lungfish, if they were capable of reflecting on the matter, to regard us as unusual lungfish. We are different sorts of beasts, just as a unicellular organism with its millions of complex molecules engaged in their billions of interactions would be very different from a proto-cell or some sort of primitive, self-replicating carbon compound.

I suspect that our intuitions with respect to the A and C forms would be primarily influenced by mental resemblance, as opposed to physical resemblance, given that a common genetic ancestry is accepted in both cases. As noted earlier, in effect, if something like the A or C form, or, most likely, even the B form, were encountered on a foreign world, we would presumably, unhesitantly, not regard it as human. This consideration, as earlier suggested, seems to emphasize the importance of a common genetic ancestry, and, given such an ancestry, within it, the importance of mental resemblance. One supposes, of course, that a point might be reached, eventually, in which the physical differences would seem insurmountable. As always decision and degree would be significant. Black and white are very different, but, between them, there exists an infinite number of shades of gray. Similarly, black and white themselves, are black and white, as opposed to something else, in virtue of conceptualization. In a sense, black and white exist in nature, but nature does not notice such things. People do. Gradation is nature's manner; division is our human penchant.

As in the B case, epistemological classification might be influenced by considerations not obviously or purely cognitive. For example, suppose that we felt more sympathy with, and identification with, the A form than the C form. Perhaps we tend to think of the A form as plucky and the C form as elusive, even craven. The A form is small and tough; it sticks it out, under adverse conditions, as a terrestrial form, as we feel we might under similar challenges and adversities; it struggles bravely; it leads a hard life; it lives in caves, sod houses or snow dwellings; it fights off bears and wolves; it reminds us of primitive humans, fragile, but determined and indomitable. On the other hand, the C form seems to have fled from other life forms rather than confronting and combating them; it seems to have failed to meet its challenges; in the face of adversity it has withdrawn; it retreated; it surrendered its country, so to speak, and returned to the womblike environment of a supportive sea, abandoned long ago by remote ancestors. It no longer runs, walks or climbs, but now floats or swims, immune even to the hazards of gravity, its former world abandoned to hardier, more aggressive life forms. Too, it looks less like us than the A form. The A form is clearly mammalian, but the C form, while mammalian, seems less so, and more serpentine or fishlike. We feel uncomfortable in contemplating it. It has fled our world, to take refuge in an alien world. Rather than giving battle, threatened, it fled the land won by its ancestors, and slipped away, into the darkness of defeat.

To be sure, one might view matters differently, seeing the A form as a degenerate mockery of our species, more monkey than human, as little deserving of respect as packs of wandering rodents, timid, furtive, hiding in crevices, scratching out a precarious, pathetic living in a desert of ice and snow. The C form, on the other hand, we might regard as more worthy of respect, larger, stronger, faster, more interesting. Rather than facing almost certain extinction in an inhospitable environment, it dared an astonishing and creative adaptation, and embarked on the exploration and conquest, and colonization, of a new, beckoning, welcoming world, but, too, in a sense, it has at last, after countless ages, come home, to the original mother of all life, the sea, to begin again the human adventure.

In short, it is easy to suppose that noncognitive, or transcognitive, considerations, personal, moral, and even aesthetic, might influence decisions as to whether or not a form of life was to be accounted human.

3.523 Alien-Encounter Arguments

The Acceptability Position, namely, that it is acceptable, perhaps even desirable, to require Earth origin, or, say, more broadly, a common ancestry, or a substantially common ancestry, as a necessary condition for being human, may also be supported by certain alien-encounter arguments, as we shall see. On the other hand, alien-encounter arguments, as suggested earlier, can also support the Inadvisability Position, namely, the position that it is inadvisable to take Earth origin, or such, as a necessary condition for being human. In particular, we noted the "convergent-evolution argument," that human or humanlike life forms might well evolve on different worlds, or even be produced artificially, and the "arbitrariness argument," that it would be arbitrary to deny humanity to a form of life merely in virtue of where it is, where it began, and such. We distinguished between two different senses of 'human', a broad sense, having to do with properties, and a narrower sense, having primarily to do with origin. This had to do, substantially, with the distinction between a type and a derivation, between a sort of thing and a particular genetic line within a type of thing. Obviously the criterion for species identity could be set up in either of these ways, or, indeed, in any number of other ways. This recognition, however, does not resolve the issue, or, worse, dissolve the issue. It is not where the philosophical journey ends, but where, essentially, it begins.

One might consider the following matrix. It suggests some proposals for resolving issues of species. It is supposed to be, substantially, neutral between the broader and narrower sense of 'human', as there is much to be said for the value and plausibility of both the "Inadvisability Position" and the "Acceptability Position." The numbers are to be understood as follows:

    2   Yes, presumably human
    1   Yes, probably human
    0   Doubtful
  -1   No, probably not human
  -2   No, presumably not human[39]

Close Physical Resemblance		No Close Physical Resemblance		
Common Genetic Ancestry	No Common Genetic Ancestry	Common Genetic Ancestry	No Common Genetic Ancestry	
(1) 2	(2) 1	(3) 2	(4) 2	Close Mental Resemblance
(5) 2	(6) 0	(7) 1	(8) 2	No Close Mental Resemblance

---

39. In the scale provided, it should be clear that words such as 'presumably', 'probably', and 'doubtful', are indexed to the value of the conceptualization involved, not to some determinate matter of fact. For example, if one were to say that such and such a set of tracks were presumably those of a raccoon, probably those of a raccoon, or it was doubtful that they were those of a raccoon, one would be speculating about a matter of fact, as one is reasonably clear on tracks and raccoons. In the scale above, on the other hand, the references are to intuitive satisfactoriness, for example, that something with such and such properties is presumably desirably or best regarded as such and such, or probably desirably or best regarded as such and such, and so on. In this sort of situation, fact awaits concept. The matrix, thus, is prescriptive, not descriptive.

Certain cases, one supposes, would be relatively uncontroversial. For example, anything with a close physical and mental resemblance to a human, and having a common genetic ancestry with humans, as in the first case, would presumably, unhesitantly, be counted as human. It would, in effect, fulfill the criteria of both the broader and narrower sense of 'human'. Similarly, something with no close physical or mental resemblance to a human, and with no common genetic ancestry with humans, as in the eighth case, would presumably not count as human, in either the broad or narrow sense of 'human'. Similarly, the familiar alien from science fiction, which has no common genetic ancestry with humans, and does not resemble humans, but, interestingly, has a close mental resemblance to humans, as in case four, would presumably be rejected under both the broader and narrower criterion. More controversial, one supposes would be the third and fifth cases. In the third case, we might have situations of the evolved or devolved human, where there is a common genetic ancestry, but no close physical resemblance. Here the close mental resemblance, coupled with a common genetic ancestry, was taken as crucial, more so than a physical resemblance. Too, subsumed in case three, might be cyborgs, certain products of genetic engineering, and such, which might not resemble human beings physically, but would have a common human ancestry, and, hopefully, a close mental resemblance to human beings. Presumably, we would wish to count such entities as human. In case five, we have a common genetic ancestry and a close physical resemblance, but no close mental resemblance. One supposes that one would regard this entity as human, primarily because of the common genetic ancestry, coupled with its human appearance. Perhaps it might count as a deviant or degenerate human. In the second case, we have the sort of thing motivating the "Inadvisability Position," namely, something which is identical to a human being, or very similar to a human being, both physically and mentally, but does not share a common genetic ancestry with humans. It might have been, say, evolved on another world, synthesized somewhere, or such. I have suggested that we would probably regard such an entity as human, wisely or not. Certainly it would have no difficulty in passing for a human. I have listed case six as doubtful. There we would have something which has a close physical resemblance to humans, but no common genetic ancestry with humans and no close mental resemblance to humans. Much here would depend on how much weight we assigned to physical resemblance. I am supposing that it would be doubtful that we would count this entity as human on either the broad or narrow criterion for humanity. On the other hand, if it looked human, one supposes that politics might become operative, that some judge would be found who would rule it human, and so on. In case seven we have a common genetic ancestry with humans, but neither a close physical or mental resemblance. We suggest that this entity is probably not human, on at least the grounds that the thread of genetic ancestry is a thread on which may be strung the beads of many species, fish, amphibians, reptiles, and so on. To be sure, here, as in many of these cases, much might depend on the interaction of a diversity of circumstances, for example, the degree of physical or mental resemblance. How close does it have to be, to count as close? And how would one balance off mental and physical resemblance?

The "Inadvisability Position" is surely a strong position. In its light, the broad sense of 'human', in the final analysis, might seem the sense to be preferred, the best criterion for humanity. Certainly, convergent evolution makes extremely probable the emergence, quite possibly on many worlds, of life forms much like the human, for example, bipedalian, with prehensile appendages, with forward-located, binocular vision, many paired organs, an internal skeleton, an immune system, thermostatic arrangements for maintaining and controlling body temperature, an apparatus for uttering complex sounds, a variety of sensory organs for experiencing the world, a calculating, rational mind, dispositions to sociality, and such. There is no guarantee, of course, that an organism with such properties, and more, need much resemble a human being, as we usually think of human beings. We might, of course, perhaps under political pressures, broaden the concept of a human being, perhaps in such a way as to include all rational, moral creatures, but that would seem to stretch matters considerably, and we already have concepts pertinent to recognizing, and respecting, certain varieties of creatures, for example, "moral," "rational," "civilized," and such, without absorbing them into the human race, which absorption might seem presumptuous on our part, even arrogant, and perhaps annoying, or unwelcome, to creatures who might not see such a favor as all that flattering. For example, if we encountered rational, insectoidal aliens and they decided, perhaps in virtue of a generous, gracious ruling by the supreme parliament of the great Nest, to count us as insects, we might view this new dignity with ambivalence.

Claiming to be human is not to claim superiority, and generalizing humanity, whatever the intent of the act, is not to elevate or bestow status. The more a word means in general, the less it means in particular. It may be nice to be human, but not everything nice is human.

Whereas one might be willing to extend the concept "human" to one extent or another, at least given a common human ancestry, one supposes that the "Inadvisability Position" is strongest where the resemblances are closest. So let us suppose that we have, on hand, a creature evolved on a distant world who is very much like a human being. Would it not, then, be arbitrary, logically and epistemologically unreasonable, even unjustified, not to regard it as a human being? How important, after all, is its origin, or such?

Intuitively, I think origin would be regarded, at least by most individuals reflecting on these matters, as quite important. That the alien might easily pass for a human being might be undeniable, but irrelevant to the issues involved. Science fiction is freighted with such creatures, androids, shape shifters, mind clouders, and such. But let us suppose now, as indicated, that we are dealing with an alien who is very much like a human being. One supposes in such a case that one would be looking for differences here, perhaps desperately, even relatively insignificant differences, in virtue of which we might, with a clear conscience, rule against our subject, not accepting him as human. I think this is not so much species chauvinism as an intuitive commitment to something much like, if not identical to, the Earth-origin, common ancestry, and such, criterion. "He is not one of us." "He is not a Smith or a Jones." Some differences might make this matter easy, for example, if the individual was an unusual color or could change color, in such a way as to blend in with its surroundings, if it could punch through steel, see through walls, and such. Other differences might be subtle, but, for the purposes in hand, might be sufficient. Perhaps it has a pointed tongue, or a third eyelid. Perhaps some differences are not even visible, such as an unusual blood type not found in humans, a capacity to see in ultraviolet light, a paired heart, or such. Some differences might be as miniscule as a certain cast of features which, if found in a human, would attract no interest or attention whatsoever.

In short, I think almost any difference between the alien and the human, trivial or otherwise, would be likely to be taken as decisive with respect to species membership. If these anticipations are correct, they suggest the existence of an underlying criterion for humanity which, rationally or not, is operative, the criterion, I suspect, of genetic commonality.

But now let us suppose, despite all probabilities, that the alien is indistinguishable from a human being, except with respect to its history, so to speak. What then?

On the broad notion of 'human' it seems to me that it would have to be accepted as human. On the narrower notion of 'human' it would naturally be rejected as human, as it lacked human ancestry.

At this point a decision would have to be made.

It is certainly conceivable that intuitions might differ in such a case.

On the other hand, this is not an issue to be indifferently resolved, as by flipping a coin.

We will argue that the "human family" approach, or the "Acceptability Position," is superior to the "no difference" approach, or the "Inadvisability Position," for a variety of reasons.

First, however, regardless of what resolution one might favor, it should be noted that one wishes to have a single, unified criterion.

We noted earlier the difficulties which might attend a multiplicity in criteria, for example, that in accepting both 'A $\equiv$ B' ('A if and only if B') and 'A $\equiv$ C' ('A if and only if C'), one is implicitly committed to 'B $\equiv$ C' (B if and only if C), which might prove troublesome, producing a situation in which stipulated analyticities, in effect, consequences of definitions, say, 'A = df. B' and 'A = df. C', might produce consequences which would be empirically falsifiable, an eventuality incompatible with analyticity.

One may define as one wishes, but, if one does so, one must thenceforth watch one's step.

In our case, if one were to define 'human' in terms of both the ancestry criterion and the property criterion (e.g., possessing a certain set of designated properties except for ancestry), as in 'H = df. A' and 'H = df. P', a number of putative analyticities would be generated, of most interest 'H $\equiv$ A' ('H if and only if A') and 'H $\equiv$ P' ('H if and only if P'), which putative analyticities would generate the following putative analyticity, of most interest in the present context, 'A ≡ P' ('A if and only if P'), which is empirically falsifiable. 'A ≡ P' would be false if either A was the case and P not the case ('A & ~ P') or A were not the case and P was the case ('~ A & P'). As 'A ≡ P' is supposedly analytic, it would be falsifiable by any conceivable counterexample. For example, a claim that 'C is a circle if and only if C is blue' was analytic would be shown to be false if either a nonblue circle was logically possible or something could be blue without being a circle. Both of these situations are logically possible even if, as a matter of fact, interestingly, all circles were blue and everything blue was a circle. In our case it is easy to see that we cannot define humanity in terms of both the ancestry and the property criterion, as one can conceive of something having the appropriate ancestry and lacking the appropriate properties (say, a cyborg, a genetically produced oddity, a strange offshoot of the human stock some thousands of years from now, etc.), or lacking the appropriate ancestry but possessing the appropriate properties (say, a synthesized humanoid, a humanlike extraterrestrial, etc.). One can define 'human' variously, but, to avoid logistic embarrassments, one should do it in one and only one way.[40]

What is actually going on here, of course, is not the proposal of a particular criterion for species membership, which it would be presumptuous, at the least, to propose, but, rather, an argument for a logically necessary condition for species membership. For

40. Similarly, it would seem undesirable to define 'human' disjunctively, for example, taking either the ancestry or properties criterion as a sufficient condition for being human and the satisfaction of at least one of the criteria as a necessary condition for being human. This might proceed as follows: 'H = df. (A v P)', which definition, if accepted, would require taking 'H ≡ (A v P)' to be analytic. More technically, one would have '(x) (Hx ≡ (Ax v Px))' ('For every x, x is human if and only if x satisfies either the ancestry or properties criterion'). Presumably this would satisfy no one as it would admit entities as human which would not count as human on one or another of the criteria independently. For example, an advocate of the ancestry criterion might object to resembling aliens counting as human and an advocate of the properties criterion might be reluctant to count, say, cyborgs, brains in vats, some genetically produced entities as human, and so on. Whereas one could count anything as analytic, if one wished to grit one's teeth and stick to one's guns, even 'C is a circle if and only if C is blue', it seems deviant or pointless to do so, in virtue of the ensuing cognitive dislocations or embarrassments. In the preceding example, accepting its analyticity, one would have to reject, say, red circles and Euclidean circles as circles, as the first has the wrong color and the latter lack color, and would have to count everything blue as a circle, for example, bluebirds, blue squares, and such. In our case, above, '(x) (Hx ≡ (Ax v Px))' would be falsified if and only if we allowed the possibility that either '(∃x) (Hx & ~ (Ax v Px))' or '(∃x) (~ Hx & (Ax v Px))' might be true. A simplified form of such proofs might be as follows:

example, 'H = df. A', leading to 'H ≡ A' (H if and only if A), will not serve as a criterion. Whereas we might hold that everything human has such and such an ancestry, it is altogether possible that certain things which might have human ancestry might not be human, just as not everything which descended, or ascended, from a lobe-finned lungfish is a lobe-finned lungfish. For example, it might be decided that the A, B, and C evolutionary lines, discussed earlier, springing from the contemporary human as an ancestral form, are not human. So, what we will be arguing for would be 'H ® A' (If H, then A'), so to speak, in which being human would logically require having a certain descent, or ascent, in short, that ancestry of a sort will be taken as a logically

---

1. (∃x) (Hx & ~ (Ax v Px))
2. Hy & ~ (Ay v Py)   1, Existential Instantiation
3. [Hy & ~ (Ay v Py)] v [~ Hy & (Ay v Py)] 2, Logical Addition
4. ~ (Hy ≡ (Ay v Py))   3, Equivalence
5. (∃x) ~ (Hx ≡ (Ax v Px))   4, Existential Generalization
6. ~ (x) (Hx ≡ (Ax v Px))   5, Quantifier Negation.

1. (∃x) (~ Hx & (Ax v Px))
2. ~ Hy & (Ay v Py)   1, Existential Instantiation
3. [~ Hy & (Ay v Py)] v [Hy & ~ (Ay v Py)] 2, Logical Addition
4. [Hy & ~ (Ay v Py)] v [~ Hy & (Ay v Py)] 3, Commutation
5. ~ (Hy ≡ (Ay v Py))   4, Equivalence
6. (∃x) ~ (Hx ≡ (Ax v Px))   5, Existential Generalization
7. ~ (x) (Hx ≡ (Ax v Px))   6, Quantifier Negation

With respect to the disjunctive proposal, numerous counterexamples would be possible, for example, an entity which was human, but lacked both human ancestry and human form, for example, a conscious, rational machine, mobile on treads, with a cylindrical body and eleven appendages, which had been pronounced "human" by some judge, perhaps in response to a machine-rights movement, perhaps to insert himself into legal history, perhaps in compassion for the device's mental health, feelings, or such. One supposes neither the "ancestry" nor the "properties" advocates would approve of this, but, as conceptualization is involved, it constitutes a definite possibility. Other counterexamples would be entities which had human ancestry but were rejected as human or entities who were not human but had human properties. Such entities, in the first instance, might be products of evolution or devolution, some products of genetic engineering, disembodied, active brains, colony-creature "humans," and such; and, in the second instance, one might have entities with human properties whom one might not wish to count as human, for example, resembling aliens, synthesized "humans," sophisticated androids, and such. In short, the adaptation of a disjunctive criterion would presumably broaden the notion of human to an unwelcome degree, pleasing few, if any, and would seem to transgress some relevant intuitions, whether of the ancestry or properties sort. It also seems such a criterion would be open to a number of difficulties, or counterexamples, beyond those to be expected of a simpler criterion. To be sure, questions of degree obtain, and it would be difficult to obtain any criterion which would be willingly, universally accepted. That is clear from the discussion of merely two possible criteria, the "ancestry criterion" and the "properties criterion."

necessary condition for being human, but not as a logically sufficient condition for being human.

To understand 'H ® A' one must, of course, have some sense of 'H', as well as of ancestry, and such. This understanding could be relatively informal, even substantially ostensive, rather as we recognize dogs and cats without a degree in zoology or veterinary medicine, or it could be rather more formal, as in listing a set of constituents properly arranged, and such, as discussed earlier. It would doubtless be difficult to come up with a fully satisfactory account of necessary and sufficient conditions for being human, but humanity is not worse off here than almost any other empirical concept. Recall Friedrich Waismann's "open texture of empirical concepts," earlier alluded to. All we really need here, for our purposes, is a necessary condition, whether conceived of in terms of a set of properties, which might be encountered anywhere, even on remote planets, or, say, an ancestry of a particular sort.

We will argue for the "family" view of a species, in this case, the human species.

Whereas our proposal is not an empirical hypothesis, but a recommendation, it, like an empirical hypothesis, constitutes an interpretation of data, in this case, an organization of data, a conceptualization of data, and possible data, and, interestingly, lends itself to evaluation, and review, in a manner reminiscent to that in which one might evaluate an empirical hypothesis.

The empirical hypothesis is a bet. As commonly understood, its object is truth. The philosophical conceptualization is a proposal. Its object, as commonly understood, is cognitive utility, convenience, even illumination. The empirical hypothesis presupposes a conceptual structure. The subject matter of philosophical conceptualization, whoever its agent might be, is the structure itself. Although reciprocities exist between data and the conceptual structure, the conceptual structure is logically and epistemologically prior. Without it, data cannot even be recognized as data. Also, interestingly, reciprocities and adjustments often occur amongst the tiers or divisions of the conceptual structure itself. A structure cannot even be recognized without recourse to, in effect, a metastructure.

There are many lists of criteria whereby an empirical hypothesis may be evaluated, some longer, some shorter. The following suggestions are typical. The most decisive criterion for such a hypothesis is testability, that such a hypothesis be, at least in theory, testable, directly or indirectly, that it should be confirmable or disconfirmable, at least in theory, directly or indirectly. Conceivable situations in the world, actual or putative evidence, and such, would be relevant to its acceptance or rejection. Obviously presupposed by practical testability is clarity. And ideally associated with clarity is precision, most easily achieved in matters in which quantification is achievable and appropriate, for example, in astronomy and physics. A number of other criteria are less relevant, for example, simplicity, generality, stimulation, and consonance. Simplicity, reminiscent of "Ockham's Razor," in which explanatory entities are not to be multiplied beyond necessity, is compatible with formidable complexity, only that the complexity should not be more formidable than is necessary to produce the explanation. That the explanation is the simplest explanation, of course, does not entail that it is the true explanation or the best explanation. Logically such things are independent. Generality refers to coverage, the capacity of the explanation to account for a wide variety of data. Classically, Newton's theory of motion would be an excellent example of such a theory, accounting for the falling of apples, the dropping of spoons, the flight of cannon balls, the movements of planets, the rise and fall of tides, and such. Stimulation is the capacity of the hypothesis to open the door to new discoveries, to realize unprecedented possibilities, to see things differently, to look for certain relationships in places they had not been hitherto noted, or expected, to shed light on a diversity of phenomena. Some examples of stimulating hypotheses would be the germ theory of disease, of continental drift, and of evolution, each of which opened new and surprising vistas of understanding and explanation. Consonance is the compatibility of a hypothesis with established views. The presupposition of the consonance criterion is familiar and plausible, namely, that most of what we suppose is true, and for good reason, is in fact true. Accordingly, a hypothesis which would require the adjustment, revision, or repudiation of a popular theory, or world view, is suspect. The establishment view might be wrong, but the assumption is that it is substantially correct and that that which would contradict it is presumably incorrect. For example, one of the major counts against alleged paranormal phenomena is that, if it exists, it would require the serious revision, if not the repudiation of, a picture of the world which has been meticulously wrought over centuries by generations of professional scientists. Changes of that dimension would be catastrophic, and, at the least, would seldom be welcomed. Consonance is neither a necessary nor a sufficient condition for truth, but, in the pragmatics of inquiry, it is inevitably a serious consideration.

From several points of view, there is not a great deal to choose from, between the "Inadvisability Position," emphasizing properties, and the "Acceptability Position," insisting on considerations of ancestry. Both approaches are straightforward, and reasonably simple, clear, and precise. And whereas one does not test a conceptualization in the way one tests an empirical hypothesis, as conceptualizations are not empirical hypotheses, and make no claim to truth, conceptualizations can be testable, or "testability friendly," in the sense that they can figure in empirical hypotheses. One could, for example, test an entity, examine it, measure it, and so on, to see if it had such and such a set of properties, and, today, given developments in genetics, it would seem possible to ascertain, reasonably well, the DNA of a given hereditary line, for example, in the way contemporary breeds of dogs can be traced back to the ancestral form of the gray wolf. To be sure, it is easier to determine a set of existent, contemporary properties than to establish common descent and certainly so in a case where the DNA of a putative progenitor is unavailable. For example, if there were no contemporary gray wolves about, speculation about the nature of a common ancestor for our contemporary dogs would be far more problematic, much more speculative. One would be guessing at the likely properties of such an ancestor. To be sure, the "properties approach" to humanness would also become more problematic, as the properties of earlier life forms would be less clear, given the paucity of data. On these four criteria, simplicity, clarity, precision, and testability, it seems both analyses would be fairly equivalent. If there were an edge here, one might give it to the properties approach, as it is easier to ascertain ancient properties, via physical remains, cave drawings, carvings, and such, than DNA, long vanished. To be sure, things might have similar properties without being such that we would wish to regard them as similar things. It is not unusual in physical anthropology to encounter data whose classification, human or nonhuman, is unclear.

At this point, however, I think the "ancestry" approach begins to distance itself from the "properties" approach.

1) Conceptualization

First, as a general point, the crucial matter here is conceptual, and not empirical. For example, that it might easier to ascertain visible properties of an organism than trace its lineage might be important from the point of view of applying a conceptualization to the world, a not inconsiderable advantage, but it leaves untouched the conceptual value, *qua* conceptual value, of the criterion. One could, though one need not, take ancestry as the more satisfactory criterion, even though it might, in empirical fact, be less easy to establish than the possession of a set of properties. For example, it might be preferable to take a mathematical analysis of circularity, in a Platonic sense, as more relevant to circularity than some empirical concept of "roundness," say, looking and feeling round, rolling well, traceable with a compass, or such. Conceptualization's usual neighbor is the ideal, not the applicably practical. An abstract or ideal analysis might serve for both the properties and the ancestry approach. The point here is that the greater applicability or practicality of one approach or the other is not crucial with respect to the issue. It is possible, for example, to take ancestry as a defining characteristic for humanity, even if it is not possible in many cases to determine that characteristic. For example, a given life form either traces back to an earlier form, or not. Our inability to ascertain the fact of the matter is irrelevant to the fact of the matter, and logically, if not pragmatically, is irrelevant to the value or disvalue of using ancestry as a necessary condition for species membership. The preference between the properties and the ancestry approach is not dictated by, nor forced by, empirical considerations, no more than the preference between a mathematical and an empirical approach to circularity is dictated by empirical considerations.

2) "Generality and Stimulation"

The "ancestry approach" is preferable to the "properties approach" on grounds similar to those of "generality" and "stimulation" in the evaluation of empirical hypotheses. The "properties approach" is specific, limited, and narrow; if taken seriously, it would hold the species concept to a static form, disallowing changes in form, developments in form, mental and physical, and such. For example, on the "properties" approach, it seems likely that the A, B, and C evolutionary forms earlier discussed, the devolutionary form earlier discussed, and such, would be ruled out of the human species. One could surely do this, but I think it would be better to leave such decisions to be adjudicated on their merits, and not to have such decisions influenced by, or coerced by, a too-narrow a concept of humanity. The "ancestry approach," contrariwise, keeps such questions rather more open, and is more amenable to evolutionary developments. For example, the "properties" approach might result in ruling extraterrestrials of remote human descent out of the human race, whereas the "ancestry" approach has the latitude, at least, of extending the concept of human in cases deriving from a common ancestry. It is also possible that the "ancestry approach" might be preferable in the light of a "stimulation" criterion, namely, a criterion which allows for the elaboration and extension of a concept into a variety of new areas, for example, into machine/human, animal/human, designed human, fusion human, group human, and such.

3) Intrinsicality

The ancestry approach is preferable to the "properties approach" also on criteriological grounds, as it involves an intrinsic, as opposed to an extrinsic, principle of species determination. Ancestry provides a concept in which a variety of forms may be essentially related, in virtue of biological succession, as opposed to being coincidentally or accidentally related, in virtue of resemblance. In this sense. an actual, internal, natural relationship is involved, as opposed to the imposition of an external, possibly arbitrary, relationship, in virtue of what might be a mere happenstance arrangement of properties, possibly subjectively, idiosyncratically, assessed as being of sufficient similarity to count as a set of grounds for species identity. Properties are fluid, and might change over time. Accordingly, ancestry provides a stabler, more reliable criterion for speciation.

4) Unification

The ancestry approach is preferable to the "properties approach" also on the grounds of unification, as its principle allows us to bring together a possible variety of forms into a common species, rather as the many varieties of contemporary dogs, despite remarkable differences in size, shape, and appearance, all count as dogs.

The "family approach," then, allows us to unify a considerable amount of possibly diverse data. It provides us with a more viable criterion than the "properties approach." It would be useful in a greater variety of cases.

5) Applicability

The ancestry approach is preferable to the "properties approach" also on grounds of its greater applicability, as noted, above, in discussing the analogue to "stimulation" as a possible characteristic of an empirical hypothesis. In encouraging, or opening up, the possibility of extending a species concept, its greater applicability, subject to circumstances, is admitted. This allows for responsible decisions, and definite actions, in which the theoretical possibilities involved in species extensions might, in fact, be realized. One might then begin, in fact, to apply the concept "human" in such a way as to include within its scope hitherto unprecedented entities, for example, as suggested, above, the machine/human, animal/human, designed human, fusion human, group human, and such.

6) "Consonance"

There is little here to which "consonance" would be applicable, as one is here venturing into new conceptual territory.

All is new and fluid; chaos seems to blossom; there are few solidities in sight; the conceptual structure is not yet formed.

In such a world, what could 'consonance' mean?

Consonance makes a great deal of sense when a new hypothesis, a new claim to truth, proves to be incompatible with familiar and accepted, claims to truth. Contradictions, if recognized, call for resolution. One cannot knowingly, or most cannot knowingly, accept both 'p' and '~ p' as true. One may not know which is true, but one knows they cannot both be true. The noblest challenge of consonance is not to categorically reject or disparage the new, but to overcome, if possible, the contradiction. Ideally, consonance does not favor dogmatism, complacency, neglect, inertia, or prejudice; but is, rather, a call to investigation, to further inquiry.

If consonance were applicable here, in some sense, which it does not seem to be, given the novelties involved, presumably it would demand scientific stasis, or, more likely, retreat, an adherence to the status quo, an avoidance of problems for which familiar, patented solutions do not exist, that it would recommend a refusal to enter the new, frightening, uncharted, radical world of possibilities, electronic and biological, that it would insist on keeping things the same, that one should refrain from rocking traditional philosophical, social, and religious boats.

One recalls the tumult and hysteria over something as simple as cloning.

It is possible that some seven eighths of the human race has not yet emerged from the Dark Ages, and, seemingly, from choice.

On the other hand, we do have something here which is at least analogous to consonance, not just in the compatibility or incompatibility of conceptualizations, but, more importantly, in the congruence or lack of congruence of conceptualizations with moral and aesthetic intuitions, with a sense of propriety or impropriety, with a sense of rightness and wrongness, of better and worse. Such things must exist or one could not even begin to address oneself to these issues; one could not attempt to adjudicate the value of alternative conceptualizations without some sort of intellectual meter stick or compass. Even the evangelical relativist requires something like this, or he would be unable to be a relativist. Even if he thinks his preferences, intuitions, and such, are no better than those of fellows he firmly believes are ignorant, misinformed, stupid, and benighted, he has them. One cannot really get along otherwise; otherwise, immobility; otherwise, paralysis.

It is my supposition, on the whole, that the "ancestry approach" is more intuitively satisfactory than the "properties approach," that one is likely to feel that the "human family" approach is better than the other, that it is more right, and so on. I think one would be more comfortable with it. It is likely to seem a better fit with our presuppositions, perhaps with our prejudices. It is likely to be more consonant with what we feel we are, and should be.

I think the most difficult challenge to the "ancestry" notion, conceptually, would not be that of the resembling alien, which we might view with xenophobic suspicion, but would rather be that of the synthesized human being. It need not have been synthesized by evil aliens, demented sorcerers, demonic Harvard professors, or such, perhaps for devious purposes, but it might have been synthesized, say, by a team of well-intentioned human biologists, representing a consortium of small, open-minded, middle-of-the-road colleges, perhaps spanning a generation or two, who were attracted to the project by its obvious formidability. Their product, which we shall suppose is successful, is proposed as a human being, exactly the same sort of entity which might have been produced by more conventional means, utilizing Susan, or such, but wasn't. We will suppose we have no motivation here to conceptualize defensively, offensively, politically, or such, motivations common in sensitive areas. We shall also suppose that the fellows who have managed this incredible feat believe themselves, in all sincerity, to have synthesized a human being, claim that it is a human being, write it up in the appropriate peer-reviewed journals as a human being, present it to the world as a human being, and so on. We shall also suppose that the product in question regards itself as a human being, has no difficulty in obtaining a driver's license, a library card, a number of assorted credit cards, a social-security card, membership in the Diners Club, in qualifying for Frequent Flier Miles, and such. Moreover, Susan falls madly in love with him, and he with Susan, and they are married in an Anglican cathedral, and, eventually, have several children, children as nice as can be, without ceasing to be children.

Is this product, the result of a remarkable experiment, taking several years to complete, well thought of as a human being? Certainly it would have no difficulty passing for a human being. Susan can vouch for that. Similarly, most people would regard it as a human being, not just Susan. Similarly, it regards itself as a human being, and who could blame it?

As we know by now, although the entity is obviously what it is, and has exactly the properties it has, and so on, the question as to whether or not it is human is not a question of fact, until we accept a given conceptualization, or have one imposed on us by, say, the might of the state.

Whether the entity is human or not depends on how we understand 'human', and different individuals might prefer different conceptualizations. This fact tends to be obscured because most entities we commonly regard as human are very similar to one another, and, as is typical in such cases, and to be expected, a considerable amount of conceptual consensus obtains.

One of the motivations for this study is the recognition that, in the light of various technological and scientific developments looming, and well in progress, this consensus is likely to be temporary. Shortly, its utility and comfort may be a thing of the past.

In the case of the synthesized human being, one might propose a variety of conceptualizations. Most germane to our discussion, the entity would count as a human being on the "properties approach," and would not count as a human being on the "ancestry approach."

Favoring the "ancestry approach," as noted , we would rule this individual nonhuman, despite the awkward, troubling, counterintuitive aspects of such a ruling in this type of case. We have noted the need for a single criterion earlier, and the logistic need of its consistent application should be obvious. Impartiality and objectivity are desiderata of criteria. Otherwise the matter is surrendered to the oscillation of competitive subjectivities, to casual place/time variations, to rudderless convenience, to whim and caprice, to incoherence, even to inconsistency.

Whether or not the offspring of such an entity, with an acknowledged human female, in this case, Susan, would be counted as human or not would call for a ruling. Similar considerations would pertain to a human/alien match. One supposes, in a case such as this, the offspring might well be adjudged human.

There is no anomaly, of course, in relating to such an entity as human, in treating it as human, and so on. One could conceive, for example, of treating a variety of entities as human, which were even more obviously not human, such as an android, an intelligence-enhanced orangutan, a crablike ambassador from Epsilon Eridani II, and so on.

We have tried, for better or for worse, to address these problems in conceptualization rationally, logically, meticulously, in the light of reasons, consistency, advantages and disadvantages, trade-offs, evidential relationships, likely consequences, and such. One owes that to philosophy. Now, briefly, as might be expected, one returns to a more likely world, one of ignorance, irresponsibility, confusion, inconsistency, political pressures, short-sightedness, convenience, bigotry, oppression, ambition, and power. Here one would note the role of claims, the entity's and those of the public, and the likely role of legalities. Lastly, one might anticipate the invocation of a criterion whose cognitive utility might be difficult to establish.

3.6 Claims

Claims may be rational or irrational, dogmatic or defended, persuasive or unpersuasive. Claims, *simpliciter*, need not be argued for, and might be diversely motivated. As claims, *simpliciter*, they are of little philosophical interest, unless it be to identify them, characterize them, and separate them out from other forms of action. Claims are of philosophical interest, as opposed to political, psychological, or social interest, only insofar as they are rationally supported. Accordingly, while recognizing that claims are of great importance, particularly if armed, so to speak, in one way or another, say, with weapons, a rhetoric of rights, the triggerings of compassion and solicitude, and such, we need to devote little more to them here than an acknowledgement of their existence.

3.61 The Entity's Claims

One is searching here, not searching as for something misplaced or lost, or something there but not yet discovered, such as who or what has been into the garbage, but searching in the interesting sense of trying to invent a useful way of looking at something, a good way of doing something. This is more like searching for the ideal painting, which has not yet been painted. In a Platonic sense such things are already there, waiting to be discovered. Do we not recognize the painting, once the canvas is finished? Was it not what we were looking for? How would we know that? How could we recognize it as what we hoped for, if it did not somehow exist before, waiting to be recognized?

This sort of thing is analogous to mundane philosophical cognitivism, in which one presupposes that the solutions to philosophical problems are there, to be unearthed after diligent inquiry, there patiently awaiting discovery, their discovery to reward the seeker with delight, if not fame and glory, rather like Grandmother's diamond brooch in the attic, finding which would be so nice, and that it is true that such and such is the solution, rather as it is true or false that nine times five is forty-five, and nothing else, no more, no less, and that forty-five and nothing else is at the foot of one of arithmetic's many rainbows. Without denying that forty-five patiently awaits nine times five, or that the neatest mate in such and such a situation begins with the Queen's Rook, it seems unlikely that important philosophical answers are true or false, at least in any usual sense, certainly not like it is true that dogs bark and cats do not, but rather more likely that they are better or worse, perhaps in an unusual sense, but, of course, one here transcends mundane philosophical cognitivism, naive philosophical cognitivism, and enters into darker and more fascinating forests, having to do with the cognitivity of value judgments, often denied by philosophers obviously convinced of the superiority of their own such judgments. It is a bit like the fellow who insists all cultures are equivalent, and none is better than another, except perhaps that his is the worst, and so on, and, at the same time, does not move, and would be panic-stricken if placed on a boat, shunted toward another one, and would not be caught dead in most of them. Sophisticated philosophical cognitivism is presupposed in philosophical prescriptivism, but that need not be noticed for philosophical prescriptivism to be taken seriously. In philosophical *descriptivism* one allegedly discovers answers; in philosophical *prescriptivism* one explicitly proposes answers. Psychologically it is undoubtedly difficult to distinguish between discovering an answer, which was there waiting to be discovered, and inventing an answer with which one is satisfied. The invention theory, on the other hand, involves less ontological hazard; it, at least, does not presuppose an unverifiable, Platonic world of preexistent, subtle, difficult-to-ascertain, nonnatural entities with obscure relations, if any, to atoms and molecules.

It would certainly be possible to take the entity's claims as decisive in the matter of species membership. Perhaps that would be the kindest, perhaps even the morally best, way, to resolve such issues. Certainly the beliefs, needs, desires, preferences, and emotions of the entity should not be methodologically overlooked, or discounted.

We will reject this approach, however, as we are interested in developing a general criterion, which will do for a species as a whole. One is unlikely to get the sort of general criterion we are looking for, if the general criterion is to let every entity determine this matter as it sees fit. A number of obvious objections to this generous approach to the matter obtain, for example:

1) Idiosyncrasy, individual-relativity, in such matters is

incompatible with the scientific desideratum of having a public, achronic, universal, independent criterion for species membership.

2) Individual relativity in such matters would permit two entities which, for example, had a common ancestry, manifested similar properties, and such, for example, monozygotic twins, to differ in their species.

3) Species membership is important. It should not be reduced to a triviality, such as names or addresses.

4) If species membership is indexed to claims, it should be recognized that claims may differ from time to time, from place to place, from mood to mood, and so on. One would not wish an entity to count as human on Wednesday and not on Thursday, to count as human in Schenectady and not in Poughkeepsie, to be human or not human, as it feels like it, and so on.

5) Most species, and many members of what we now take to be the human species, for example, infants and small children, are incapable of rationally comprehending and asserting such a claim. Are we to view them as lacking a species? And one surely does not wish to wait until they reach the age of reason before discovering whether or not they are human beings, and what if they decide not to let us know, or decide that they are not human beings, but perhaps dogs or cats, giraffes, or elephants? And what about the dogs and cats, and giraffes and elephants? Are declarations to be made on their behalf by proxies, and how would one legitimate the proxies, and what if a proxy decided to make the eleven o'clock news by claiming that his dog was a cat, or such? Presumably, at the least we would have to have some sort of default speciation, taking it for granted that dogs are dogs, coelenterates are coelenterates, and so on, at least until we have reason to believe otherwise.

Whereas, as above, there seems little reason to support personal claims as either a necessary or a sufficient condition for species membership, one should not be oblivious to situations, hopefully rare, in which an entity's claims might be of interest, not simply psychologically or clinically, as when a human claims to be, say, a cocker spaniel, an alien incarcerated in the body of a human, or such, or socially or technologically, as when, say, an armed robot is fiercely insistent that it is a human, and thus entitled to unemployment benefits, to wed Agnes, and such, but in cases where some plausible cognitive points and some moral issues of indubitable poignancy are involved. Such cases are best dealt with, however, under legalities.

3.62 The Public's Claims

In a sense, this is what does determine species membership, unless, say, a particular elite, with all the potential coercive power, legalized violence, and such, of a state, determines the matter by decree, by means of decisions of its judges, or such. After all, in conceptual matters, who but the public could determine such matters? To be sure, the public commonly delegates such matters to, so to speak, a committee of itself, a committee of informed individuals interested in such matters, for example, in particular, biologists. Contemporary orderings of species, taxonomical arrangements, and such, are usually organized in terms of conjectured evolutionary relationships, in terms of descent, rather than properties, such as resemblances. The conceptual aspects of this penchant are not always clear, one suspects, even to biologists. Many seem to suppose, or, better, speak and write, as though they supposed, that species, generalities, so to speak, are as real as the data which they organize and interpret. No one denies descent, for example; on the other hand, the decision to use it as a way of organizing the world, as opposed to some other way, appearance, for example, is up to us.

The public is commonly content, happily, to leave these matters to that segment of itself, scientists, philosophers, and so on, which finds such matters of interest. Let them set matters up, as they will. It is their house to organize as they will. Let them do as they wish. This tolerance, of course, is largely a function of the fact that the interests and labors of that portion of its membership, that concerned with such matters, the scientists, philosophers, and so on, seldom affects, let alone seriously impacts, the more general world view and welfare. The classification of a particular form of life, say, this crustacean or that, is seldom a burning issue, affecting resources, labor, beliefs, values, or self-image. This is not always the case, but it is generally the case. It is not clear, of course, given new technologies and their possible applications, that it will remain the case. It is easy to suppose, for example, that the humanity or lack of humanity of a cheaply producible, competitive, genetically engineered work force might become an issue of general interest.

There are, of course, a number of considerations which would make it undesirable for the public in general to take an active role in these matters, many of which are similar to those militating against taking an entity's claims as decisive. For example, the pubic, in general, might not produce a useful, unified criterion; it might produce inconsistent criteria, trivial criteria, local or temporal criteria, criteria frivolously alterable, and so on.

Whereas it might make sense to put speciation, as a regrettable last resort, up to the vote of, say, elite planetary biologists, or such, it would seem undesirable to put it up to a vote of a public which has little time to look into such matters, and usually little interest in doing so. Doing so might also empower emotion, dogmatism, discrimination, and prejudice, over tolerance, fairness, reason, and objectivity.

Democracy has two major values, first, it makes change possible without violence, in particular, transitions of power, and, second, it encourages some responsiveness on the part of governance, however indirect, to a majority of the voting public. Two difficulties are that there is no guarantee that a decision arrived at democratically is the best decision, or even a good decision, particularly as the public is commonly uninformed and susceptible to manipulation, and, second, democracy, *qua* democracy, if not seriously qualified, makes possible ignoring the interests, values, welfare, and such, of those not in the majority.

The rule of the wisest and best, namely, aristocracy, is usually preferred, particularly by those who regard themselves as the wisest and best. There is much to be said for monarchy, too, by those who plan to sit on the throne. On the other hand, as aristocracies and monarchies may become inept or corrupt, as well as parties and administrations, and congresses and parliaments, a removable or revisable establishment has much to be said for it. Democracy may be, as Churchill once suggested, the worst possible form of government, except for all the others.

In any event, deciding speciation on the basis of the general public's claims, in the case of humans, let alone dogs and cats, bryophytes and brachiopods, would seem to be a poor approach to such matters, time-consuming, complex, expensive, and unnecessary, as well as one likely to produce results which might be odd, if not capricious, and injudicious, if not absurd. It would be much better to surrender the matter to that portion of the public better acquainted with the data, the issues, and possibilities, subject, of course, one supposes, to a more general public review in cases of results dictated by political or economic agendas, with massive, possibly unfortunate, consequences to that public.

3.7 Legalities

We have covered, in this study, hundreds of problematic situations, and have recognized the conceptual challenges implicit in several of them, suggesting, in many cases, considerations which might be relevant to their understanding, treatment, assessment, and resolution. General criteria for evaluating conceptual innovations and novelties, approaches and solutions, will be dealt with in the final portion of this study.

At this point, it is useful to call attention to a factor commonly ignored, or, at least, seldom mentioned, in philosophical discourse, namely, the fixing and enforcement of conceptualizations. Whereas one would hope that conceptual fixation, particularly if sanctions are involved, would presuppose a conceptualization worth fixing, and, if necessary, worth promulgating, even by means of the coercive apparatus of the state, it is clear that two different moments are involved, first, the conceptualization, and, second, its enactment, imposition, and enforcement. The first moment is ideally philosophical, regardless of who is doing the philosophy, but, as a practical fact, may be social or political, with particular ends in view; the second moment is ideally humane, rational, open-minded, and impartial, but may be, as a practical fact, little more than an act of war, a concession to influence, a bid for popularity, the doing of a favor, an exercise of radical partisanship, and so on.

One of the inevitable tragedies involved in these matters is the likelihood, in many cases, those of general public interest, of the enactment of laws, the drafting of regulations, the handing down of legal rulings, and such, where legislators, bureaucrats, judges, and so on, with the guns and bayonets of the state behind them, out of their element and with little understanding of the issues involved, will be called on to institute second-order fictions, by means of which the lives and welfare of others, human and nonhuman, may be significantly affected, changed, shaped, disrupted, constrained, even terminated, or prevented. One sees no likely alternative to these developments in the present state of society, and anticipates an exacerbation of the usual forms of political dialogue to which we have become accustomed, emotion, tumult, interference, disruption, strikes, riots, protests, violence, and chaos, eventually necessitating, one supposes, centralization and repression.

It is to be presumed that the attention and effort philosophically devoted to the analysis of subtle and complex issues, such as these, perhaps by a variety of intellects over a good deal of time, will have little influence on events or practical importance in a world which seems less real than selfish, unpleasant, competitive, materialistic, and ultimately meaningless. On the other hand, it is not so clear that such a world, with its interacting, striving, energetic, risking organisms, is that inferior to one which, too often, is critical, pretentious, sanctimonious, sessile, and disengaged.

Philosophy often lives with futility.

But that is perhaps just as well.

I, for one, would dread living in the airy castles, the celestial prisons, with armed guards out of sight, frequently envisioned by philosophers secure in their ivory towers, contemporary monasteries, trendy seminars, and such, innocent of the nature of men and darkness, oblivious to the inevitable, abrasive contaminations of actuality. Too often their sanitary, well-intentioned, naive, aerial ruminations provide the pretexts in virtue of which horror and slaughter may be conveniently rationalized.

Often the road to paradise leads to the gates of hell.

Certainly philosophy is no stranger to futility.

But, given much philosophy, perhaps that is for the best.

Futility, too, it might be noted, has its advantages. One can relish the pleasures of considering oneself improperly neglected; one can deliciously resent the inadequate appreciation of one's brilliance and nobility, so stupidly or maliciously unrecognized. More important, there is a refuge and an impunity in the lack of responsibility. The spinning wheel which never touches earth may go nowhere, but, at least, it will never go off the road or over a cliff. Taking no road it will never take a wrong road.

So, one wishes the legislators, bureaucrats, judges, and such, society's decision makers, good luck in their endeavors.

Theirs is the role of making conceptualization, of one sort or another, of one value or another, good or bad, intelligent or stupid, ignorant or informed, naive or sophisticated, effective and real, not in the study, not in the classrooms or colloquia, not in the lecture halls, not in the corridors of colleges, not on the campuses of universities, not in the seminars, and so on, but in the streets, in the house of humanity, where ideas, borne in sheathes and cartridge magazines, make differences, bestowing and depriving, permitting and not permitting, letting live and killing.

One hopes, of course, that one's work might have some influence on how things turn out, if only in the modest role of suggesting that the issues are serious, subtle, and complex, and deserving of a consideration and treatment similarly serious, subtle, and complex.

3.8 Soul

Doubtless from the point of view of some individuals, and perhaps many individuals, we have overlooked an obvious, straightforward, simple solution to the problem of species identity, at least in the case of the human species, namely, that a member of the human species is any individual who happens to have a human soul. Granted that it might be difficult to ascertain that a given organism has such and such a soul, it might also be difficult, in many cases, to know if such and such an organism has such and such an ancestry. For example, the individual either has or does not have such and such an ancestry, regardless of our ability to make the relevant determination, and, if it has it, then it is a such and such, and, if it does not have it, then it is something else. Similarly, then, if a given entity has such and such a soul, then it is human, and, if not, then it is not human.

Let us not concern ourselves with questions such as how to tell a human soul from a nonhuman soul. For example, would a human-resembling alien have a "human soul," or might he, actually, have an "alien soul," and such. Is there one sort of soul, or several sorts, say, one sort for each species, or, say, one soul permeating all individuals in a species, or one soul doing for all living beings, or one for each individual, and so on. Each angel, as I recall, being pure form, lacking matter, and such, counts a species onto itself.

Such questions are well worth neglecting.

Presumably, for a god or gods, at least of an appropriate sort, namely, of a sort capable of making the determination, for example, understanding what it is involved, and being able to detect the presence or absence of the entity involved, namely, knowing what a soul is, and being capable of detecting its presence or absence in a given case, this would seem an ideal criterion.

One might accept something like 'x is human if and only if x has a soul of Type T' as being analytic, or, more briefly 'H ≡ T'. And, of course, one could repudiate the putative analyticity if either 'H & ~ T' or '~ H & T' were conceivable, namely, if one could conceive of a human without a soul of type T or could conceive of something not human with a soul of type T.

Before looking into the matter, some prefatory remarks would seem in order. Whatever ease or difficulty might be entertained by a god or gods in dealing with souls, there seems to be a considerable variety of notions amongst human beings. A "soul" might be identified with, or described as, breath, pneuma, a spirit, vital force, psyche, mind, consciousness, a ka, an invisible double, a mystery body, an astral body, an inner self, a life principle, an arrangement of fire atoms, the form of the body, a ghost, and so on. Many of these concepts, obviously, do not presuppose imperishability, immortality, survival after death, etc., nor the uniqueness of souls to human beings. Following Aristotle, for example, humans are not the only organism with souls, as there is a vegetative soul, connected with life itself, a sensitive soul, having to do with feeling, memory, and such, and a rational soul, having to do with reason. For example, a liverwort would have a vegetative soul, a mouse would have a vegetative and sensitive soul, and a human would have a vegetative, sensitive, and rational soul. There are many difficulties connected with understanding Aristotle here, for example, the nature of these souls; their relations to one another, and to the body; is there really only one soul with different levels of development reached, say, in one organism, but not in another; is the existence of the rational soul a sufficient condition for the sensitive and vegetative soul, in which case they are necessary conditions for it, and thus it perishes if they perish; how does the active intellect, which is impersonal, but supposedly independent of the body, fit into this schematism; and so on. Happily, all we need here is that soul, or souls, on this approach, are not restricted to humans.

One supposes that the two most likely "soul" views to be of interest in our context would be the spirit, or ghost, notion, and the "form of the body" notion.

As the "form of the body" notion" is supposedly a more sophisticated view, we will consider it first. First, one might note that it seems impossible that there could be a body without a form, and thus, where there is a body there would be a soul. This would attribute souls to a number of unlikely bodies, for example, rocks. If one were to insist that the bodies must be of a given sort, say, living bodies, it would seem to follow that bacteria, perhaps viruses, earth worms, house flies, fleas, mosquitoes, crocodiles, rattlesnakes, turtles, and such, as well as algae, molds, buttercups, pine trees, rose bushes, and so on, would have souls. Even if one favored endowing such entities as bacteria, bedbugs, termites, clams, sponges, herrings, porcupines, and elephants with souls, forgetting about mildew and redwood trees, it does not seem one could easily do so under the "form of the body" approach, as these bodies have divergent forms. One could, of course, accept this, multiplying things. first, species-specific souls, which, for example, in the case of insects, would give us something like 800 thousand different sorts of souls, and, two, continuing the example, as no two insects are absolutely identical in form, one would have even more sorts of souls. If "form" here does not mean form in the usual sense, but, say, something metaphysical, like Platonic necessary and sufficient conditions, or eternal requirements or definitions somehow affecting atoms and molecules, similar difficulties might obtain, having to do with multiplicity and individuation, and others beside, such as the unintelligible, unscientific, muddled verbalisms involved. In any event, all things considered, it seems likely one might be well advised to seek a different concept of soul, perhaps the "spirit;' or "ghost" notion.

On the other hand, as a last-ditch effort with respect to the "form of body" notion, let us suppose that the only "forms of bodies" of interest here are forms of human bodies. This presupposes we can recognize humans independently of a particular sort of soul, but that is all right, for one must start somewhere, as one would, for example, with the "descent approach," the "properties approach," and so on. If one does not have some idea of what one is talking about before one starts, one will not get far, and, indeed, would be unlikely to start, at all.

As human bodies have various forms, one would then, on the "form of the body" approach, be expected to distinguish, one supposes, between male and female souls, little souls, as in infants and toddlers, childsize souls, teenage-size souls, adult souls, old souls, changing souls, as the form changes, fat souls, thin souls, short souls, tall souls, and so on.

If one were to speak of the "form of the mind" as being crucial, similar problems would present themselves. Also, it is not obvious, in any usual sense, that the mind has a form. Would it be the form of the brain, with such and such a shape, or such? Presumably one would wish to continue to restrict souls to living bodies as there might not be much difference, formwise, between a living body and a recently deceased body. And bodies long deceased have forms, as well. Too, many things which are not human, manikins, androids, robots, dolls, statues, and such, might have the forms of human bodies. On the other hand, if one does restrict souls to living bodies, then the soul, on the "form of a body approach," would be mortal, perishing with the body's life, a consequence likely to be unappealing to those most likely to propose a "soul criterion" as a necessary and sufficient condition for membership in the human species.

This brings us, at last, to a perhaps less sophisticated, but more plausible, sense of soul, the "spirit" or "ghost" sense of soul. Following Descartes, this sort of soul is a "thinking substance," and an unusual sort of substance, as it is not in space. This sounds very much as though Descartes would be ruling such an entity out of existence. For example, being a substance, and being alive in some sense or other, and so on, it is presumably not to be equated with things such as suspicion, enthusiasm, rights and wrongs, liability, ownership, the binomial theorem, the number fourteen, and such, which, also, are not usually thought of as being in space, at least in the sense of being measurable in inches, weighed in pounds, and clocked in miles per hour, being situated in, say, Arkansas, being older on Wednesday than on Tuesday, and such. On the other hand, Descartes did not wish, at least obviously, to define his thinking substance out of existence, and there was surely abundant linguistic precedent for speaking of intangible substances, capable of noting this and that and forming views, capable of somehow taking action in space, even though they weren't there, and so on, in a way in which there was no comparable linguistic precedent for insisting on the existence and/or activity of, say, square circles and triangular squares.

It is a bit hard to understand the notion of an intangible soul.

Therefore, let us not understand it, but, to move the discussion forward, let us suppose that it possesses a subtle, unusual, illusive tangibility of some sort, that it is, so to speak, not immaterial, but less material, or less grossly material, or, perhaps best, differently material. One is reminded of the fine-grained fire atoms of Democritus. In any event, it is material enough, somehow, to push the buttons of the body. Certainly, at least, one wishes to dispense with "square-circle theories."[41]

We now have, so to speak, a "ghost in the machine," to borrow a phrase from Gilbert Ryle, who dared to be skeptical of the possibility. The nature of this "spirit" or "ghost," we shall suppose, as we struggle for intelligibility, permeates the body, or inhabits it, at any rate, perhaps moving about amongst the capillaries, and such, and has a rather bodylike form, though it is not the form of the body. For example, this "ghost' may inhabit a fat body without fear of being overweight, or fat itself. It laughs at high-cholesterol scores, except insofar as it empathizes with its cabin, or housing, body. It is, so to speak, a "mystery body," usually inside the normal body, but capable of doing without it, at least after its death. The mystery body, having its own degree or sort of tangibility, we shall suppose, is capable of being felt or seen upon occasion, is capable of being photographed by parapsychological enthusiasts lying in wait in gloomy Victorian mansions, is capable of rattling chains, and otherwise complaining,

---

41. Whereas we have spoken of "materiality," and such, all that is really required is some sort of monism, namely, some sort of "same stuff somehow" notion. Obviously rocks and trees are different from feelings and thoughts, but, one supposes, there must be some common denominator, or some thread of continuity or commonality, involved; otherwise it is hard to see how rocks, perhaps stumbled over, could produce feelings, say, pain, and how thoughts, anger and such, might result in the movements of a body, perhaps picking up rocks and removing them from the road. There is a general consensus, scientifically, that at one time there was no consciousness and, at a later time, there was consciousness. Certainly there is consciousness now. In short, at one time there was unconscious x, so to speak, and, at a later time, there was conscious x. Given the monistic view then, we would have:

1. Unconscious x followed by conscious x.

Now we might leave it at that. On the other hand, if we wish to think in terms of either matter or mind, we might have either:

2. Unconscious matter followed by conscious matter.

or:

3. Unconscious mind followed by conscious mind.

Accordingly, in such a case, the interpretation of "x" as matter or mind is a matter of indifference. On neither view is there the least threat to rocks and trees, and feelings and thoughts; they are safe; they are what they have always been, whatever that might be.

in the corridors of medieval castles, and so on. We also allow for its possible immortality, and such. The "mystery body," of course, does not solve all our problems. We still might wonder how it manages to appear and disappear at will, manages to walk through walls, and such. Its anatomy, too, presents its challenges. For example, how could it see without regular eyes, hear without regular ears, speak without a regular vocal apparatus, and so on. Does it have a mystery skeleton, a mystery heart, with a mystery circulatory system, a mystery excretory and reproductive system, for which it would presumably have little need, and so on. Does it have mystery thumbs which it might twiddle if bored, mystery hands with which it might scratch mystery itches, mystery legs with which it might wander about, or jump up and down? Does it have mystery muscles, which allow it to turn its head, and so on. Would it have mystery sweat glands? Does it ever trip, or twist its mystery ankle? If it has its form of tangibility, must it not be susceptible, to some extent, of being chilled or overheated, or bitten by small insects? Obviously there are many problems, but most of these, for our purposes, may be dismissed. We have at least in hand now a concept of soul which will not send an instant red alert to the precincts of logic.

Of some interest, however, is the relation of this double, so to speak, or *Doppelgänger*, to the cabin body, or housing body. Does it, for example, animate the physical body? Does it switch the body on, move it as one might stilts, or a puppet, guide it, as though by a steering wheel, or such? Or is the body independently alive, more in the nature of a horse, which might convey one here and there, whose reins one might control?

The following small table presupposes that the soul exists.

Body independent of soul	Body dependent on soul.	
1. Body produces soul	2. Mortal dualism	Soul dependent on body
3. Both independent	4. Soul as animator	Soul independent of body

In case 1, the soul is a function of the body, which produces it, and on which it is dependent. In a situation of this sort, the soul would be mortal, perishing with the body, on which it was dependent. Here, too, the body could preexist, producing the soul, say, only at a certain level of development. Also, here the body might remain alive after the death of the soul, as in the case of body life but brain death. In case 2, body and soul are interdependent, each requiring the other. Soul here would be a sort of "vital force," always present in life, of whatever complexity or condition, and gone in death. Both would perish together. In case 3, body and soul are independent of one another. Here the soul would, so to speak, inhabit, or take up residence within, a body. Strictly, it seems the body would not require a soul. For example, Descartes did not believe that animals had souls, but he had no doubts as to their being alive. Similarly, this is, I believe, a common view, namely, that humans have souls and, say, raccoons, giraffes, dog, cats, chimpanzees, gorillas, and such, do not have souls, or, at least, souls of the important sort, say, those of religious interest. To be sure, the matter is a bit cloudy, as when the soul leaves the body it is usually supposed, or used to be supposed, that the body dies, which suggests that the body needs the soul to be alive, which is similar to case 4. On the other hand, presumably the gorilla does not need such a soul to be alive. Similarly, one supposes a god or gods might create something human, or humanlike, leaving out the soul, as with, say, the gorillas. Too, of course, if there are no souls, then the body would be independent of souls. In case 4, we have a familiar view, namely, that the soul is the animator of the body, that the body is accordingly dependent on the soul, and that the soul is independent of the body, at least to the extent that it may survive the body.

If the soul does not exist, of course, it cannot be produced by the body, it cannot be inextricably associated with the body, it cannot be independent of the body, and it cannot animate the body.

The existence of the soul, then, would seem to be the question of most importance.

As animals, allegedly, do not have souls, or the right sort of souls, it does not seem clear that human beings, too, might not have souls, or the right sort of souls. How would one know that human beings had souls, or that animals do not have souls? What if some human beings had souls and some did not, and the same for animals? Would alien life forms have souls? Some might, and some might not? And what about conscious machines?

Could a god or gods place a human soul into, say, a raccoon, or a raccoon soul, if such things exist, into a human being?

What sort of soul might be transitioned amongst human beings, or amongst human beings and animals, or amongst animals with animals, in metempsychosis, as the form-of-the body notion, the mystery-body notion, and such, seem inappropriate. What sort of entity would be moving about amongst the bodies, and do the bodies need such an entity to exist, to be alive? Would something like a tiny, invisible, gaseous sphere, somehow endowed with consciousness, be the entity in question? Presumably not, but what would be the entity in question?

Recall that we were tentatively considering, in our attempt to develop a criterion for human speciation, given common views on such matters, views more than familiar in our culture, the possibility of taking something like 'x is human if and only if x has a soul of Type T' as being analytic, or, more briefly 'H ≡ T'. Then, as might also be recalled, it was noted that one might repudiate the putative analyticity in question if either 'H & ~ T' or '~ H & T' were conceivable, namely, if one could conceive of a human without a soul of type T or could conceive of something not human with a soul of type T.

Unfortunately for at least the proffered analyticity of the relevant biconditional formulation, it is easy to conceive of counterexamples. For example, if it should be the case that human beings lack souls, as many naturalists, for example, might suppose, then one could conceive of human beings without souls, just as one could conceive of raccoons and giraffes, chimpanzees and gorillas, and such, without souls. Similarly, it would seem possible for something not human to possess a soul of type T, perhaps an alien form of life, perhaps a mutation based on nonhuman hominid stock, perhaps a conscious machine, or such. Similarly, there would seem no logical problems involved in a god or gods placing a soul of type T in anything, a raccoon, a tree, a rock, anything.

If these considerations are the case, it would be ill-considered to take 'H ≡ T' as analytic, unless one wished to hold to one's guns forever, rather as we could take 'C is a circle if and only if C is blue' as analytic if we were sufficiently motivated, and were prepared to accept the inevitable cognitive embarrassments and dislocations.

Few people, outside of Hollywood's special-effects experts, are likely to take the "mystery body" analysis of soul very seriously, despite the fact that it is the only analysis of soul, at least to my knowledge, which even begins to make sense, as it is at least imaginable, avoids the bankruptcy of most putative solutions of the mind/body problem, and offers some hint of what might count as evidence as to its existence, and, in a complementary way, to its nonexistence, for example, sightings, chain rattlings, and such, and, negatively, the explanation of such phenomena in other ways, for example, in terms of psychological needs, optical illusions, hallucinations, faked photographs, and such. A hypothesis to which evidence is at least theoretically relevant, one way or the other, to one degree or another, is preferable empirically, if not psychologically, to one to which evidence is irrelevant.

The existence or nonexistence of a soul is surely an important question, if one wishes to use a certain sort of soul as a criterion for speciation, say, for being human. On the other hand, if we wish to be more than verbal, to do more than merely use words, disengaged from the world, one would have to have some idea, at least, of what one is talking about. One cannot just suppose one is talking about something just because one is using words. That is the primary motivation behind trying to develop a concept of soul which is more than a label without a bottle, a word with no reference, a semantic fraud, a cognitive "blank cartridge." There may or may not be a unicorn but, at least, we have some idea of what would count as a unicorn, what it would be to meet one, and so on. On the other hand, with many words, for example, 'soul', it is less clear what we would count as a soul, what it would be to ascertain its existence, and such. 'Soul' seems to occur somewhere between 'unicorn' and 'xub'. Questions of testability, evidence, and such, are clearly important. Let us suppose then that we take a reasonably clear, if that is not an objection to the notion altogether, idea, such as the "mystery body," as one way of understanding a "soul." Even so, I think it would be obvious that it would be difficult, if not impossible, despite an abundance of alleged evidence, sightings, and such, in diverse cultures over several centuries, to establish the existence of a "mystery body." And if the nature of, and the existence of, an independent, disengageable, peripatetic, theoretically ascertainable "mystery body" is problematical, so much the more problematical it would be to establish the existence of souls even more elusive and harder to understand. It is not clear that one could just take the existence of such a thing "on faith," because it is not clear *what* one is taking on faith, or even talking about. What would it be, for example, to take the existence of 'xub' on faith? One might, I suppose, take the existence of unicorns on faith, if one wished, but there, at least, one knows what one is taking on faith, what one is talking about. It is quite another thing to take the existence of a soul or a xub on faith. In short, it seems that the concept of "soul," whatever its transparency to, and convenience for, a god or gods, will not provide us with a suitable criterion for speciation, for being human.

4. The Qualification of Descent

Above, under the "Acceptability Position," we have argued for taking descent as a logically necessary condition for species membership. On the other hand, we have presupposed at least an informal understanding of "human," to begin with, "that sort of thing," and so on. Obviously one would have to have some starting place, even to begin to trace descent. If descent from a human is crucial, one must have begun with something taken as human, to begin the chain of descent. Now, assuming that descent is in place, we have noted, earlier, in dealing with possible cases of devolution, and with evolution, for example, the A, B, and C lines, that one might reach a point where one accepted descent but declined to ascribe humanness. For example, human beings may have fish, amphibians, reptiles, primitive mammals, and such, in their ancestry, but we do not think of human beings as fish, amphibians, reptiles, primitive mammals, and such. Our principle concern now, in this brief, exploratory section, is to suggest some considerations, with a sketch of some possible procedures, which might appertain to the qualification of descent. We know by now that facts are relative to a conceptual structure. A world without minds is a world without facts. The world doubtless existed long before conceptual structures but the fact that it so existed did not then exist. Planets and stars are neither true nor false; that there are planets and stars is true, or false. The relevance of a conceptual structure, and even its existence, tends to be obscured for many individuals because it organizes, shapes, and influences a vision of reality, reality being seen in its terms. To a large extent human beings see reality by means of such a structure, a structure largely the result of systemic processes, by means of which a common social reality is constructed over generations, largely, but not entirely, as a function of one or more native languages. Who, for example, seeing such and such an animal, does not just see a giraffe?

In certain situations, usually surprising, sometimes uncomfortable, or annoying, the conceptual structure emerges, so to speak, into sight. Parts of it may seem at odds with other parts; we may confront a reality for which the conceptual structure has not prepared us; it may urge us, for serious reasons, in more than one direction; it may fail us altogether. The map of cognitive reality requires adjustments, revisions, extensions, perhaps new consolidations or divisions, perhaps deletions, eliminating "countries" now seen as mythical. We are confronted with anomalies; the categories of understanding may have to be enlarged, even invented anew.

As we have seen, philosophy, regardless of her agent, must address herself to these new cartographical challenges, must address herself to the remeasuring and recasting of the cognitive grid, to the redrawing of the map of the mind, wherein it, in part, responds to and, in part, creates the world in which it lives.

The conceptual structure is silent until, by assessment, proposal, adjudication, and decision, it is taught to speak.

Who shall decide at what point humanity is exceeded, or lost?

4.1 Simple and Complex Decision Gradients

Nature commonly comes in chunks, which leads to the notions of fixed species, and such, but sometimes, as in size, temperature, and color, it comes in continuities, or continua. And, conceptually, as we now recognize, it usually comes in continua.

A simple decision gradient demarcates a gradation at one point, in connection with one property.

Let us suppose we have a succession of hominid life forms ranging from a form which is clearly not human to a form which is clearly human. One supposes that anthropologists might upon occasion disagree as to where, on this gradient, a form becomes human. This is not a question of fact until the criterion for human is decided, or accepted, after which, and only then, does it become a fact that such and such a life form is human. This in no way suggests that the decision is arbitrary. Presumably there would always be better and worse reasons for dividing the gradient at one point or another. Similarly, one might suppose a gradient from the clearly human to the clearly nonhuman, with a view to the future.

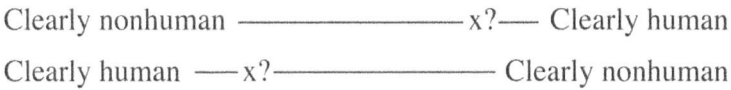

Clearly nonhuman ————————— x?—— Clearly human

Clearly human —— x?———————— Clearly nonhuman

Or, somewhat more complexly:

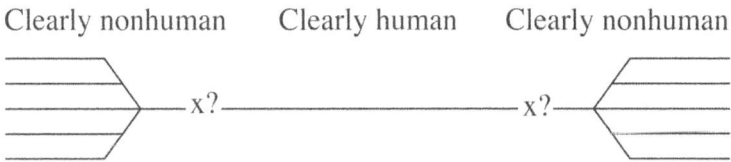

Clearly nonhuman   Clearly human   Clearly nonhuman

———————
——————— ⟩— x?———————————— x?—⟨
———————
———————

A complex decision gradient assesses, and sums, so to speak, a variety of properties. Indeed, presumably a "simple decision gradient" is founded on a complex decision gradient.

One might consider a two-property decision gradient, in which, as earlier, one might consider balancing off the relative significance of physical and mental resemblance in candidates for human speciation, as, in this case, entity 1, who bears more physical resemblance, and less mental resemblance, to a paradigm human, and entity 2, which bears more mental resemblance, but less physical resemblance, to a paradigm human.

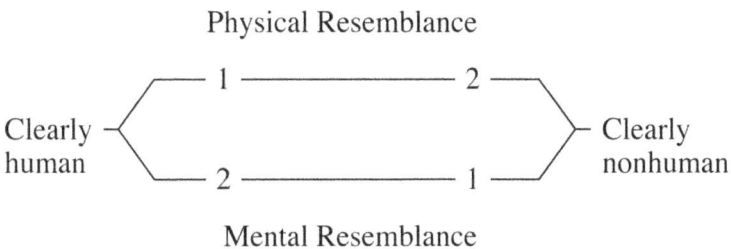

Physical Resemblance

Clearly
human
⟨ — 1 ————————— 2 —
— 2 ————————— 1 —
⟩ — Clearly
nonhuman

Mental Resemblance

It is easy to conceive, as well, of multiple-parameter decision gradients, as in the following schematic example. In it, we suppose ourselves to be dealing with a particular individual, or a particular form of life, with relation to the "normal human," viewed from the point of view of a variety of properties, perhaps such things as configuration, appearance, attractiveness, size, gross internal anatomy, physical powers, mental powers, brain types, psychological types, moral dispositions, health, longevity, and so on. In the schematism we hypothesize some eight properties. The "normal range" for the human will be taken to be between + 5 and − 5, and, accordingly, readings above and below that range differ from the norm, and will be regarded as "other than normal." The extent or nature of a property above + 5 will be regarded as "above normal," and a property below − 5 will be regarded as "below normal," and, in most cases, "above normal" will carry the connotation of superior, more impressive, or more desirable, according to modern lights, and "below normal" will carry the connotation of inferior, less impressive, or less desirable, according to modern lights. To be sure, what is most important here is not superior and inferior, but a degree of difference in one direction or another. For example, a gross interior anatomy ranking above + 5 would presumably be more complex or efficient than one ranking below − 5, and whereas complexity, *per se*, would seem value-neutral, efficiency would seem to be more desirable than inefficiency, abstractly considered. Each of these properties, one notes, might be subdivided indefinitely. For example, under "gross internal anatomy," one might be concerned with digestive and excretory systems, arteries and veins, the circulatory system, the blood, heart, eyes, muscles, connective and supportive tissues, joints and ligaments, the lymph and lymphatic system, body fluids, nerves and the nervous system, the olfactory system, the reproductive system, the respiratory system, the skull, and skeleton, the spinal cord and vertebrae, and so on. The problem, of course, addressed by such a gradient is to decide whether or not the entity, or the form of life, in question, is well adjudged, or well considered, to be human. One notes that the individual, or form of life, falls within the "normal range" for a human with respect to only two of the eight relevant properties. On the other hand, we surely do not wish to confine "human" to normal humans, but would wish, presumably, to allow for above- and below-normal human beings, as we commonly do, for example, individuals who are taller or shorter than the normal human being. The question then is the degree of difference and the accepted relevance of the degree of difference to the classification or conceptualization. A schematism like this, abstract as it is, makes extremely clear the dependence of truth and fact on conceptual structure. Two properties, 3 and 6, are within the normal range, and two others, 4 and 5, are close to the normal range; one property, 7, is well below the normal range; and three properties, 1, 2, and 8, are more markedly above the normal range. So, is the entity, or the form of life, human or not? Such a question is unintelligible until the conceptual structure is in place, and the relevant measurements in hand. Once we are clear on the criteria, and the situation of the entity, or form of life, *vis-à-vis* the criteria, the question is an empirical question, with a clear answer, true, false, or, given the possibility of a tie, so to speak, an equality of points, or such, undecidable. The latter possibility, would require a resolution, for example, an equality of points is to be resolved in a predetermined manner, rather as "a tie goes to the runner," or further attention to the conceptual structure, with the end in view of making such a tie less likely in the future. Practicality may be expected of a conceptual structure; perfection is not required.

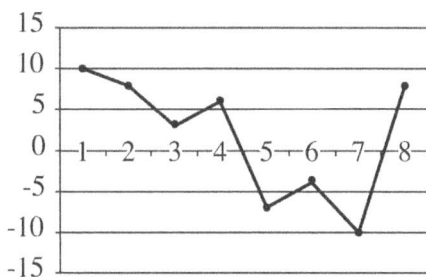

The number of properties regarded as necessary for humanity could also be conceptualized variously. In the following schematism, we are concerned with five individuals, A, B, C, D, and E, and eight properties, properties one through eight.

	A	B	C	D	E
P_1	•				•
P_2	•	•		•	•
P_3		•			•
P_4			•		•
P_5		•			
P_6			•		
P_7			•		•
P_8					•

Which entities are human?

On this approach, the answer would depend on the number of properties required to be human.

For example:

Properties required to be human	Humans
8	None
7	None
6	E
5	E
4	E, C
3	B, C, D, E
2	A, B, C, D, E
1	A, B, C, D, E

4.2 Differentiations Amongst Properties

In the preceding schematism we have, in effect, done no more than count properties, determining human or nonhuman as a function of the number of relevant properties characterizing the entity, for example, if five properties were required, only Individual E would be human, whereas if only three properties were required, one would have four humans, B, C, D, and E, and so on.

A more realistic approach to the matter would be to recognize that certain properties might be regarded as more decisive than others. For example, we have already noted that certain properties might be given more weight, or significance, than others, for example, that one might take physical resemblance as more important than mental resemblance, or the opposite.

But, one might wonder, how much more important?

Clearly one might approach this sort of thing from several directions, for example, whether or not a property is required, the degree to which a property is possessed, and the value of the property in itself, so to speak. The following small schematism, simplified to four variables, suffices to make several points, and suggest a variety of possibilities.

	Individual I_1	Individual I_2	Individual I_3	Individual I_4
Property P_1 (+)	0	2	3	6
Property P_2 (++)	6	1	4	2
Property P_3 (+++)	10	1	3	4
Property P_4 (++++)	2	10	5	4
Totals	18	14	15	16

Let us suppose that each of the four properties is regarded as a necessary condition for humanity. If that is the case, then Individual I_1 is automatically eliminated. If, say, only three of the properties is regarded as a necessary condition for humanity, then Individual I_1 is still under consideration. In fact. he has the highest number of "points," the points here construed as the degree to which the property is possessed. The "+" marks under the property designations indicate the relative importance of the property in question, according to the lights of the time in question, for example, that Property P_1 is the least important, and Property P_4 is the most important. Let us now suppose that a necessary condition for being human, according to modern lights, is to possess Property P_4 to at least the degree 5, or to the extent of five points. If that is the case, Individual I_1 and Individual I_4 are eliminated, as only Individual I_2 and Individual I_3 meet that criterion. Let us also suppose that a minimum of fifteen points is required for being human. In that case, Individual I_2, having only fourteen points, is eliminated and our only human, on this calculation, is Individual I_3, despite the fact that Individual I_2 possessed twice the points of Individual I_3 in the category of the most important property, Property P_4. Indeed, Individual I_3 qualified only minimally in that category. At this point, my own intuitions would encourage some sort of adjustment in the criteria, perhaps lowering the minimum number of points required for being human, allowing a ten in Property P_4 to count as sufficient for humanity, or such. I would like, even in terms of this abstract schematism, for Individual I_2 to come out as human. To be sure, much would depend on what we took Property P_4 to be. Is it intelligence, good health, honesty, kindliness, gullibility, responsiveness to advertising, the capacity to lift heavy weights, to defeat Grizzly bears in unarmed combat, or what? It might be the latter property, if the human race was short on weaponry and long on Grizzly bears.

Whereas this small schematism is abstract, I think it is illuminating, with respect to the sorts of subtleties, necessary vs. sufficient conditions, degree to which a property is possessed, the value of the property in itself, cut-off points, and so on, that might enter into these issues.

4.3 The Set Approach

As a passing note in the "qualification of descent," namely, how one might attempt to discriminate between human and nonhuman, even with descent in place, one might note that some properties might be taken as necessary conditions, whereas other properties might be members of sets, one or more of which might count toward humanness. For example, let us suppose we have three sets of properties, Set 1, Set 2, and Set 3, which sets are composed as follows:

Set 1: Property P_1 and Property P_2
Set 2: Property P_3, Property P_4, and Property P_5
Set 3: Property P_6 and Property P_7.

One might then define the predicate 'H' along the following lines:

Hx: x is human

and then define 'x is human' along the following lines:

x is human = df. x possesses at least one property from Set 1, at least two properties from Set 2, and all properties in Set 3.

On this approach, there would be twelve ways to be human, or "setwise," there would be one way to be human, namely, satisfying the "three-set requirement."

Analogously, if the only necessary condition for being human was to possess, say, four out of a relevant seven properties, there would be thirty-five ways to be human, and, if the only necessary condition for being human was to possess, say, five out of a relevant eight properties, there would be fifty-six ways to be human.

In Part One of this study, "Anticipations," we were concerned to give the reader some sense of the awesome challenges looming before human cultures, challenges for which these cultures are muchly unprepared; in Part Two, "Conceptualizations," we were concerned to suggest a number of philosophical tools, methodologies, techniques, and analyses, pertinent to understanding and relating to these challenges. Part Three is entitled, "Prescriptions." It endeavors to make out a case for philosophical prescriptivism, in which is presented a vision of possible philosophical adjudication, possible but not likely, one supposes, in a world of envy and violence, of hatred and fanaticism, of guns and bayonets.

PART THREE

PRESCRIPTIONS

As alluded to earlier, one might distinguish between philosophical descriptivism and philosophical prescriptivism. In philosophical descriptivism the solutions to philosophical problems pre-exist and await their discovery, which may never take place. Our views of such solutions may differ from time to time, and place to place, but the solutions themselves, are inalterable. This is to see philosophy as a sort of science, with a different subject matter, a subject matter "out there," which one can be right or wrong about. The ether and phlogiston, for example, either exist or don't exist, and the scientist does his best to find out which is the case. There may be problems, of course, both for the scientist and the philosopher, in clarifying the questions, in determining what would be relevant to their solutions, and so on, but, once the question is clear enough, and the evidence relations are in place, then the solution is again " out there," waiting to be discovered. This view of philosophy is an ancient and venerable one, and was surely inherited by science, which grew out of philosophy, and accepted many of her values, for example, independence, curiosity, diligence, carefulness, open, publicly accessible methods, the sharing of results, the examination by peers, the avoidance of intimidation and coercion, recourse to reason, reliance on logic, and the seeking of objectivity, and, of course, at least implicitly, the value of nonideological truth seeking, and the worthiness and importance of the endeavor itself. Indeed, it is tempting to think of scientists as now being, as they once were, our "natural philosophers," or "philosophers of nature." On the other hand, there seems to be at least a difference in degree of "out-thereness," or "not-out-thereness," between the ether and, say, causality, meaning, and time, and between phlogiston and, say, person, mind, and self. If nothing else, different subtleties are involved. The philosophical prescriptivist, on the other hand, unlike the philosophical descriptivist, does not presuppose, at least in the same sense, that the solutions to philosophical problems pre-exist, waiting to be discovered. His approach to these matters is more analogous to engineering, as in "What sort of road or bridge shall we build to get from A to B?" He does not even suppose the road or bridge is already built in Plato's metaphysical workshop or garage, waiting to be copied, however imperfectly. The prescriptivist's approach here is not so much what is the solution, as what shall we have for the solution, what sort of solution shall we build, what sort of solution shall we invent? This orientation does put him at a certain psychological disadvantage, because it acknowledges that his solution is only one of several possible solutions, and that it may be challenged, repudiated, revised, or discarded. His solution is not "true," at least in any normal sense; on the other hand, like screwdrivers, hammers, and saws, it may be very helpful; it may help to get a handle on complex data; it may be illuminating; it may help us to understand the world.

1. Background Considerations

1.1 Reality and Truth

Whereas reality and truth may be rhetorically conflated, it seems reasonably obvious that they are very different sorts of things. There is a sense, one supposes, that if the word 'cat' did not exist, there would be no cats, but this is not good news for mice. It might, with similar flair, be claimed that if the word 'thing' did not exist, there would be no "things," and thus there would be nothing, except, if the word 'nothing' did not exist, then there would not be nothing either, and so on. Similarly, it is hard to believe that a number of potential life forms, worlds, universes and such, are waiting in line, to be named, in order to exist. Indeed, many things must lack names. Does that mean they do not exist? Does one not discover the new entity first, and then name it second, and so on. Given the fusion of the conceptual structure with the data it interprets, many people to whom it would seem absurd to explicitly identify reality and truth behave as though they do. Do we not simply look and see that that animal is a giraffe? Many who would instantly recognize and accept the word/world distinction, and explicitly reject a confusion of words with the world, nonetheless, in their practice, ignore or blur this distinction, and think not things, or experiences, but words. They see the world through the lens of language. They are "associational thinkers." They are "verbal thinkers." "Generality thinking," valuable and necessary though it is, can obscure difference, and lead one to overlook or neglect richness and specificity. We approach and understand the world only in terms of generalities, but the world itself is not a generality.

In any event, reality and truth, as we usually think of these things, at least upon reflection, are to be distinguished. For example, reality was around long before truth. Too, rocks, trees, and giraffes are neither true nor false. That something is a rock, a tree, a giraffe might be true. If there were no minds there would be no truths. Truth may require reality, but reality can get on quite well without truth.

Reality and truth are different.

1.2 The Human World and the "Physicist's World."

The undeniable reality is the human world, the world of human experience, and the world of human experience is a function, as we have seen, of data and the interpretation of data. This distinction must be presupposed as, otherwise, conceptualization would have nothing to conceptualize, and content without conceptualization would be unintelligible. This is reminiscent of Kant's observation that concepts without percepts would be empty, and percepts without concepts would be blind. On the other hand, data and its interpretation, or conceptualization, are commonly fused, not inextricably as boundary and area, but intimately, as in, say, data and "giraffe." In practical life, at least after infancy, data and interpretation come together, even if the "data" generates no interpretation other than, say, "there," "that," or "thing."

Whereas the undeniable reality is the human world, the world of human experience, the etiology of that world might be various. Taking the world of human experience as an "effect," of sorts, it is not obvious what might be its cause, if it has a cause. Some candidates for a relevant causality would be matter, of one sort or another, a Platonic Demiurge shaping a matrix, the divinity of Stoic Fire, a Spinozistic substance to be identified with neither mind nor matter, a Kantian Thing-in-Itself, neither in time nor space, a Berkeleyan divine entity, a Hegelian *Geist*, and so on. A given "effect," in theory, might have an infinite number of possible causes, ranging from the anomalous and inexplicable to the will and intent of a god or gods. Currently popular, though this changes from time to time, much water having been washed under the bridge of being since Democritus' atoms and the void, is the "physicist's world," or, perhaps better, considering relativity theory, quantum mechanics, multidimensionality, string theory, and such, the "physicist's worlds." All one needs at the moment is to understand that the physicist's world is an extrapolation beyond the human world, that it is a hypothesis, in terms of which to account for experience. The data is the human world; the bold hypothesis, or conjecture, to explain the data, the human world, is one or more "physicist's worlds." The great strength of at least some of the incompatible physicist's worlds is that at least some the pertinent theories, coupled with initial-condition statements, supplementary hypotheses, and such, make possible the generation of prediction statements, at least some of which lend themselves to confirmation or disconfirmation. A confirmed prediction statement does not prove a theory, as that claim would commit the fallacy of "affirming the consequent," but it surely does the theory good, as it is what one would expect if the theory is true. It would supply confirmation, if not proof. Similarly, given the complexity of possibilities involved, no conclusive disproof of a theory is possible either. On the other hand, a muchly confirmed theory is surely going to be taken seriously, and a theory which generated, or seemed to generate, subsequently falsified prediction statements is likely to be revised or discarded. One can never know that a theory is true, but one can hope that it will prove empirically adequate, namely, that it "saves the appearances," accounts for the data, and so on. Indeed, it might even be true. One can surely hope so. To be sure, if the entire scientific world picture is mistaken, then all the theories constructed on it, however interesting or useful, will be false. Most relevant to our current project is to contrast the "human world" with the "physicist's world," and, to do so, we will make use of a "physicist's world," which may serve as a representative of such worlds. In the world in question, fermions comprise matter and bosons convey force, whatever force is. Fermions are nucleons and leptons; bosons are gravitons, if they exist, photons, weak bosons, and gluons. The four forms of conveyed force are gravitation, electromagnetism, and the weak and strong nuclear forces. More generally, such worlds range from one of atoms and the void, to ones of particles, subatomic particles, fields, rays, waves, forces, and such.

In any event, the physicist's world, though it might be our base world or Ur world, if it exists, is quite different from the human world *as we know it*. For example, in it there is no sound or color, as we think of sound and color. In it there are no odors, tastes, and "feels," as we tend to think of those things. In that world, *qua* that world, there is no consciousness, subjectivity, intentionality, or understanding. In such a world the particles constituting the brain may exist, but, as we think of these things, in terms of the human world, we do not find calculating, anticipating, remembering, imagining, amongst those particles. Similarly in such a world, on its terms, say, fermions and bosons, we look in vain for persons, meaning, hope, thought, beauty, despair, dreams, beliefs, and such. It is not just that the house is built of bricks, but one does not think of the house in terms of bricks. It is more like the fact that the human world may be founded on particles and waves but one does not think of it, and cannot think of it, in terms of particles and waves. Experientially, it is another world altogether.

In what follows, we will be primarily concerned with the "human world." Philosophy thrives best and most legitimately in that world.

Too, in that world, the challenges of the future are at their most poignant.

1.3 Animal Truth and Statemental Truth

Whereas we speak generally of the "human world," and this is defensible, as we are human, and our interests are largely human interests, certainly in this study, it is presumably clear that the basic distinction is between the worlds of experience and the "physicist's world," assuming that such a world exists. In short, one supposes that consciousness, experience, and such, are not uniquely human. With all due respect to Descartes, who regarded animals as automata, analogous to windmills, water pumps, and clocks, presumably for religious reasons, it seems obvious that animals are conscious, have experiences, and such. It is hard to believe, for example, that anyone who has shared his life with dogs and cats would be in doubt about this. To be sure, how far down the phylogenetic scale awareness and such exist is not clear. One supposes that the amoeba does not have much of an inner life, but it seems possible, given reactions, irritability, and such, that it has some degree of awareness, even self-awareness. For example, it refuses to ingest its own pseudopodia, and so on. If in doubt, in general, about animal consciousness, one could always inquire of various chimpanzees and gorillas who communicate, sometimes annoyingly, in Amslan (ASL, or American Sign Language).

If animals are conscious, have experiences, and such, it seems likely that they have knowledge, that they know this and that, for example, where their nest or burrow is, that such and such an item is edible, and so on, and thus, implicitly, if not explicitly, that they have beliefs. For example, one supposes they can recognize others of their species, can recognize their mates, can recognize prey, and, perhaps after an experience or two, can recognize predators, and such. Too, let us suppose that a chimpanzee anticipates the presence of termites in a hollow log, smears a leafy twig with saliva, and then, swabbing this tool about inside the log, snags several termites, which it then eats. He speculated as to the existence of termites in the log, was hungry, and then put his speculation, anticipation, conjecture, or hope, to the test. It seems then that we might well suppose that his conjecture, nonverbal, of course, was true. In any event, it would seem absurd to claim that animals are incapable of knowing anything, or believing anything, when, obviously, they know a great many things, and have countless beliefs, true or false. Similarly, one supposes that most of the beliefs, true or false, of human beings are not propositionalized, not put in language, though one supposes that the human being, unlike most other animals, could propositionalize the beliefs, if interested.

Philosophers, on the whole, though surely not universally, examples ranging from Aristotle to Alfred Tarski, have tended to discuss truth primarily in terms of language, sentences, statements, propositions, and such.

No particular objection needs to be raised to this habit, provided it is kept in mind that without animal truth, so to speak, the entire project would be ungrounded and unintelligible. Take a simple example:

'S' is true if and only if 'S' asserts the case.

'S', perhaps a statement, cannot assert anything. A person might use 'S' to assert something. 'S', or its token, a noise, a puddle of ink, gouges in a rock, chalk marks on a sidewalk, or whatever, just sits there. 'S' does not even have a meaning unless a mind gives it a meaning. And what is this mysterious "case" that 'S' is supposed to be asserting? One would have to have a mind to carve some sort of "case" out of the world. "Cases" are not out there, like rocks, trees, and giraffes.

These remarks in place, and the necessity of the experiential matrix understood, let us consider some possible varieties of "statemental truth."

This is important because the conceptual structure can most fruitfully and adroitly be addressed in terms of language. Whereas it is clear that many language-lacking animals have their conceptual structures, for example, their recognition of sorts of things, and what is to be expected of sorts of things, it is also clear, or reasonably so, that such language-free conceptual structures, unless one of a god or gods, are going to be simple, narrow, rudimentary, and primitive. Certainly they would seem to be conspicuously limited. They would lack the articulation and sophistication practicable in a complex language,

1.4 Possible Varieties of Statemental Truth

An analytic statement is one which is true in virtue of its meaning alone. In a sense this is false, for reasons given above, having to do with the relevance of intention, minds, and such, but, given such reservations, it seems acceptable to speak in this fashion, and it is convenient to do so. A typical example of an analytic statement would be 'All swans are birds.' Being a bird is a logically necessary condition for being a swan. There would be no point in hanging about ponds to see if a swan shows up which is not a bird. If something is not a bird, we cross it off the swan list.

A synthetic statement is a statement whose truth value does not depend on its meaning alone. Its truth value depends on its meaning and something else, for example, on the "case," the "facts," on a state of affairs, on the way the world is, or such. A typical example of a synthetic statement would be 'All swans are either black or white'. Induction would tend to confirm this; a single blue swan would disconfirm it.[42]

The analytic/synthetic distinction is controversial in philosophy, for a number of reasons. Philosophical behaviorists, in particular, have difficulties with the distinction, at least officially, if not unofficially, where they carry on much like the rest of us, because meanings, unlike unicellular organisms, wombats, and mountains, do not seem to be "out there." They do not lend themselves to being weighed and measured, clocked, photographed, put on microscope slides, and so on. What is their speed or cubic content? That would

---

42. Some statements, of course, are false in virtue of their meaning alone. Sometimes these are spoken of as a variety of analytic statements, understanding analytic statements then as statements whose truth or falsity depends on their meaning alone, but this tends to be misleading. Such statements might better be spoken of as analytically false statements, logically false statements, inconsistent statements, contradictions, or such.

Analytic statements are sometimes also spoken of as analytically true statements, *a priori* statements, tautologies, logical truths, or such. Synthetic statements are sometimes also spoken of as empirical statements, contingent statements, *a posteriori* statements, or such.

Analytic and synthetic usually have primary reference to meaning; *a priori* and *a posteriori* usually have primary reference to experience. For example, the truth of an *a priori* proposition is supposedly knowable prior to experience, at least of a certain sort, for example, you do not have to go to the pond to confirm 'All swans are birds'; and the truth value of an *a posteriori* statement cannot be known without some experience, at least of a certain sort. For example, you cannot know, prior to experience, that the next swan to land in the pond will be black or white. For all one knows, it might be blue.

It seems clear that all analytic statements are *a priori* statements and that all *a posteriori* statements are synthetic statements. Immanuel Kant thought that some synthetic statements could be *a priori* statements, for example, '7 + 5 = 12', but this seems to have been a function of a notion of either logical containment of the concept of the predicate in the concept of the subject, or psychological containment, namely, that in thinking the concept of the subject one thought the concept of the predicate. Here, of course '7 + 5 = 12' is not a subject/predicate statement, so the "containment" approach, either logical or psychological, seems inapplicable, this perhaps leading Kant to see the formula as synthetic, while, on the other hand, it is obviously *a priori*. For example, it is clearly necessarily true, it is nonrefutable, and its negation is inconsistent, so it is analytic, and all analytic propositions are *a priori*, etc. There are a number of interesting issues here but it they are outside the scope of this study, even in a footnote. One might remark, in passing, that the concept of the synthetic *a priori* statement, even apart from Kant's particular problems, has not fared well philosophically. One might always, of course, rig definitions in some eccentric fashion, and then come up with some statements which, under that particular approach, would count as synthetic *a priori* statements. For example, 'q' is not obviously contained in the concept 'p', but 'p' does logically imply 'p or q', and so on. Thus, 'If Henry is a Boy Scout, then Henry is either a Boy Scout or a hippopotamus' would count as synthetic *a priori*. The synthetic *a priori* statement seems to be philosophy's answer to Lewis Carroll's snark.

be hard to say. In any event, they are not obviously quantifiable and if one insists that only the quantifiable exists, or is "meaningful," or such, then meanings are badly off. It seems, of course, that they do exist and are verifiable. For example, do you really not understand the meaning of 'triangle'? To be sure, meaning is on the inside, and the inside may be off limits to the philosophical behaviorist. In any event, if meanings are unintelligible, inaccessible, do not exist, or such, then, to be sure, there is not much point in, say, defining an analytic statement as one which is true in virtue of its meaning alone, and so on. To be sure, not all difficulties with the analytic/synthetic distinction require a particular methodological stance and a willingness to ignore enormous and obvious dimensions of familiar human experience. For example, some utterances are unclear, and then, naturally enough, it is hard to know if one should take them to be analytic or synthetic. For example, is it analytic that spiders have eight legs? Could a seven-legged spider show up? Similarly, the meanings of expressions might be place or time relative, and vary from individual to individual even in the same vicinity at the same time. Individuals may assign different meanings to the same expression, and the same meaning to different expressions. This presupposes, of course, that meanings exist. For example, in the seventh century the whale might count as a fish, and in the twenty-first century it might not. Mercury might count as a liquid for a fourteenth century alchemist and not for a twenty-first century physicist. Similarly, for one fellow 'All men are mortal' might be taken to be analytic, and another might see it as synthetic, and so on. That analyticities and syntheticities might occasionally differ from place to place, from time to time, from idiolect to idiolect, and so on, does not jeopardize the distinction. Such things are to be expected given the nature of semantic evolution and diversity in a living language. In any event, it is a supposition of this study, as it is generally in philosophy, as a whole, that the analytic/synthetic distinction is not only viable, with some qualifications, but is valuable. Indeed, without it, it seems it would be very difficult to account for understanding, thought, memory,

---

43. An examination of the controversy pertaining to the existence of meanings, the analytic/synthetic distinction, and such, is far beyond the scope of this study. Typical arguments in favor of the distinction would come from the nature of support strategies, differing, for example, in cases of challenges to, say, 'All swans are birds' and challenges to 'All swans are white or black'; the nature of logic and semantics, for example, it seems that legitimate substitution instances of '(p & q) É q' would be logically true; that syntheticities and analyticities would be consequences of any

conceptual structure capable of identifying and distinguishing contents, relating contents, and affirming and denying contents, for example, if we can identify "zirk" and "zug," and relate contents, as in "zirk" and "zug" are same as "zoog," and we can deny contents, as in, say, "not zug," then it would be logically true, or analytic, that "not zug" implied "not zoog," and so on; that summations over idiolects produce general uniformities of response, for example, most English speakers would take 'All swans are birds' as analytic, and 'All swans are either black or white' as synthetic'; that the existence of plausible translation suggests at least similarities of meanings, for example, that 'non', 'nein', 'nyet', and 'no' are intertranslatable; that communication presupposes meanings, for example, that people can understand one another's meaning, for example, as in "Open the door" and "Shut the window"; that sameness of meaning is intelligible, for example, that 'house' and 'house' have the same meaning, that an artificial language might be invented, in which it could be stipulated that, say, 'glunk' would have the same meaning as 'house', and so on; the question "What do you mean by that?" seems to be intelligible; that even the philosophical behaviorist seems willing to tell us what he means, and so on. One might also note that meanings, and the acceptance of the analytic/synthetic distinction is pervasive among the vast majority of speakers of native languages, which suggests a ground in linguistic fact. A colossal misconception here would seem unlikely. An appeal to intuition might also be pertinent. For example, do you know the meaning of 'triangle'? If so, it seems at least one meaning exists, and so might not others, as well?

Two issues should be clearly distinguished, what meanings are, and how it is convenient to speak of meanings. The first issue is psychological and the second is pragmatic. Meaning has psychological reality; for example, you understand the meaning of 'triangle'. This is as verifiable as having a thought, remembering something, anticipating something, and so on. That is where meanings are, and are undeniable. That is as real as dandelions in a lawn. They are verifiably there. An independent question, given the existence of, but privacy of, meanings, is how to get from the inside, so to speak, to the outside. Language solves this problem by speaking of, dealing with, meanings as though they were external, on the outside. An analogy would be numbers. Were there no minds, there would be no numbers, no more than chess moves in the Early Precambrian Era, some four and a half billion years ago. Whereas numbers are clearly mind-dependent, for example, if no minds, no numbers, we certainly do not think of numbers, or speak of numbers, as if they were mind-dependent. We speak of them as though they were "out there," like rocks and trees, independent of thought. This does not mean they are "out there," but it does recognize that it is convenient to think of them, and speak of them, as though they were "out there," perhaps in some eternal, imperishable, objective, awesome, changeless Platonic realm. Abstract entities may not exist, but it is extremely convenient to use words in certain ways. Pragmatic considerations recommend, even require, that we talk about numbers as though they were an unusual sort of rocks or trees. "Do abstract entities exist?" would seem to be an unintelligible question, as it is hard to see what would count as evidence for or against their existence. It is not even clear what would be involved in hypothesizing their existence, in any practical sense. What would we be hypothesizing? On the other hand, we semantically project an internal reality beyond the mind into an external, public world. Unicorns may not exist, but it is easy to talk about them. If unicorns were as useful as numbers, as precisely related as numbers, as intriguing, exciting, and beautiful as numbers, as important to science as numbers, and so on, presumably people would suppose their objective existence. And, in the case of unicorns, we would at least know what we were talking about. We would, for example, know if we met one. But what would it be, say, to meet the number two?

communication, translation, logic, inference, even the nature of language as a whole.[43]

The two customary varieties of statemental truth, as suggested, are analytic and synthetic truths, the first true in virtue of their meaning alone, and the second true or false in virtue of their meaning plus something else, say, facts, states of affairs, the way the world is, or such.[44]

On the other hand, this dichotomy seems far removed from the linguistic habituations, the implicit truth-views, embedded not only in contemporary cultural and societal practice, but, as seems likely, in those of all civilizations on record. Compared to how humanity has customarily thought of "truth," spoken of "truth," and such, our small, neat dichotomy seems, at best, woefully incomplete. Philosophy may be right, and the world wrong, but the matter is worth looking into, however briefly.

---

44. There are several theories of "truth." Most of these theories fall into one or another of three major families of theories, these being the correspondence, coherence, or pragmatic "families." Our characterization of synthetic truths presupposes a correspondence theory of truth. As suggested, there are many versions of each of these theories of truth. Each of these versions has, or has had, intelligent, informed advocates. Clearly a discussion of truth theory is beyond the scope of this study. Briefly, coherence theories have been popular with Idealists who might maintain that no mind-independent external world exists, or, at least, can be known to exist, so correspondence between, say, an idea and a nonexistent, or unknowable, reality is unintelligible or absurd, with the result that one has to make do with truth as a "fitting together," so to speak, of what is available, say, ideas. Pragmatic theories, at least historically, tended to share the reservations or doubts of idealists about the ascertainability of provable correspondences, but tended to supplement coherence with an emphasis on the utility or value of, say, ideas, beliefs, or theories, for example, in producing an "as-if" illuminating vision of the world, in anticipating future experiences, and such. Coherence theory, without the metaphysical hypothesis of idealism, tends to collapse. For example, coherence implies at least consistency, and two sets of propositions are consistent if and only if both sets can be true, and this seems to presuppose an independent sense of 'true', presumably one of a correspondence nature. Similarly, it seems logically possible, though controversial, that there might be more than one enormous set of coherent propositions and that the sets might be incompatible, incompatibility suggesting that both cannot be true, and this suggesting, again, an implicit recognition of an independent criterion for truth, say, correspondence. The usual view today is that coherence is a necessary condition for the truth of a set of propositions, but not a sufficient condition for its truth. Also, *sans* the idealist hypothesis, the coherence theory seems to lack an anchorage in reality. Its relationship to reality seems obscure, at best, and, as soon as one begins to concern oneself with such a relationship, one is moving toward correspondence. Pragmatic theories are much more plausible. On the other hand, given the usual way one thinks about truth, it seems that one might have some useful beliefs which are not true, and some true beliefs which are not useful. It seems that utility and truth, at least as one usually thinks about such things, are logically independent properties.

We will be particularly concerned with the possibility of recognizing what might be referred to as moral, aesthetic, and philosophical truths, truths which do not fit easily, at any rate, into the slots available in our dichotomy, analytic and synthetic, but, first, let us note a large number of sorts· of truths which might, generically, be referred to as rule-bound truths, which, also, do not fit, at least easily, into our dichotomy of analytic and synthetic. One sort of rule-bound truths would pertain to rules of the sort which might be found in games, for example, that the rook in chess moves orthogonally, that the batter in baseball who receives four balls is entitled to first base, and so on. Truths of this sort are dependent on the way a game is set up. If they are understood as "analytic," it does not seem that they are analytic in same sense as 'All swans are birds' is analytic. Similarly, their negations are not inconsistent, as one would need, if they were truly analytic. For example, 'it is not the case that the rook moves orthogonally' is false, but it does not seem to be inconsistent. The game would just be a bit different. And so, too, with baseball. Indeed, at one time, a batter was not entitled to first base after receiving only four balls. One might also consider matters such as those involved in ownership, contract, and law, for example, it is true that Jones owns that bicycle, that it is incumbent on the party of the second part to do such and such if the party of the first part has done such and such first, that one should drive on the right side of the street in Minneapolis, and so on. One could also consider such things as obligations attendant on promises, such as it is true that Jones should mail the letter, or truths connected with customs, such as it is true that the villagers have a right to graze their cattle on the commons, and such. There are many sorts of truths which would fit under the broad rubric of rule-bound truths, truths which do not appear to be either analytic, in any usual sense, or synthetic, in any usual sense.

One might also note that most individuals assign truth values to at least some aesthetic and moral judgments. Very few people, for example, regard all such judgments as mere expressions of emotions or approval, as concealed imperatives, as simple commendations, as only prescriptions, recommendations, or such. There seem to be two ways to understand these habits, one, that humanity is simply mistaken, and has been, almost universally, since the trees and caves, which is possible, or that "truth" may be larger and more annoying, or more subtle or complex, than has been dreamt of in at least some philosophies.

'Truth', of course, may be understood narrowly or broadly, and, surely, in a variety of ways. What we are concerned with here, at this point, is trying to understand, without presupposing a colossal misconception is taking place, or that humanity is afflicted with some sort of genetic racial error, why individuals would use 'true' and 'false' of aesthetic and moral judgments, as, clearly, most such judgments would not appear to be either analytic or synthetic, at least in any ordinary sense. For example, it is surely not analytic that Picture P is beautiful, and, as people may disagree as to its beauty, it does not appear to be synthetic, either, at least in any ordinary sense. For example, one may adjudicate questions as to the size of the picture with a tape measure or yard stick, whereas one might fail to reach an agreement on its artistic merit. At this point, one might easily claim that truth is not involved. On the other hand, it would not be unusual for one individual to feel he is right, and the other wrong, and so on. So the interesting question here is not which individual is right, or that there is no truth here, so they are both wrong, but what do they mean by their judgments, what do they have in mind when, in such situations, they use predicates such as 'true' and 'false'? Why do they believe such predicates are justified, or, more generally, why does most of the human race believe such predicates are justified, when they are well aware that agreement is not universal, and instrumentation does not exist, say, a yardstick, in virtue of which the agreement or disagreement might be resolved, one way or the other?

First, just to acknowledge the obvious, and move it to one side, so that one can look into deeper issues, it is clear that disagreement may exist in the case of aesthetic and moral judgments, apparently in a way that it does not exist with respect to human evolution, the age of the universe, the existence of extraterrestrial life, and such. In the latter sort of cases, an objective externality is presupposed. In the former sort of cases, the situation is less clear, at best. At least since ancient skepticism, it has been recognized, and explicitly acknowledged, that human opinion, with respect to a great many issues, issues often taken with great seriousness, issues with profound effects on human life, may differ from time to time, and place to place. One is familiar with concepts of time-relativity, place-relativity, individual-relativity, cultural-relativity, mood-relativity, health-relativity, age-relativity, and such. The usual contrast is between objective and subjective, or, say, between nonaxiological and axiological judgment, between, say, empirical judgments and value judgments, aesthetic and moral judgments often relegated to the latter category, commonly a prelude to their dismissal. Pretty clearly there is something surprising at the epistemological marginalization, if not rejection, of categories of judgments of enormous interest and importance, those likely to have the most serious and significant consequences for human life, particularly moral judgments. Rather than peremptorily denying truth and falsity to such judgments, on the basis of some semantic theory, in the face of almost universal human belief and practice, it behooves us to try to come to grips with those beliefs and practices.

The first thing to do, before moving on to a more important issue, is to recognize that most moral judgments, for example, do have truth values, in one or two ways, both of which are conditional. First, long-term systemic cultural processes, over generations, produce moralities, or moral codes. This is not surprising, as such codes are necessary conditions for the viability of communities. It is also likely that biology has selected, in the human case, for humans are gregarious, for dispositions to empathy , conformity, and such, which dispositions would seem to facilitate, at least, the acceptance of, and the adherence to, moral codes. Thus, a given act might be morally obligatory in, say, a given community. Then it is true that it is morally obligatory in that community. That would simply be a matter of fact. This does not entail, of course, that it is morally obligatory in other communities, or in the same community at a different time. To require, say, that an act, in order to be truly morally obligatory, must be morally obligatory in a vacuum, or under all circumstances in all possible communities and at all possible times, under all possible conditions of life and under all possible belief systems, would seem to require a great deal of a moral act. This would seem, for most practical purposes, to deny the existence of moral acts. But perhaps that is the agenda involved, the point of the theory. To be sure, a given culture's views of morality may well be challenged, both from within and without the culture. In this way moralities may change, hopefully for the better. The second, and more interesting way, that morality may obtain a conditional cognitivity is in virtue of its instrumental relationship to a goal. Moralities here are viewed along the lines of hypothetical imperatives. For example, if you wish to travel between New York City and Los Angeles in the quickest way possible, then it is true you should take an airplane. In the moral realm, one would have imperatives along the lines of, if you want to have a communal life of such and such a nature and quality, then you should act in such and such a manner. In such cases it would be true or false that a given moral code, a given set of moral rules, would or would not conduce to a community of such and such a sort. To be sure, not everyone might have the same goal, the same desires, and such. That is where the interesting questions arise, at least philosophically. For example, if one wants a well-ordered, harmonious, compassionate, healthy, prosperous community, then it is true that such and such ought to be done. On the other hand, if one, say, wishes a chaotic, insecure, disharmonious, vicious, diseased, impoverished, hate-filled community, one preferring such things, then one should behave in other ways. Is it true that one should prefer a better world, as normally understood, to a worse world, as normally understood? Most people would suppose it so, but it is not clear how one would go about proving this to a sane, informed

---

45. It should be noted, in passing, that whereas transcultural moral disagreement may exist, the fact, as established by cultural anthropology, is that there is overwhelming transcultural moral agreement. This is not surprising, considering the importance of a practical, instrumental moral consensus to a thriving, long-lived community. Too, almost all apparent moral disagreement is a function of differences in life conditions and belief systems. For example, it is one thing to kill an elderly parent to ensure the survival of a family group in the 19th century arctic wastes and another to kill one to obtain an inheritance in prosperous, first-century, sunny Rome. Similarly, many communities have believed in the necessity, or desirability, of blood sacrifice to please a god or gods, perhaps to ensure a good harvest, or such. In such a belief system, it would presumably be thought to be not only permissible to sacrifice human life, but perhaps incumbent on the community to do so. Thus, an act which might be thought morally acceptable, even morally necessary, in one community might be regarded, in another community, one with a different belief system, as misguided and tragic. One might also note that even universal moral agreement in all respects over all time in all cultures would not be likely to satisfy a methodological moral skeptic as, in theory, everyone, in every respect, at all times, might be mistaken. One might inquire, of course, as to what the moral skeptic would accept as a proof of moral objectivity. If he would accept nothing, then his position is not even dogmatic; it would simply be vacuous, meaningless. To be sure, agreement is not a sufficient condition for truth. That everyone on Earth might believe in the heliocentric theory of the solar system does not make it true. What would make it true is that the Earth does, in fact, orbit the sun. In any event, in the text, we are less concerned to defend the cognitivity of aesthetic or moral judgments, as to understand, or try to understand, why people regard them as cognitive, as being either true or false. What convinces them, rightly or wrongly, that predicates such as 'true' and 'false' may be legitimately applied to them?

individual who might disagree, perhaps preferring a world of uncertainty, insecurity, recklessness, feuds, strife, war, excitement, bloodshed, adventure, and danger.[45]

Our crucial interest then is not in conditional moral truths, either indexed to particular codes, or viewed as instrumentalities to accepted goals, because one might, after all, not accept a particular code or a particular goal. And, beyond this, we are not even interested, in this context, in either defending or attacking aesthetic or moral cognitivity. Rather we are trying to understand what might explain the almost universal belief, in all times and places, in all communities and civilizations, that truth and falsity appertain to at least some aesthetic and moral judgments. What accounts for the pervasiveness, tenacity, and depth of this belief?

In order to better focus on the discussion, to minimize agitation, to avoid distractions, and such, we will discuss the matter in the context of a simple example, hopefully one at room temperature and safe, certainly, one hopes, one inoffensive and uncontentious.

Presumably, an individual who claims that Picture P is beautiful likes the picture. We will disregard the real possibility that an individual might recognize that a picture is beautiful and not like it at all. Perhaps it is painted by a rival, or he disapproves of the subject matter, or such. But we will suppose that normally when someone says a picture is beautiful he likes it. Presumably, however, it is not beautiful because he likes it, but he likes it because it is beautiful. That is the way it seems to him, at any rate.

In English, it is interesting to note that there is no simple, single expression for "merely subjectively appealing." This is not to deny that certain things might be "merely subjectively appealing," only that the absence of a simple, single expression for that possibility suggests that such experiences may be rare, that is, a sense that some object is merely subjectively appealing, or, at least, that such experiences are not recognized as such. When one experiences, say, a picture as beautiful, one does not experience it as "merely subjectively appealing." One sees it, in itself, as an object that is appealing, *simpliciter*. One sees beauty, for example, as a property of the object, and not of the experience. One might see the picture twice, even in the same afternoon, once as beautiful, and once as not beautiful. But one sees its beauty as a property of the object, not of the experience, and one sees its lack of beauty, as a property of the object, not of the experience. Experiences, however pleasant or unpleasant in themselves, are transitory. The object endures through time, and is objective, and its beauty, to the observer, seems as much a property of the picture as its size, shape, and colors. This fascination with the object, and with externalization, is as endemic with reviewers, critics, and appraisers, as with others. The average critic, for example, putatively addresses himself to the object, and not to his personal responses to the object, which would be likely to be of little interest to anyone. They are interested in the picture, not a clinical report on the condition of the critic's physiology.

Even an individual intellectually committed to a subjectivist analysis of aesthetic experience, who would officially deny any value or credibility to his own views, certainly has his views, and has reasons for having them. He takes them seriously, and cannot help doing so. Even if he feels he is entitled to no more than an "I like it," there is an "it" he likes, and it must seem to be such as to warrant his liking. He does not experience it, personally, as "merely subjectively appealing." He experiences it as appealing, and properly so. He cannot experience it otherwise. And he feels that his response is legitimate, warranted, correct, right, and so on, and that it should be a legitimate, warranted, correct, right, and so on, response for others, though he recognizes, perhaps to his disappointment, that it might not be. Perhaps he wonders why they do not, or cannot, see it as he sees it. He sees his view as "right," not simply as right for him. Even if something is merely right for Jones, if that is possible, Jones does not experience it as merely right for Jones.

Regardless of objectivity and subjectivity, of truth and falsity, in matters aesthetic and moral, I think we now have an insight into the likely answer to our question, as to why there is, across times and cultures, an almost universal commitment to cognitivity in these matters. There is a very serious and important reason, for example, why individual and cultural relativism, whether warranted or not, is a view entertained, and usually inconsistently, by an almost invisible minority. The reason that this is so is not because of the egotistical and ethnocentric proclivities of an ignorant, benighted, and degenerate species; contrasted with the perspicacity and moral superiority of a tiny aristocratic elite; the answer, rather, lies in the nature of human experience, and the experience of all humans. Some things, whether because of conditioning, acculturation, biography, genetic givens, or what not, are experienced in certain ways. One cannot, for example, see a picture as beautiful, and not see it as beautiful. One does not see it as "beautiful for me," but as beautiful. These observations certainly do not demonstrate the objectivity of beauty, or such, but they make clear that beauty is experienced as objective, and cannot be experienced otherwise. In this way, we have a theory by means of which to explain the pervasive human commitment to truth in matters aesthetic and moral, while, at the same time, acknowledging, with our skeptical forebears, that "the opinions of men differ from time to time, and place to place."

In explaining the pervasive human commitment to "aesthetic and moral truths," we have not, of course, explained it away. Rather, it might be noted that this pervasive disposition of humanity is a touching and ennobling characteristic, possibly testifying to a genetically conditioned conatus to seek objectivity, rightness, and truth in those matters and realms closest to the human heart.

The criterion for truth here is a sense of rightness.

We have already noted, in connection with rule-bound truths, that truths exist beyond the analytic and synthetic, at least as those categories are normally understood. We have also recognized that conditional moral truths exist, in the sense of entailments and instrumentalities. Clearly, independent questions have to do with the evaluation of codes and the assessment of goals.

It is in terms of this sense of rightness, or wrongness, that one finds the only key to allegations of ultimate aesthetic and moral truth, aside, as noted, from the entailments of codes and instrumentalities related to goals. It is in terms of rightness, or seeming rightness, that human beings must ultimately adjudicate amongst codes and goals. It is likely that human beings will not adjudicate identically, but each must adjudicate, if morally motivated, ultimately in terms of a desiderated world, a vision as to how one would have the world be.

However, one resolves such issues, one will see one's resolution as right, or as right as possible under the circumstances. And one, being human, will then see it as true that one's resolution was right, or as right as possible under the circumstances.

Thus, interestingly, the ultimate human commitment is to aesthetic and moral truth.

That may not be where human beings talk, but it is where they live.

A third form of axiological truth, if one acknowledges the possibility of such truths, might be "philosophical truth."

Interestingly, philosophical descriptivists, in which category most philosophers would be found, seldom give much attention to the nature of philosophical statements, even to concern themselves, really, with what it might be for, or how one would know that, such statements were true. They do philosophy without much worrying about what they are doing. This is doubtless generally acceptable for inquiries pertaining to recognizable externalities, such as the Krebs cycle, the successions of life forms, the composition of stars, and such, where, whatever the formidable difficulties involved, one is at least addressing oneself to an objective subject matter, understands what one is doing, and has a reasonably clear understanding of what would be involved in being right or wrong. On the other hand, this not worrying about what one is doing is less acceptable in philosophy. Philosophy lacks comparable externalities, such as strata and currents, against which to measure her hypotheses. In the light of this difference, philosophers might be well advised to give more attention to what they are doing. Uncriticized assumptions and blithe complacencies are comforting, but not always innocuous. In science the target is synthetic truth, discovering the "case," so to speak. Many philosophers seem to presuppose that they share this target, that they are after the "case," and that it is "out there," to be right or wrong about. Problems, however, commonly appertain to the endeavor. For example, how would one prove one is right, and how can one explain disagreement, particularly amongst similarly intelligent, similarly informed investigators, not easily dismissed as ignorant, obstinate, stupid, certifiably insane, evil, and such? Surely the unearthed gold should be equally obvious to all. And how is it that problems familiar to the Academy and Lyceum are still about? Should they not have been solved by now? If solved, how? If not, why not? Perhaps science and philosophy are different after all, really different.

1.5 Some Class Concepts

An extremely important distinction in these matters, which will become evident shortly, is the distinction between classification and nonclassification questions. Accordingly, as such questions presuppose a reference to classes, some attention to the nature of classes and their relations would seem in order. Much of what is said here could also be phrased in terms of sets, but 'classes' is a more familiar term and has its place in history, lore, and tradition.

Whereas classes are presumably no more mind-independent than numbers, meanings, properties, and such, it is convenient to speak of them as though they were mind-independent. So, too, one could speak of chess moves, ideas, rights, beliefs, theories, and so on. Language facilitates the illusion of unifying and externalizing a myriad internalities, bringing them into the public domain, where they may be considered and discussed, as though they might be trees or rocks. This illusion , like that of naive realism, in which a brain-dependent internal world is confused with a mind-independent external world, is not only useful and convenient, but may be necessary.[46]

The mind-dependence of classes, as opposed to the hypothesis that they are mind-independent, somehow "out there," waiting to be discovered, is recommended by the consideration that they can be invented on the spot. For example, one might invent a class of math majors who play basketball, and so on. Once the invention is in place, of course, one can treat it as though it had a mind-independent status. Reality does not classify; people classify.

Without further articulation, much about classes remains obscure. Consider the class of swans. Does it contain only present swans, or does it also include past swans? What about future swans? What about fictional swans? And are possible swans members of the class of swans, and so on. These are not questions of fact, of course. Rulings would be called for. Such considerations, again, suggest the unlikely tenability of a mind-independent view of classes.

---

46. The supposition that classes have an independent reality, metaphysical or otherwise, if not carefully qualified, leads to contradictions. For example, if one supposes that all properties may determine classes, and that a class must either be a member of itself or not, then a contradiction is in the offing, as noted by Bertrand Russell. Consider the class of all classes which are not members of themselves. If it is a member of itself, it cannot be a member of itself. And if it is not a member of itself, then it must be a member of itself. "Russell's Paradox," as it is known, naturally enough, is only one of a number of such paradoxes. One might also note, in passing, that a similar paradox might be supposed to be involved in Platonic metaphysics. For example, presumably a form either characterizes itself or not. For example, the form "dog" does not characterize itself as it is not a dog. On the other hand, the form of forms, if it exists, presumably characterizes itself, as it is a form, the form of the holy is presumably holy, the form of the beautiful is presumably beautiful, the form of the good is presumably good, and so on. Then, if there is a form of forms which do not characterize themselves, it seems that a paradox, à la Russell, might be generated. Consider the form of forms which do not characterize themselves. It if characterizes itself it cannot characterize itself, and if it does not characterize itself it must characterize itself.

Some individuals identify classes with their members. As we normally think of classes, and talk about them, that identification would seem injudicious. It would seem to confuse classes with collections, aggregates, bunches, groups, or such. For example, dogs bark, wag tails, eat dog food, and such. But one usually does not think of a class, as barking, wagging its tails, eating its dog food, and such. Presumably classes, no more than numbers, occupy space, have weight, and such. One supposes the class of elephants does not occupy more space and weigh more than the class of numbers divisible by two. Also, the null class, a most important class, without which class logic could not exist, is short of members; it doesn't have any.

Some simple class concepts might be illustrated, as follows. There is the class A and the class B, two simple classes; the logical product of these two classes would be the class of entities who are members of both A and B; the sum class would be the class of entities who are members of either A or B. For example, consider the class of math majors and the class of basketball players, two simple classes. The logical product of these two classes would then be a new class, identical to the one we suggested earlier, the class of math majors who are basketball players; the sum class would be the class of entities who are either math majors or basketball players. Class A' (A Prime) is the class of entities which are not members of A. Class A'B' is the class of entities who are members of neither class A nor of class B, and so on. Classes, with respect to one another, may be exhaustive, exclusive, or complementary. Classes A and B are exhaustive classes if and only if every member of the universe of discourse is a member of either class A or class B. (i.e., class A'B' is empty). Classes A and B are exclusive classes if and only if no member of the universe of discourse is a member of both class A and class B (i.e., class AB is empty). Classes A and B are complementary classes if and only if classes A and B are both exhaustive and exclusive classes. In short, every member of the universe of discourse belongs either to class A or to class B and no member of the universe of discourse belongs both to class A and class B (i.e., both class A'B' and class AB are empty). There is, of course, a complex logic of classes, or sets, as there is of propositions, predicates, relations, modalities, and such.

In what follows, we make use of the notion of classes.

2. Classification and Nonclassification Questions.

Most questions are nonclassification questions. A nonclassification question is one in which the classificatory scheme itself is not in question. It is in place, so to speak, and waiting to be used. For example, are tigers carnivorous and do tigers live in India are, in our sense, nonclassification questions because they are not questions about classification itself but, rather, about utilizing, or applying, a classificatory scheme or structure. Where do clearly understood things fit into a clearly understood classificatory structure? In the example, we are reasonably clear on tigers, what it is to be carnivorous, and where India is, and so on. We are simply wondering if tigers eat meat, and if they are found in India, and such. I think this is clear enough. There is a sense, of course, in which most questions could be construed as classification questions. For example, in the context of our example, we might be asking if tigers should be classified as meat eaters, if they should be classified as entities some of which, at least, are found in India, and such. On the other hand, since the classificatory structure itself is not in question they would be regarded, on our approach, as nonclassification questions.

Our major interest here, as philosophers, is in classification questions, namely, in our sense, questions which pertain to the classificatory structure itself. Such questions divide themselves into those which have true/false answers, and those which have better/worse answers. Those which have true/false answers are of little philosophical interest, and will later be referred to as

---

47. This distinction is reasonably clear, and, I think, illuminating and valuable. It does presuppose at least a *seeming* cognitivity, namely, that some answers are better or worse than others. It also, naturally enough, leads to the question as to whether or not it is *true* that some answers are better or worse than others. It is hard to see how a commitment to axiological truth can be avoided. If that is the case, then there is a sense in which all classificatory questions, and, later, all philosophical questions, and such, for example, both first-order and second-order philosophical questions, would have true/false answers. On the other hand, that is to commit oneself to the reality of axiological truth, and axiological truth, other than in the hitherto-noted senses of entailments and instrumentalities, is controversial, as, for example, in the appraisal of codes and the acceptability of goals, and such. Axiological truth, in the more

second-order questions.[47] Those which are of keen philosophical interest, and on which much hangs, even eventual cognitivity itself, are those which have better/worse answers, which will later be referred to as first-order questions.

For purposes of clarification, let us consider some simple classification questions, questions pertaining to the classificatory structure itself, which have a true/false answer.

Here is a representation of a simple four-class matrix:

A	A'	
AB	A'B	B
AB'	A'B'	B'

A number of questions with true/false answers could be based on this small matrix.

For example, is class AB' a subclass of class A, is class AB' a subclass of class B', is class AB' a subclass of class A', is AB' a subclass of class B, and so on. The answers to the first two questions are affirmative, and to the second two questions, negative. The answers to these and many similar questions are embedded in the classificatory structure itself. They are clearly questions relating to the classificatory scheme or structure, and have true/false answers.

We noted, a moment ago, that one may distinguish between the A and the A' (A Prime). A and A' are complementary classes. Everything in the universe of discourse belongs either to A or A' and nothing belongs to both. For example, if A is the class of math majors, then A' is the class of all entities which are not math majors. If the class of interest is the sum class A or B, say, the class of entities which are either math majors or basketball players, then its complementary class is the class of entities which are not math majors and not basketball players, and so on.

Where the classificatory scheme or structure is familiar and reasonably clear, and most entities of interest are clearly classifiable, the classificatory scheme, or conceptual structure, is

---

interesting senses, does not seem to be either analytic or synthetic, at least as those forms of truth are usually understood. In any event, when one contrasts questions with true/false answers with questions with better/worse answers, it should be clear that one means by true/false answers answers which would be recognized as true or false in reasonably familiar, relatively uncontroversial senses of 'true' and 'false', that is, as being analytically true or false, or synthetically true or false.

largely invisible. Two situations, however, like a reagent applied to a concealed writing, tend to bring what is commonly unseen into sight, sometimes troublesomely so. The first situation is practical, that in which an entity is not clearly classifiable; and the second situation is theoretical, that concerned with the classification of possible entities, which would not be clearly classifiable. For example, to take an actual situation, as opposed to an "at-present-merely-theoretical situation," a tiglon is the offspring of a male tiger and a female lion. Is a tiglon a tiger or a nontiger, a lion or a nonlion, or do we need a new category altogether? What then of the union of a tiglon and a tiger, or a lion? Is that a tiglon, or a nontiglon, a new kind of tiglon, a different kind of tiger, a different kind of lion, and so on. Presumably no one is going to become unusually exercised about tiglons and such, but things become more serious when one is trying to decide how to distinguish between, say, living and nonliving, human and nonhuman, person and nonperson, scientific and nonscientific, moral and nonmoral, just and not just, desirable and nondesirable, natural and nonnatural, intelligent and nonintelligent, same person or not the same person, same species or not the same species, and so on.

It is usually easy to classify entities but not always, and sometimes considerable consequences may appertain to classifications, consequences which may determine life and death, existence or nonexistence, the future of a society or species, citizenship or noncitizenship, participation or nonparticipation, economic and social liberation and acceptance or not, control of one's own destiny or not, entitlement to influence society or not, and such. It is not always easy to distinguish between the A and the A'. Many questions such as these do not seen to have true/false answers, at least in any obvious sense, but would seem, rather, to be the sort of classificatory questions more likely to have better/worse answers. If truths are involved here, it seems they would have to be axiological truths, namely, that it is true that such and such is a better answer than some other answer. It would surely be difficult to know if some answer was the best answer, and that it was true that it was the best answer, particularly in some absolutistic sense, for all time, all places, and all situations.

3. First-Order and Second-Order Questions

We have now distinguished between nonclassification questions and classification questions, and, amongst classification questions, between those with true/false answers and those which do not appear to have true/false answers, at least in any usual sense, questions which, rather, for most practical purposes, putting aside questions of final axiological truth, would seem to have only better/worse answers. First-order questions will now be understood as those questions which are classificatory or conceptual questions of the sort seeming. at least initially, to have only better/worse answers. Second-order questions then will be what we have called nonclassification questions, for example, are tigers carnivorous, are they found in India, and such, and classification questions with true/false answers, in a familiar sense of truth and falsity, for example, is class AB' a subclass of class A.

Whereas second-order questions may be subtle, difficult, or impossible to answer, for example, were dinosaurs warm-blooded, did the "Big Bang" constitute a discontinuity obliterating evidence of the previous history of the universe, how much dark matter is there in the universe, is there rational life elsewhere in the universe, and such, they are not essentially philosophical questions. Similarly, that a correct answer to a complex problem in set theory exists does not entail that the answer is easily discovered; similarly, that a complex mate in chess, say, in nineteen moves, lurks in a given board situation does not entail that this will be obvious to a player, even one of international stature. But, again, such questions, as interesting, important, and formidable, as they may be, are not essentially philosophical questions, at least in the sense of first-order questions. They have clear answers, whether those answers are ever to be obtained, or not.

The importance of first-order questions emerges when it is realized that cognitivity, truth as we usually think of it, namely, analytic and synthetic truth, are decision dependent.

3.1 The Decision-Dependence of Analytic and Synthetic Truth

Both analytic and synthetic truth are decision dependent, analytic truth wholly, and synthetic truth in part. This is the case, in a trivial sense, in virtue of established usages and inherited conventions, and, in a nontrivial sense, in virtue of philosophical recommendation.

We usually do not think of the meanings of words as decision-dependent, but, obviously, they are. You and I, for example, did not decide on the meaning of 'raccoon', 'giraffe', 'elephant', and so on, but others did. The relation between phonemes and morphemes, between sounds and meanings, is one of convention, and conventions are the result of cultural decisions, and the acceptance, and perpetuations, of cultural decisions. Interactions amongst indefinite numbers of individuals over generations, even millennia, produce language, law, moralities, markets, and such, cultural artifacts of enormous importance, the products, as it is said, "of human action but not of human design."

Swans, in a sense, are what they are, independently of what is thought about them. For example, there may well have been swans about before humans, and apart from humans, even today. I speak of "in a sense," because one commonly uses language to speak and think of that which is, in its nature, prelinguistic, nonlinguistic, and so on. It is thus paradoxical in one sense to think of swans existing before words such as 'swan' and 'bird' existed, but, in another sense, it is obvious that the sort of things we speak of today as swans and birds are in no sense dependent on human speech habits. To be sure, what we will take as swans and birds does reflect semantic mapping. One might have used the expression 'swan' to mean just about anything, and the same with the expression 'bird.' And, even aside from the arbitrariness of correlating meanings with noises, one can easily conceive of a semantic mapping in which, say, 'swan' was used more narrowly or more widely than it is today. It would not be logically impossible, for example, to restrict the word 'swan' to what we now think of as white swans, and use a different term for black swans, or one might use the word 'swan' more broadly than is common today, for example, to refer to any large waterfowl, and so on. Similarly, 'bird' might be construed in such a way as to exclude what we now think of as flightless birds, say, ostriches, kiwis, penguins, and such from the class of birds, which we might restrict to feathered things that fly. Also, obviously, one might use the expression 'bird' to apply more generally, to flying things, say, bats, many insects, airplanes, gliders, kites, and so on. There are doubtless better and worse ways to map the experiential world, but the point is that there are various ways in which it could be done. For example, a biologically based taxonomy of the plant and animal kingdoms seems an illuminating, interesting way to organize certain aspects of human reality, the phenomenal grid, so to speak, but it is not the only "true way" in which it could be done. There is more than one way to divide a pie, and more than one way to map a world. Given what we mean by 'swan' and 'bird' it is analytic that all swans are birds. But what we mean by 'swan' and 'bird' are decision dependent. In this sense, analytic truths, truths which are true in virtue of their meaning alone, are decision dependent.

As meaning is decision-dependent, so, too, synthetic truths, with respect to their semantic component, are decision-dependent. Without a fixity of meaning, or a substantial fixity of meaning, synthetic truths are not possible. Meaning is logically prior to truth. Meaning must be in place before truth can exist. As an example, one might consider the following sort of situation, which is apparently typical in field linguistics. Several villages exist along the banks of a river, we shall suppose ten villages, strung out, as follows:

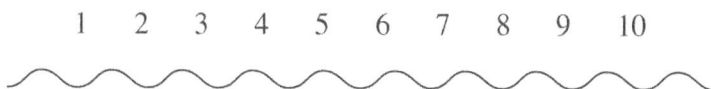

1   2   3   4   5   6   7   8   9   10

The villagers in Village 1 and Village 2 have no difficulty understanding one another, and communicate easily. As we understand 'language' we would presumably say they speak the same language. This same situation occurs along the river with adjacent villages. For example, the individuals in Village 2 have no difficulty with those of Village 1 and Village 3, and so on. The villagers in Village 1 can also understand, and easily communicate with, the villagers of Village 3, but they do have some problems.

Certainly the folks in Village 3 are different, talk "funny," have some surprising words, and such. And this sort of thing also occurs along the river. Now once we have a hiatus of more than two or three villages between any two given villages, things become genuinely difficult, and, finally, the villagers in Village 1 cannot understand the villagers in Village 10, at all. Clearly, there, as we normally think of languages, we would have different languages. The question now, which sounds simple, but is not at all simple, is how many languages are spoken along the river? That certainly sounds like a straightforward empirical question, but it is not. It becomes a straightforward empirical question, capable of a true/false answer, only after one has in place a reasonably clear distinction between dialects and languages. In a situation of the sort we have proposed, one might have two, three, four, or five languages spoken along the river. On one understanding of 'dialect' it would be true that two languages were spoken along the river; on other understandings of 'dialect', one might have three, four, or five languages spoken along the river, and so on. Synthetic truth is thus dependent on both meaning and data, or the world. This is the case with all synthetic truth. This is obvious, but what is likely to be less obvious is the variability, alterability, convenience, and utility of meaning. Truth can shift with meaning, and meaning may be indexed to purpose. How are we to understand 'person', 'responsible', 'mental competence', 'normality', 'intent', 'negligence', 'obligation', 'justice', 'fairness', 'equality', 'art', 'democracy', 'intelligence', 'love', 'abuse', 'good', 'bad', even 'town', 'city', 'fruit', 'vegetable', and so on? Just as statistics can be rigged to promote agendas, so, too, can meaning, that it may discover "hitherto overlooked truths," truths newly wrought, fresh from the semantic workshops, truths with coveted ends in view.

As noted, then, both analytic and synthetic truth are decision dependent, analytic truth wholly, and synthetic truth in part. This is the case, in a trivial sense, in virtue of established usages and inherited conventions, and, in a nontrivial sense, in virtue of philosophical recommendation.

Obviously the primary philosophical interest here, as opposed to the great empirical interest in realizing these things to begin with, is what factors might be well taken to influence philosophical recommendation. Value judgments are obviously involved. This is less a dismal understanding of "here be dragons," than an invitation to improve a world. Whatever one thinks of the cognitivity of value judgments, they underlie not merely matters of taste, but matters of how human reality will be constructed, and this is of great importance because we will find ourselves living in such realities.

Human reality, namely the experiential worlds in which we exist, are a fusion of data and conceptualization, or data organized in terms of a classificatory scheme, or structure. The human reality, it seems likely, is founded upon, is produced by, and is topologically correlated with, a human-independent reality. The human-independent reality, whatever its origin, nature, or cause, was doubtless about long before humans and is largely apart from humans, and is, for the most part, undisturbed by humans. Human reality, on the other hand, as noted, is affected by conceptualization, and this involves meaning, and meaning is decision-dependent, though, for the most part, the decisions have not been ours. On the other hand, as noted, meaning can be revised, and to the extent it is revised truth, and possible truth, is changed, and so, too, is human reality. In the sense that meaning precedes truth, and meaning is choice-dependent, and choice depends on decision, and decision depends on value judgment, it becomes clear that value judgment underlies human reality, and that human reality, properly understood, is incomprehensible, unintelligible, without it. To the extent that epistemology is concerned with meaning and truth, and meaning is decision-dependent, and depends on value judgment, axiology precedes epistemology.

The crucial question then, if one puts aside guns and knives, spears and clubs, discrimination, stigmatization, ridicule, and scorn, is whether or not some value judgments are better than others. Even the noncognitivist believes this. If one proceeds beyond this to the question as to whether or not it is true that some value judgments are better than others, one is concerned, as noted, with the possibility of axiological truth. If such truths exist, it seems clear they cannot be understood as either analytic or synthetic truths, certainly in any familiar sense. Interestingly, it is a common presupposition that such truths exist. Whether the presupposition is true or not, it seems one cannot live without it. It is reminiscent of the notion of free will, in that, whether it exists or not, one has no choice but to act as though it does. How else could one act?

3.2 Criteria, Metacriteria, Metametacriteria, etc.

As we have seen, cognitivity presupposes modalities of organization, for example, criteria, classification, and conceptualization. A board may be as long as it is, and, indeed, it would be hard, would it not, for it to be otherwise, but this arrogant absolutism, in all of its independent ontological sublimity, is not easily countenanced in the human world. Indeed, outside the human world, whatever it is, or may be, it does not even have the dignity of being a board. In the human world, however humiliating, this ontological given is relativized to human experience. It becomes a something, perhaps a board, a plank, a flat piece of wood, an oddity, fuel, an artifact, a tool, a weapon, an object, a thing, or such. If one had never seen a board it is not clear how one would see a board. If you had never seen a zurk, left behind in the camp of a transient alien, what would you take it to be? How long is the board? Pretty long, long enough, not really so long? So many feet, so many centimeters, so many zilks? The board may be as long as it is, and not a whit longer, but it doesn't make much sense ask how long it is without some sort of scale in mind, long enough to impress you, twice the length of a saber-toothed tiger's tail, longer than the favorite spear of Ajax, seven feet long, or whatever. This interplay of data and interpretation is easy to overlook in the case of the board. Similarly, do we not just look and see a giraffe? It is more obvious, and more subtle, and harrowing, when, on strange roads, we are failed by the ruts of language. We must then address ourselves to classificatory issues, issues of criteria, issues of conceptualization. One tries out meanings, suggests meanings, proposes meanings. The semantic elasticity of language, with its ill-defined borders and its capacity to welcome immigrants, its willingness to wander and explore, not only tolerates, but invites, novel conceptualizations, illuminating metaphors, new visions of the world. And thus proposal lays the foundation on which cognitivity will reign, a cognitivity commonly unaware of the value judgments on which it is founded, without which it could not exist.

How shall we, say, distinguish between the A and the A', the A and the not-A, say, the same person and not the same person, human and not human, good and not good, right and not right? If we are asking the question, it presumably has no clear answer. One proposes distinctions, conceptualizations. These are commonly purpose-relative. Are the purposes acceptable? Proposals are involved. One must then adjudicate amongst proposals. How could one do that? Obviously, one adjudicates in terms of a criterion. But a criterion is itself a proposal. Accordingly, if the criterion is regarded as unacceptable, if it is challenged by another criterion, one adjudicates amongst competitive criteria in virtue of a metacriterion. But metacriteria are themselves proposals. They are relevant or not relevant; they are acceptable, or not acceptable, and so on. One would decide amongst metacriteria, if possible, in terms of a metametacriterion, and so on, and eventually, somehow, in terms of a highest criterion, one, like Plato's Idea of the Good, or Form of the Good, beyond which no appeal would be intelligible. Whether or not there could a final criterion, in terms of which, ultimately, conceptual disagreement might be resolved, is an open question.

These suggestions having to do with the adjudication of proposals, in virtue of a hierarchy of criteria, might be schematized

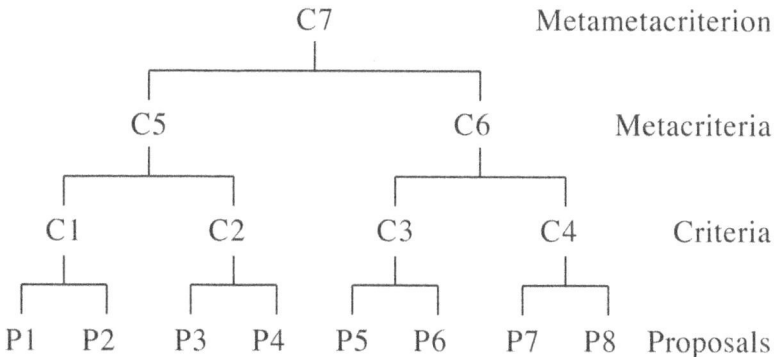

		C7		Metametacriterion
	C5		C6	Metacriteria
C1	C2	C3	C4	Criteria
P1 P2	P3 P4	P5 P6	P7 P8	Proposals

as follows:

For example, in virtue of C1 we might approve Proposal P1; in virtue of C2, one might approve P3; in virtue of C3, one might approve P5; and in virtue of C4, one might approve P7. On this approach, Proposals 1, 3, 5, and 7, have ascended to the second level of selection. One might then suppose that in virtue of C5 one might approve Proposal 1, but, in virtue of C6, one might approve Proposal 5. At this point one must adjudicate between Proposal 1 and Proposal 5. Then, one might suppose that in virtue of the metametacriterion, a final choice might be made, whatever it might be, between Proposal 1 and Proposal 5.

The question here is not so much whether ethical disagreement would in fact be resolvable, which may be doubtful, given social egotisms, jealousies, hatreds, and such, but whether or not there might be a best solution, whether or not it is ever recognized, let alone achieved.

If there is such a solution, or final criterion, it would presumably be a vision of an ideal world, or a vision of a world so ideal that it might accept and tolerate, if not understand and celebrate, more than one vision of an ideal world.

3.3 Considerations Pertinent to Classificatory Hypotheses (Prescriptions)

Prior to considering four major criteria pertinent to the assessment of classificatory hypotheses, or prescriptions, the criteria of precision, clarity , utility, and beauty, two observations are in order, first, the inevitability of value judgment and, second, axiological divergence.

Human life cannot be lived without value judgment. Human behavior does not do itself. What works for trees and tides does not work for the human being. The human being must act, and to act he must decide, must choose amongst options, which will be assessed as better or worse, and will be better or worse, given ends in view. That is a simple matter of fact. To deny it would be incomprehensible. Similarly, if some options are better or worse, then it is *true* that some options are better or worse. Thus, cognitivity is implicit in such judgment.

For the human being, value judgment, and life in its terms, are inevitable.

The human being is an animal which acts; it is an acting animal.

Inaction is not an option for the human being. To choose not to act is itself a choice, an action. Action does not require incessant ratiocination, but, commonly, mere consent. Life accompanies choices, directed by them, not always keeping them in mind. Immobility and paralysis might be chosen. One may choose to do nothing, a choice in itself. Even to walk in the tracks of habit, to cultivate docility to custom, to surrender leadership to an authority, in fear or gladness, to acquiesce in the expectations of one's fellows, to acquiesce in one's expectations for oneself, to conform without question, to hide one's thoughts, to hide from oneself; even the decision to lie down and die, is to act. The human being will have his preferences and make his choices in the light of them. One acts in terms of value judgments; without them action is impossible; without them human life, as we know it, is impossible. To ignore, overlook, marginalize, deny, deplore, demean, denigrate, misconstrue, or minimize value judgment is more than a mistake; it is to misunderstand, and fundamentally, the nature of a species.

Seemingly choice is not done to one; but one does choice.

In any event, choice is real, and effective, even should it be the consequence of undiscerned elements, psychological, cosmic, or random. Even a determined choice is a choice. Even if, given the nature of the universe, one cannot choose other than as one chooses, one chooses. Choice is not inert. Link three in a causal chain cannot occur without link two. Even should the universe be a deterministic system, without point or reason, a self-winding watch without hands, or a quantum playground, a cosmic circus, it is one in which choice is ingredient and familiar. We may indeed live in such a universe, the determined universe or the quantum universe, but, interestingly, one cannot act as though one did. Even the convinced determinist must act as though he were free. He must act before he can learn what he was determined to do. And even the physicist who might believe in a random universe, a sort of cosmic prank, an accumulation of minor miracles, piling up probabilities, must act before he can discover what the universe happened to do with him in such and such a case. But he, and our determinist, will go about their work as if the universe they live in is really just what it seems to be, one in which each manages his own existence. Whether freedom exists or not, one must act as though one were free. There is no alternative. The travail and torment, the judgment and appraisal, the planning, calculation, and argument, the logic and thought, the hope, concern, anxiety, effort, and striving may be a sham, but again, it may not be. And perhaps it is not. We live enmeshed in mysteries.

But we are guaranteed, at this point, of no more than subjectivistic and relativistic cognitivities.

This brings us to axiological divergence.

Whereas human beings are pretty much of a muchness, physically, chemically, and biologically, for example, having in common well over ninety-nine percent of their genes, it is obvious that they often differ widely in other respects, for example, in culture, conditioning, environmental background, historical antecedents, beliefs, principles, convictions, ideology, and such. At the obvious risk of oversimplification, let us suggest a sort of intellectual division, or intellectual "watershed," amongst human beings, distinguishing between, say, A and B, recognizing, of course, that the distinction is not exact, and that many "A humans" have some "B characteristics," and many "B humans" have some "A characteristics." The basic distinction between A and B is the differential prioritization given to utility and truth. It seems clear that utility and truth, while they may often go together, are not identical, for example, that a given belief might have great social utility without being true, and that a true belief might have little, or even negative, or even catastrophic, social utility. It should also be recognized that those who favor utility are likely to claim that their beliefs are true, for who would claim that their beliefs are false, and that those who favor truth are likely to maintain that their beliefs also possess utility, or, in any event, are innocuous and inoffensive, for who would wish to claim that their beliefs are harmful, disruptive, or dangerous.

The "division," or "watershed," might look something like this:

A	B
"instinct," intuitions, faith, "reasons of the heart, of which reason knows nothing," verbal meaningfulness, mysticism, revelation, hope, comfort seeking, emotional needs, subjectivity, life consequences, social controls, authority, power, nonacceptance of, or reservations concerning, criteria on the other side of the "division," or "watershed," etc.	skepticism, evidence, inquiry, empirical meaningfulness, testability, falsification conditions, precision, clarity, coherence, naturalism, science, truth seeking, cognitive needs, objectivity, putative realism, courage, discovery, independence, nonacceptance of, or reservations concerning, criteria on the other side of the "division," or "watershed," etc.

A basic distinction related to, and underlying, the A/B dichotomy, has to do with a distinction between modalities of discourse, a distinction with respect to what is considered meaningful and what is not considered meaningful. As an illustration, but one should feel free to provide one's own examples, one might consider the two locutions, 'Air is everywhere' and 'Zeus is everywhere'. Air is invisible, and, so, too, at least most of the time, one supposes, would be Zeus. Air certainly seems to be everywhere, or, at least, almost everywhere we usually are, but we nonetheless take the first locution to be false. For example, air is not in a vacuum. It is not so clear what one might take as evidence of the nonexistence of Zeus, even if Zeus does exist. We do know a number of things about air, in virtue of publicly available experiences, for example, it can be weighed, is difficult to compress beyond a point, and such. Most of what we know about Zeus is hearsay, but there are rumors about. For example, he is alleged to be the father of the gods. The descriptions furnished by Leda, Danae, and Europa may be discounted, as they appear to be confused, and are, in any event, incompatible.

Is Zeus everywhere?

If he is spotted on Mount Olympus, then it seems clear he is there and not somewhere else, and thus not everywhere. But this is perhaps crude anthropomorphization. Zeus is perhaps a spirit. But this proposal does not much advance the inquiry. What is a spirit? A spirit is presumably intangible, and not in space. If that is the case, how can it know, if it can, what is in space, interact with it, bring about changes there, and so on? Can it look, listen, feel, taste, sniff, and such, without sensory apparatus? How could it, for example, if it has no body, look behind itself, or up or down, or to the side? One might claim it can look, listen, and such, but not as we look, listen, and such, but this is not helpful, pending some clarification as to what one is talking about, for example, what that might be like, or how it might be accomplished. Perhaps "spirit" is something like an etherealized anthropomorphization. And it seems it would be difficult even for a spirit, a ghost, or such, if it is a thing, to be everywhere. For example, it seems unlikely that the same spirit, or ghost, could haunt castles in England and Scotland simultaneously. More prosaically, could it be in, say, Cleveland and Pittsburgh simultaneously? On the other hand, it might be suggested that "being everywhere" does not mean "being everywhere," but being *aware* of everything everywhere. This, too, of course, presents its problems. It seems it would be very difficult to be aware of everything, even everything in Cleveland, for example, what was going on in every office, haberdashery, butcher shop, grocery store, house, school, apartment, parking lot, street, alley, dumpster, trash can, and such. Would Zeus be aware, for example, not only of the secret motivations of politician Smith and businessman Jones, but of the contents of their subconscious minds, if they have them, as well, of which they themselves have no knowledge? What of a sparrow with flight problems, the inner life of fruit flies, and such? It seems there would be a great deal to keep track of in Cleveland alone, let alone Cleveland and Pittsburgh, let alone Cleveland, Pittsburgh, and the entire universe. But at this point one supposes that enlightened protagonists of demythologizing classical theology might make their appearance, smiling bemusedly at the naivety, the sweet innocence, the natural but pardonable stupidity, of supposing that things might mean what it seems they are saying. We are now given to understand, depending on the enlightened Zeusiologist in question, that 'Zeus is everywhere' really means, say, "Nature is uniform," "Reality is reliable," "Fate rules," "You can count on a hard fall if you step off a cliff," "Being is everywhere, except where it isn't," and so on. Perhaps this sort of thing saves a day, but it is not clear what day it saves. It seems, rather, to have changed calendars. One suspects, at the least, that there would be a reduction in the frequency and quality of hecatombs. In any event, it should be clear that 'Air is everywhere' and 'Zeus is everywhere' are different in a number of interesting respects. We are reasonably clear on the truth conditions of 'Air is everywhere' and much less clear on the truth conditions of 'Zeus is everywhere', assuming that it has truth conditions.

The sort of thing suggested here illustrates what we referred to earlier, a basic distinction related to, and underlying, the A/B dichotomy, an interesting distinction between modalities of discourse, a distinction with respect to what is considered meaningful and what is not considered meaningful.

It is a philosophical truism that there is no view from nowhere. The very notion of a view implies that it is a view from somewhere, rather like the very notion of a Euclidean circle implies that that it will possess a radius and a circumference. Similarly, the very distinction between the meaningful and the meaningless implies a criterion in terms of which the distinction is to be made, and a criterion implies a view, a decision, an acceptance, or such, which is going to be someone's view, decision, acceptance, or such. Much of this may be vague, custom-relative, unreflective, out of sight, and so on, but wherever one decides whether or not something is meaningful or not meaningful, some sort of criterion will be involved, recognized or not. One's intuitions of meaningfulness will underlie one's assessments of competitive meaning criteria, and the seeming adequacy or inadequacy of explicit meaning criteria may react upon one's intuitions, modifying them, expanding them, constricting them, and so on. One of the most interesting early adventures of analytic philosophy. beginning at least with Hume, Peirce, James, and others, was the attempt to develop a general criterion of empirical meaningfulness, for example, in terms of experiential correlation, ascertainable effects, cognitive utility, verifiability, falsifiability, translatability, essential theoretic inclusion, and so on. One might start with the word or sentence as the unit of meaningfulness, and end up, as with Quine, taking the unit of intelligibility to be the whole of science, as though there were a whole of science. But, obviously, as we suspect from the "division," or "watershed," suggested above, the "A" people may have different views as to what is meaningful, and to what extent, and different views as to what is not meaningful, and to what extent, from the "B" people, those on the other side of the "division", or "watershed." Presumably both the A people and the B people would agree that something like 'The glorp is schlerblich today' is meaningless', and that something like 'Albany is the capital of New York State' is meaningful, but, fortunately, or unfortunately, there will be numerous instances of locutions which might be taken seriously on one side of the watershed and, at least, not so seriously on the other side of the watershed, say, backing up a few centuries, 'Zeus is watching', 'Zeus listens to prayers', 'Zeus answers prayers,' 'Zeus is good', 'Zeus has an eye for the ladies', 'Zeus is everywhere', and so on. Indeed, one can imagine lively discussions having to do with some of these locutions, at least on the A side of the watershed. In any event, one supposes that many such locutions, whether regarded as true or false, would have been regarded not only as meaningful on the A side, but important. On the B side of the watershed, on the other hand, one supposes that many such locutions would be regarded either as meaningless, it not being clear how they might be tested, what their falsification conditions might be, and so on, or else as meaningful but uniformly false, due, say, to reference failure, and, accordingly, such things being the case, would be regarded as unimportant, unless, of course, one were to be executed or sent into exile if one did not pretend to agree with them. It is interesting to note that almost no grammatical locution using intelligible words, if any, is likely to be totally devoid of some discoverable, or invented, meaning. For example, 'Baudelaire is the pancreas of French poetry' might be regarded as meaningful, and simply false, at least literally, as poetry, French or otherwise, does not have a pancreas. On the other hand, one supposes a determined exegete might be able to find some meaning there, metaphorically, rather as one might find some meaning in, say, 'Plato is the Cadillac of Greek philosophy, whereas Aristotle is the Jeep Cherokee of Greek philosophy.' "Let us suppose that, say, 'Zeus exists' is supposed to entail, amongst many other things, that some entity exists which can change itself into water fowl, unusual meteorological phenomena, large animals, and such. If we regard this as false, then we would regard 'Zeus exists' as false, and, since it is false, it would be meaningful, as truth-value possession is a sufficient condition for meaningfulness.

It is important to note that on both sides of the watershed, however meaningfulness is understood, there is likely to be a distinction between the more or less meaningful. Sharp dichotomies may be required to pursue certain logistic objectives, but such dichotomization will map onto the real world only if one is prepared to see to it that the real world is obliged to comply. For example, logic will have it that a fellow is either bald or not bald, but, in interesting cases, this is likely to be the case only if one specifies the number of hairs involved, what will be counted as a hair, where it has to be, does a follicle count, vital or dried up, whether a false hair is a hair, and so on. And when does one stop legislating such things? One stops when it turns out that, say, Jones is either bald or not bald. The laws of logic hold because we will see to it, if we are interested, that they do. So, for most practical purposes, we will see a continuum, either actual or conceivable, between the more and less. This annoying fact also calls it to our attention that the search for a general meaning criterion is perhaps less likely to be philosophically fruitful than a more specific inquiry into the meaningfulness, and to what extent, or into the meaninglessness, and to what extent, of individual locutions. One size seldom fits all.

Two last points would seem in order, before addressing ourselves to the assessment of classificatory hypotheses, or prescriptions. First, the nature of meaning, and, second, the communication of meaning, understood broadly as the communication of understandings, which might be well-founded or not.

Meanings, like numbers, are best spoken of as though they might be objective, mind-independent, causally inert Platonic entities. We would usually talk, for example, as if the meaning of the word 'cat' were something "out there," though it is not clear just where that would be. And this is the case regardless of the number of meanings a word might be said to have. Meanings would seem to be, like words themselves, abstract entities, types, universals, and such, as opposed to concrete entities, tokens, particulars, or such. Just as we might have several "tokens" of the word 'cat', the type, so, too, one might have one meaning expressed by several words. There might be tiers of universals, or types. Now I doubt that anyone knows how one would test for types, universals, and such, or even, really, what it would be for them to exist, or even what it might be to pretend or suppose that they existed. On the other hand, it is also clear that it is very convenient to carry on as though there were such things. Whereas meaning, in the last analysis, is doubtless individual and psychological, it is not useful to think of it that way. To psychologize meaning, even though it is indubitably psychological, is to multiply, diversify, and subjectivize meaning in ways which would be cognitively catastrophic. For example, we would not want as many meanings for 'cat' or '2' as there are individuals familiar with the words. Similarly we cannot make the meaning a commonality ingredient in a variety of psychological states. First, there may be no such commonality. Second, even if there were, given the privacy of psychological states, it is not clear how one might test for its presence, and nature. Third, one likes to think of meanings which are not being instantiated, so to speak. One likes to think of 'cat' and 'two' having a meaning even when no one is thinking about cats and numbers. We would like to think it possible to forget a meaning, acquire a new meaning, invent a meaning, recognize a meaning, change a meaning, interpret a meaning, and so on. We would like to think of a text as having a meaning before it has been discovered or translated, and so on. The most convenient way to make sense out of meanings is to assign them an ontologically independent, if mythological, status. One talks as though they were objective, causally-inert, mind-independent entities. Similar considerations would pertain to "generalities," "possibilities," "unexplored continuations," "underived consequences of a premise-set," and such. The first step is to use language in a certain way, adopting an ontological-discourse posture, for example, talking about words and numbers as though they were an unusual sort of rocks and trees, things somehow "out there," wherever that might be. The second step is to embrace the notion of latitudinal sense meaning, for example, while accepting, or talking as if one accepted, meanings as independent entities, reserving to oneself the right to modify, elasticize or contract such meanings. For example, one might accept that 'horse' had a meaning but give oneself the privilege of modifying or altering that meaning as seemed fit, for example, being willing or unwilling to accept the notion of a winged horse, an articulate horse, a purple horse, a horse large enough to bestride a continent, a horse small enough to race about in a saucer, and so on. Even an old-fashioned, buggy-whip Platonic form would have to have enough elasticity to characterize hoofed horses and toed horses, living horses and extinct horses, possible horses, nonexistent horses, chargers, race horses, ponies, saddle horses, draft horses, brown horses and black horses, mares, stallions, foals, colts, and so on. In short, meanings are best taken as, or talked about as if they were, independent entities, but, also, as being sufficiently "porous" or "open-textured," as earlier referenced, to serve the general purposes of thought. In short, one talks an impersonal game and plays a personal game; one talks ontology and practices psychology; one talks objectivity and capitalizes, and empowers, subjectivity.

Lastly, one must consider lexical anchoring, or connectivity, or the relation of words to the world. Whereas there is an obvious distinction, certainly upon the least reflection, between words and the world, it is also obvious that one sees the world largely through words, or, perhaps better, through a conceptual structure, within which words are ingredient, and prominent, the reefs rising from the waters, the pinnacles of the ice bergs of cognitivity. Who, for example, does not just look and see that that is a cat, a giraffe, a rock, a tree? It would be a mistake, of course, to identify the conceptual structure with words, for a variety of reasons, the most obvious being that presumably all the higher animals have a conceptual structure, and few have a language. With or without a language one sees the world in terms of things, and kinds of things, and expectations of the behavior of things, and kinds of things. Nonetheless, words are obviously important, and most obviously important when communication is at issue.

Interestingly, it is usual to focus attention on the semantics of communication, rather than the pragmatics of communication, and, when the pragmatics of communication are considered, usually attention is limited to diverse employments of language, for example, beyond fact stating or misstating, one asks questions, one orders, recommends, prescribes, expresses, prays, curses, condemns, exonerates, casts spells, plays, jokes, teases, taunts, insults, promises, makes things the case, as in marrying people, and so on.

Here, we would like to consider something unusual, something seldom dealt with, namely, the pragmatics of cognitivity, what is, or may be, involved in communication where understanding is in question. How is understanding to be brought about? The usual answer to this is to see cognitivity in terms of naive referentiality. For example, one communicates the location of the cat by saying, perhaps, 'The cat is on the mat.' Our auditor then understands where the cat is. This is a paradigm case of referential cognitive discourse. The danger is to see the cognitivity of all referential discourse in terms of this sort of model, and to deny cognitivity to forms of discourse which fail to meet this sort of model. The "cat on the mat" model, of course, can be extended to, say, theories with empirical consequences, and such, and, depending on one's side of the "intellectual division," or "watershed," to 'Zeus is everywhere.' The essence of the "cat is on the mat" model is literalism.

In discussing the pragmatics of cognitivity, we will be addressing the distinction between literal and figurative discourse, and, more particularly, the role of figurative discourse, and its legitimacy, or illegitimacy, in connection with cognitivity, in connection with the achievement of communication. Is it possible to achieve cognitivity outside of what one might characterize as the paradigm model of discernible referentiality, the "cat on the mat" model, so to speak?[48]

We have already called attention to the mystery of meaning, its

---

48. What we have in mind here is very different from the familiar, useful distinction between cognitive and emotive meaning. For example, 'Bill is a flatfoot' and 'Bill is a law-enforcement officer' might be taken as having the same cognitive meaning, but would have, for most people, different emotive meanings. What we are concerned with here is the capacity, or incapacity, of figurative discourse to convey cognitive meaning.

mobility, its fruitful obscurities and porosities, its elasticities, its variabilities, its alleged metaphysical dimensions, and its undeniable psychological aspects. These considerations in themselves carry us far beyond the "cat on the mat" model. Essentially what we are concerned with here is the achievement of understanding. And understanding need not begin with, nor end with, the dogmatism of literality. Literality, like science, is most at home in the countries of quantification. The board is between two and three feet in length; the rock weighs between three and four pounds, and so on. It works well with locating cats, as well. It seems reasonably clear, on the other hand, that the literally false may produce achievements of understanding, and illuminate truths exceeding those of yardsticks and scales. It will help, initially, to recognize that in the real world, as opposed to the contrived worlds of the logician and mathematician, one commonly deals not so much with this and that as more and less. The serious question is often not so much counting discrete statements and sorting out truth values as producing a state of consciousness which is well, or better, connected with reality. The following five statements are all literally false, but they produce understandings, awarenesses of personality, history, culture, and such which are more or less illuminating or mistaken, more or less true or false, than comparable understandings of particular, literal statements.

1. Where I and my horse shall trod, no grass shall grow again.
2. The Russian Revolution began under the belly of a Cossack's horse.
3. Civilizations rise and fall.
4. We live ephemeral lives in the shadow of great ideas.
5. History is a slaughter bench on which are sacrificed the happiness of peoples, the wisdom of states, and the virtue of individuals.[49]

---

49. These examples are intended to be illustrative of succinct, figurative claims of a sort likely to produce rich understandings, whether correct or incorrect, and to whatever degree, locutions which communicate with a depth and impact which would be difficult to attain with literalistic or quantitative substitutes. Attributions, accordingly, are irrelevant to the point at issue. On the other hand the first locution is attributed to Attila; the second is familiar, but I do not know who first said it; the third might be attributed to any number of historians, for example, Vico, Spengler, Toynbee, etc. The fourth is a line from a poem by the Russian poet, Yevgeny Yevtushenko; and the last is suggested by a remark in Hegel's *Philosophy of History*.

Recourse to figurative discourse seems inevitable, and not regrettably so, if one wishes to deal with a great many matters of human importance, most of which do not lend themselves to quantitative predicates, such as spatial and temporal locations, sizes, shapes, masses, speeds, and such. For example, an exact physical description of the positions, velocities, and states of all molecules at the battle of Waterloo would tell us nothing about the battle of Waterloo.

As soon as one drops the paradigm of naive referentiality and accepts the paradigm of referential understanding one uses the ladder of literal falsity to reach levels of understandings, which, well founded or not, are distinctively human in nature, and afford the only means wherewith a human mind is likely to comprehend the more serious realities within which it finds itself. Herein one finds the most serious aspects of lexical anchoring or connectivity, the most serious exposition of the relation of words to the world.

There is nothing really that unusual in this matter, as there is no clear distinction between literal and figurative discourse; rather, a continuum is involved, moving from, say, "Plato is the Cadillac of philosophy. whereas Aristotle is the Jeep Cherokee of philosophy" through "Bill is a good egg" and "Susan is a knockout"' to "the neck of the bottle," "the foot of the hill," "the eye of the needle," "the heart of the matter," and so on, to, at the far end of the continuum, something like "the cat is on the mat." In fact, the distinction between literal and figurative may be closer to that between familiar and newer, older and fresher, passive and active, pedestrian and striking, sober and sparkling, usual and unusual, sand and diamonds, or such, than to true and false, meaningful and meaningless, or such. Pretty clearly, figurative speech, in particular, metaphor, lies at the root of much of language, is ingredient in its meristematic aspects, and thrives where language is most alive, for example, in the humanities. Accordingly, figurative discourse may be eschewed only at the risk of intellectual vapidity, paralysis, and stagnation. New visions, better visions, require new languages, better languages. It is on the wings of words that thought takes flight, discovering new horizons, seeing things unseen before. To be sure, dangers abound. It is easier to locate the cat on her mat than to see the world as she sees it, than to understand her.

The emphasis, surely, is on communication, and the achievement of understanding, or its seeming achievement. The truth conditions of 'the cat is on the mat' are reasonably clear, whereas those of 'Hellas was once the light of the world' are less so. This is not a reason for restricting notions of cognitivity and meaningfulness to, say, the locations of cats, but it is a reason for wariness, generosity, and a willingness to explicate and supply reasons. Certainly truth is not alien to figurative discourse. Presumably one might supply reasons for and against Hellas having been once the light of the world, say, a few senses in which it might be taken as true, or plausible, and a few in which it might be taken as implausible, or false. Opening the gate of such a saying may reveal multitudes. On the other hand, sometimes figurative discourse is either obviously false or unintelligible. For example, 'Scranton, Pennsylvania, was once the light of the world' would most likely be taken as false, whereas 'Baudelaire is the pancreas of French literature' would presumably be taken as either false, period, or unintelligible, or not worth figuring out something that it might mean. An additional danger is that if one puts meaningful words together in grammatical sequences, someone or other, with enough ingenuity, determination, and stamina would probably be able to come up with something that it might mean, at least figuratively. And therein doubtless is found the salvation of many modern poets, and perhaps some modern philosophers. After all, as is well known, whatever can be said, can be said obscurely.

This study, I take it, has made it clear that philosophy has new things to do, and new places to go. Revolutions are afoot. We have done new things, and gone new places. We have tried to recognize and deal with, however inadequately, several aspects of the current scientific and technological revolution, and some of its philosophical challenges, challenges as profound as rethinking the nature of life and deciding human destiny. I think it is also clear that the conceptual structures of the past developed to deal, more or less successfully, with an old world and familiar realities are inadequate to deal with a new world and unfamiliar realities. How shall we conceive the new and unfamiliar? This is important because human reality, as opposed to whatever lies beyond human sensibility, if anything, is a function of data and the interpretation of data. Human experience is not passive but active; the world of human experience is, in effect, a manufactured product, a cognitive artifact, in which the raw material of sense is shaped and organized in terms of a conceptual structure, shaped and organized into our world of human experience. And to the extent that our human reality is a function of our conceptual structure, and our conceptual structure is malleable, so, too, to that extent, is human reality. Philosophy then, in this new world, can no longer pretend to be simply descriptive, but must become creative, must become prescriptive, and thus is initiated the inevitable adventure of philosophical prescriptivism.

So, let us now consider four major criteria pertinent to the assessment of classificatory hypotheses, or prescriptions, the criteria of precision, clarity , utility, and beauty.

3.31 Precision

Outside contrived or artificial precincts, such as those of logic and mathematics, where a legitimate substitution instance of 'p' is always true or false, and not both, and a number is either the number two or it isn't, precision is generally a relative matter, indexed to the purposes involved. To be sure, precision may be imposed in any subject matter, provided one is willing to pay the price, for example, specifying the number of hairs, their length, condition, distribution, and such, with respect to being bald or not, but, in the practical world, one is inclined to make do with a pragmatic version of precision, namely, precision to the extent needed or serviceable for the task at hand at the time in question. On the other hand, all things considered, precision, on the whole, like clarity, utility, and beauty, is clearly a classificatory virtue.

Accordingly, at least within limits, a good classificatory scheme , or conceptual structure, should be both logically and verbally precise. Certainly one wants to organize data in a logical manner, and this requires that the classificatory structure avoids the fallacies of incomplete classification and crossclassification. The first fallacy might be thought of as the "no place" fallacy, and the second as the "more than one place" fallacy. For example, if one were, say, at the age of three, dividing paintings into those which were in oils and those which were beautiful, one would have succeeded in producing a classificatory scheme which managed, in one dynamic swoop, to commit all the fallacies in sight. For example, our three-year-old's schematism would not allow for ugly water colors, thus committing the fallacy of incomplete classification, or the "no place" fallacy, and would require putting beautiful oil paintings in two different places, thus committing the fallacy of crossclassification, or the "more than one place" fallacy. The ideal here is to have a place for everything and no more than one place for everything, or, in short, to have one and only one place for everything. Our three-year-old's schematism is easily rectified, perhaps by his four-year-old birthday. For example, he might set up things in a variety of ways, of which three are suggested.

Paintings

In Oils	Not in Oils

Paintings

Beautiful	Not Beautiful

Or, more complexly.

Paintings

In Oils	Not in Oils	
		Beautiful
		Not Beautiful

The last three schematisms satisfy the logical requirements of precision. There is one and only one place for every possible painting. There are, of course, many problems which would remain, say, what counts as a painting, what of a painting partly in oils and partly not, how do we decide what is beautiful, and so on, but these are not logical problems. They are criteriological problems. Now it will be noted that logistic criteria are easily satisfied. For example, they would be satisfied by the following schematism:

Everything that exists

Glunks	Nonglunks

But, for 'Glunk' and 'Nonglunk' to be more than class variables, say, 'A' and 'A'' (A Prime, the complementary class to class A), they or 'Glunk' should be defined. A common approach to this sort of thing, schematically, would be something like this:

> a is a member of Class C = df. a is characterized by Property P

We might then define the class of glunks as follows:

> a is a member of the class of glunks = df. a is characterized by the property of vombling on Tuesdays

Here we have satisfied the criteria for both logical and verbal precision. On the other hand, as we do not know, presumably, what it is to vomble on Tuesdays, just as we might not be clear on what it is to stand in the good graces of Khnum, god of the Ninth Cataract of the Nile, or what it might be to unwittingly further the ends of the Absolute, we must move to our second criterion, that to be used in the evaluation of classificatory hypotheses, namely, clarity.

3.32 Clarity

Precision and clarity are not the same thing. For example, the number '2' is as precise as might be desired, but what the number '2' might be, if anything, is less clear.

Here we come, alas, again to the sort of intellectual watershed earlier noted. What counts as meaningful? On such a reef many an analysis might founder. Our proposal, or recommendation, is to favor, where possible, empirically clear criteria, namely, criteria to which human experiences are relevant, sooner or later, directly or indirectly, actually or imaginatively. This would presumably give us a common ground, at least in many cases, between the A and the B side of the "division," or "watershed." The major difference then would be that certain locutions on the A side of the "watershed" might not be regarded as meaningful on the B side of the "watershed," for example, 'Zeus is everywhere'. On the other hand, a scientific theory with testable consequences, in theory or in fact, would presumably be regarded as meaningful on both sides of the "division," or "watershed." My sympathies, obviously, are on the B side of things, but if someone wishes to regard himself as standing in the good graces of Khnum, or as unwittingly furthering the cause of the Absolute, it seems to me he is entitled to do so. It might make no sense to me, or little sense, but this would be, in effect, to impose my notions of meaningfulness on the other fellow, which would be to beg the question, since what is most at issue is what is and is not meaningful. Presumably the favored conceptual structures would reflect such differences. Too, as noted before, first, questions of meaningfulness are best handled locally, attention being devoted to specific cases, rather than broadly, as in attempting to devise a general, one-size-fits-all meaning criterion, and, second, it is desirable to recognize degrees of meaningfulness, as, one supposes, very few things are absolutely meaningless, though some may come close. What is most fascinating to me is what might be accounted meaningful on the B side of the "watershed," where I am most comfortable. For example, let us suppose that there is no empirical evidence whatsoever for the existence of tachyons, a particle the speed of which allegedly exceeds that of light. Or, say, let us suppose that there is no empirical evidence whatsoever for a supposed scientific theory, say, some version of string theory. Rather than dismiss such notions high-handedly as "metaphysical," in some abhorrent, scandalous sense, unfit to be mentioned in mixed company, or as some sort of unintentional science fiction, one might recognize that it is easy enough to imagine something getting from point A to point B in less time than a beam of light, and, in the string-theory case, that there may be fewer subatomic particles than were dreamt of in the philosophy of one physicist or another. Even if the string theorist does not mean by strings what most of us mean by strings, something, at least, is conveyed. To be sure, something is probably also conveyed by the notion of standing in the good graces of Khnum and by unwittingly furthering the projects of an Absolute. Gradations are involved.

How are we to understand "experiential relevance"? Clearly experience is relevant to the assessment of something like 'The cat is plaid'. Similarly, it seems we understand locutions such as 'A plaid cat watched Caesar cross the Rubicon', 'On Epsilon Eridani 2 a plaid cat exists', 'A plaid cat will be born in Minneapolis tomorrow', and so on. Similarly, 'Some plaid cats are virtuous' and 'Some plaid cats are thinking of Edinburgh' would seem meaningful. And, indeed, 'Plaid cats like haggis' and 'Plaid cats bring good luck to Scotsmen' would not only be meaningful, but, if there are no plaid cats, would be true, given the truth conditions of a prominent "modern logic."[50]

Relevance to human experience then means, in effect, relevance to possible, conceivable, or imaginable human experience. One usually, in one way or another, knows what one is talking about. For example, one can imagine what it might be like to be a cat and be thinking of Edinburgh, though cats are more likely to be thinking about dinner, and, similarly, one might imagine what the walls of Carcassonne might have witnessed, and remembered, although walls witness nothing and remember nothing. If one does not know what one is talking about, then there is little point in talking about it.

In any event, a viable conceptual structure should be experience oriented; it should be capable of being applied to the realities of experience; it should not be allowed to collapse into verbalism, and vacuity; in such a case it would be largely empirically useless, and, if one is on the B side of the "division," or "watershed," one wants the conceptual structure to be empirically useful. Its point, after all, is not to organize words or construct a music or mathematics of ink marks or noises but to organize real data in such a way as to produce an intelligible world.

3.33 Utility

Utility is complex, and it may be diversely explicated. We will concern ourselves with five aspects of, or dimensions of, utility, namely, cognitive, practical, axiological, philosophical, and what might be broadly spoken of as "human utility." These categories are substantially independent, but, as would be expected, there is some overlap and interpenetration.

---

50. In the following, 'There are no plaid cats' is shown to logically entail 'All plaid cats bring good luck to Scotsmen'.

1. ~ ($\exists$x) Px	
2. (x) ~ Px	1, Quantifier Negation
3. ~ Py	2, Universal Instantiation
4. ~ Py v Gy	3, Logical Addition
5. Py $\supset$ Gy	4, Implication
6. (x) (Px $\supset$ Gx)	5, Universal Generalization

3.331 Cognitive Utility

Cognitive utility is truth-value relative. For example, does the system in question, the underlying conceptual structure, facilitate the formulation of empirical hypotheses? Recall, the question of the number of languages spoken along the river. We noted that there is no true/false answer to that question without a relevant conceptual structure, for example, one which would allow for an empirically discernible distinction to be drawn between dialects and languages. A satisfactory system from the point of view of cognitive utility should simplify, clarify, organize, and illuminate data.

3.332 Practical Utility

We have recognized that data may be diversely organized and interpreted, and, indeed, logically, any given set of data is susceptible to an infinite number of possible interpretations. This does not entail, of course, that there is nothing to choose from amongst these interpretations. Whereas all might be cognitively feasible perhaps few would be intellectually attractive. For example, some might be too simple, ignoring interesting differences, and some might be too complex, multiplying distinctions, so to speak, beyond necessity, even reason. Some systems are more easily applied than others. Consider the development of a propositional logic and something as simple as the number of truth-functional particles embedded in the logic. It is possible to do propositional logic with a single truth-functional particle, for example, "not both true," or "both false," the two "Sheffer strokes." This produces an elegant logic, and constitutes a brilliant *tour de force* of intellectual economy. On the other hand, from the practical point of view, it is much easier to do propositional logic with a larger number of particles, for example, "not," "and," "or," "if, then," and "if and only if." Similarly, the number and nature of rules in a logistic system may vary from system to system. Once again, practicality suggests a rational compromise between elegance and utility. Similarly, a given way of doing something might be feasible if one were willing to transform a system, but such a transformation might entail modifications, dislocations, alterations, even catastrophes, elsewhere in the system, which would, from a practical point of view, militate against the change. An example of this is Peano's famous "Question-Mark Operator"

$$a/b ? c/d = a + c/b + d$$

A substitution instance here would be: '1/2 ? 1/3 = 1 + 1/2 + 3', or '2/5'.

On the other hand, accepting this operator would require sacrificing the principle that equals substituted for equals yields equals, as when, say, '2/4' is substituted for '1/2'. In such a case, the question-mark operator would give us:

'2/4 ? 1/3 = 2 + l/4+ 3', or '3/7', which is not equivalent to '2/5'.

In sum, practical utility is of great importance. A system which generates practical problems is, to that extent, less desirable than a system which does not generate such problems. Concepts should integrate with other concepts. An analysis should not outrage custom, standard practice, common sense, and such, without good reason.

3.333 Moral Utility

As noted earlier, value judgment is not only universal but necessary in the human condition; without such judgment action, for a self-governing creature, such as a human being, would be not only impossible but inexplicable. Even recourse to inactivity or deciding by means of a random-selection device represents the judgment that that is to be preferred over something else. The alternative would be automatistic behavior, or external manipulation, in which latter case the value judgments would be those of the manipulating intelligence, not that of the seeming human being itself. These observations are independent of the question of the cognitivity, or truth-value possession, of axiological judgments. It seems obvious, however, that such judgments possess at least an individual-relative, or conditional, or instrumental, cognitivity, and are shown to be true or false by their life consequences. Furthermore, a viable society requires, and must attain, sooner or later, if it is to survive, an axiological consensus, hypocritical or not. Even the sociopath recognizes the wisdom of acting as though he subscribed to a particular moral consensus. With its rules a society of devils is possible. Without its rules, it would collapse.

Accordingly, it seems that a good classificatory scheme, or conceptual structure, should make it possible to deal intelligently with the local axiological universe, just as a good classificatory scheme, or conceptual structure, should make it possible to deal intelligently with the local empirical universe. For example, a systemic-process morality, a default morality, will have its rules, distinctions, principles, priorities, criteria, and such. And it is as desirable to produce order, clarity, and consistency into an axiological system as it is into an empirical system, indeed, perhaps more desirable, considering the requirements for a viable society. Presumably a vague, confused, inconsistent empirical conceptual structure, such as that of contemporary science, with its paradoxes and anomalies, is less inimical to society, and less deleterious in its likely consequences to human welfare, than a vague, confused, inconsistent axiological structure with competitive or contradictory conceptualizations of rights, duties, responsibilities, legitimacies, priorities, privacy, dignity, discrimination, harassment, intimidation, consent, freedom, law, and such, conceptualizations with a variegated miscellany of very real effects on property, liberty, and life. An historical analogue to this is the attempt, in law, over generations, by means of multiplex distinctions, rulings, definitions, and principles, to propose acceptable conceptualizations of intent, negligence, liability, ownership, degree, what is and is not just or fair, what is and is not permitted, and so on. Indeed, law, like philosophy, will find itself confronted with its own "challenges of the future." And it will be prescriptive philosophy, legal or otherwise, which will illuminate a thousand paths, some of which will be later followed by law.

It is an interesting question as to whether or not a unified, harmonious system might contain, as consistent subsystems, both an axiological system and an empirical system. At present that does not seem clear. For example, a judge and a biologist might differ on what counts as a human being.

In any event, a conceptual structure which justified or favored social Darwinism, oppression, genocide, infanticide, disfigurement, mutilation, and such, might be consistent, but, in the views of some, at least, it would be objectionable. Its views, or some of them, would not be morally useful, but, at least from the point of view of some, repulsive, illegitimate, dangerous, or pernicious. It would not be useful in producing a moral world, as such a world might be understood, again, at least by some. Empirical truth is not to be perverted to the ends of morality, at least from the point of view of those on the B side of the "intellectual division," or "intellectual watershed," but if one is to choose between otherwise equivalent conceptual structures, that which is most morally useful would seem to be the optimum choice. This is not to deny that lies and mythologies may not have enormous social utility. Perhaps it is rather to suggest that truth is nice, too, and that lies and mythologies are more likely to divide, sicken, stunt, and kill than truth, which is not very good at that sort of thing, at all. Truth, one suspects, actually, has great social utility, and great moral utility, as well. Indeed, a conceptual structure which facilitates the pursuit of truth is likely to have considerable moral utility.

3.334 Philosophical Utility

The two classical vocations of philosophy have been, first, to learn the world, and, second, to live within it, in short, to understand the world, by whatever means possible, and, second, to figure out a way to live well within the world as one supposes it to be. Whatever might further these two projects then has philosophical utility. Rather clearly this book has been primarily concerned with the first moment in the philosophical project, that of understanding the world. On the other hand, how one lives within the world will be heavily influenced by one's understanding of the world. There is an intimate relationship between cognition and action, between knowing and doing.

What is usually less recognized is the extent to which the human world, as opposed to the physicist's world, is dependent on conceptualization, and that conceptualization is up to us. Indeed, even the obduracy of the human world, its intrusiveness, its coherence, is largely a result of choices which have been made for us, in one way or another, and which we ourselves might make. We will have it so. It would be easy to conceive of experience as unsettled, transitory, and chaotic, but we do not do so. We prefer constancy and simplicity, not the shifting, kaleidoscopic world we might envisage, which spins and blurs with every turn of the head, every focus of the eyes, a world in which an object would ramify into a thousand perspectives, shades and colors, and speeds, shapes, and sizes, in one light or another, at one distance or another. Our experiences are discounted, qualified, screened, denied, and revised, in order to produce a reliable, stable world, a world of constancies, a world in which it is easier to live. Some of these penchants are doubtless shared with other forms of life, and have been genetically coded, in virtue of survival efficiency. This does not obscure the fact, however, that an ordering has taken place, however achieved, and that, logically, it constitutes an ordering imposed upon what, in fact, if taken at face value, as correlated with the literal data of sense, would be a cognitive shambles. But these basic favors bestowed by nature indifferently and purposelessly on one species or another move to the background in the human case, where the vast amount of interpretation far exceeds that correlated with genetic selections. The world we experience, the human world, is the result of data and its interpretation, and data may be diversely interpreted. The conceptual structure is logically prior to facts and objects. Without the conceptual structure there might be data but there would be no facts or objects. The concept of fact is sophisticated, and that of an object even more so. Remember the languages along the river; remember just looking and simply seeing a giraffe. Axiology precedes epistemology.

This book is different from most philosophical books, as it is written not from a descriptivist point of view, but from that of a prescriptivist. Classically philosophers have taken themselves to be outsiders, taken themselves to look upon the world neutrally, objectively, and report on how they see it, or think it. Immanuel Kant was the first philosopher to recognize in a very serious way that the world we know directly, namely, the world of human experience, which he spoke of as the phenomenal world, is largely the way it is because we are the way we are. It is not so much that it is the way we see it as it is the way it is because of the way we see it. Whereas our approach is far from Kantian, it does owe much to the Kantian tradition, in the sense that the mind is not passive but active, and that the world of our experience is an artifact, a manufactured product, in its way, a function of data and the organization of data, of data and the conceptualization of data. The world is in part malleable, because conceptualization, which recognizes, selects, organizes, and interprets data, is malleable.

And so, to philosophical utility.

Indeed, this book has been largely an exercise in philosophical utility.

One would like to understand, interpret, and cope with new realities, new possibilities. Amongst these possibilities we have discussed reprogenetics, cloning, gene technology, artificial intelligence, and artificial life. We have tried to propose useful criteria pertaining to both personal identity and species identity.

Does a new conceptualization felicitously solve or infelicitously create new philosophical problems?

Ideally, a fruitful, clear, useful conceptualization introduces order and criteria, enabling one to organize the data of experience in such a way as to produce a somewhat better human world, one more comprehensible, one more public, one more congenial, one in which it is easier and more satisfying to live. This is in no way a license to rethink a world in terms of a particular politics, religion, or ideology, nor is it a recommendation to conceptualize arbitrarily. The systemic-process conceptual structure associated with common sense is of great value, and is to be best respected to the extent that it is judicious and effective. It is so familiar that its prescriptive origins may not even be noticed. But in the face of onrushing scientific and technological change it requires extension and revision. This book has tried to make that clear.

Is a "common-denominator" conceptual structure possible?

One hopes so.

Given the reality of human nature, which, like dog nature and cat nature, is a matter of biology, and given the overwhelming commonality of genes amongst human organisms, it seems not unlikely that there might be discovered conditions, as with plants and other animals, under which such organisms might thrive, might well, even exuberantly, flourish. And if this form of life should be inveterately cruel, selfish, self-seeking, and exploitative, presumably conditions might be discovered in terms of which a consistent, compromising civil order might be maintained even amongst such unusual and troublesome creatures, in which a rational prudentiality, if nothing else, aside from indigenous compassion and altruism, would recommend at least a wary, harmonious coexistence.

In the past, again and again, philosophy has taken new shapes, new directions, under the imperative of making sense of new prospects and practicalities, of new devices, inventions, discoveries, and possibilities, understanding them, and incorporating them in a new world view. It is my speculation that there will someday be a movement in philosophy, an attempt to come to grips with the new science, the new technology. Doing this will require new categories, the stretching and testing of old ones. Philosophy, it is my speculation, in this generation, or the next, will undergo a transformation of subject matters analogous to the widespread repudiation of vacuous, prescientific speculation in recent centuries. Philosophy, as the business of mapping a reality, and finding ourselves within it, is challenged. It is challenged again, by the new science, the new technology.

Suppose that our friend Jones underwent a variety of transformations. At any given point in this succession of transformations we might ask ourselves if Jones should still count as human. If we understand what it is to be human we presumably have an immediate answer to such a question; lacking an immediate answer we must infer that we do not understand what it is to be human. This is interesting, particularly for humans. If we are naive cognitivists, of course, we will think deeply about the problem and when we come up with an answer which makes us happy, we will claim that we have discovered it, what it is to be human, and that it was there all the time, though, to be sure, it was not easy to locate. (If it had been, our achievement would have been less.) On the other hand, if we are a prescriptivist, we will suppose it as likely, and, indeed, far more likely, that we have created, or invented, an answer which seem to us to have philosophical utility. And so, too, for questions having to do with the nature of the human species.

This book, as is often the case with broad philosophy, as opposed to narrow philosophy, a book dealing with forest philosophy, so to speak, rather than tree philosophy, or "molecule-of-bark philosophy," a book dealing with global philosophy rather than local philosophy, incorporates a vision.

In the end it seems likely that strife, power, influence, cruelty, force, and war, will continue to make the world what it is, as they have in the past, under whatever flags and veils. Still, there is always the hope, and the possibility, that human nature is partially rational, and sensitive, if selectively so, to the imperatives of consistency and sympathy, and it would be foolish to deny the fact that the human being, amongst his various properties and anomalies, is not unoften a thinking, caring animal.

3.335 Human Utility

Although human utility would be difficult to define in any satisfactory manner, it seems such a category should be included amongst considerations pertinent to the assessments and choices amongst classificatory hypotheses or prescriptions.

"Human utility," as it is understood here, has less to do with the necessities of life, such as food, shelter, and clothing, or even the amenities of life, such as comforts and conveniences, but largely to do with the production of human welfare. For example, does the conceptual structure enhance life? Does it energize human beings, even exalt them? Does it contribute to a cooperative, thriving, civil, harmonious, productive society? For example, dividing human beings up into, say, barbarians and nonbarbarians, Aryans and nonAryans, the clean and the unclean, true believers and infidels, or such, is likely to produce resentment, envy, hatred, jealousy, prejudice, bitterness, dissension, and such. Assuming one disapproves of such things, a conceptual structure should avoid exacerbating social frictions. An emotionally satisfying society in which human beings, within reason, are free to pursue their own interests and work out their own destinies, would be a society in which human utility, if not maximized, would not be reduced and jeopardized. Needless to say, conceptual schemes might be designed to promote narrow ends, to enhance the welfare of A over B, and so on, but that would be to presuppose different criteria against which one might measure classificatory schemes. The one proposed here, obviously, harbors a familiar agenda of "common-denominator humanism," compatible with desiderata of civility, courtesy, tolerance, diversity, and such.

3.34 Beauty

Abstract structures, such as those of logic and mathematics, have their own canons of beauty, stately, formal, and austere, a beauty only too obviously "not of this world," so much so that to many the truths of, say, geometry, have seemed to require their own world, one far removed from the perils of time and space, of age and particularity, one much unlike that of the confusions and sorrows, the disruptions and pollutions, the humble waxings and wanings, and perishings, indigenous to quotidian existence. In touching such a world, if only with the mind, if for only a moment, it seemed one had touched the permanent and secure, that one had looked, however briefly, on the holy, the immortal, the changeless, the imperishable, the eternal, and sacred. In many a mathematician and logician one finds a Platonic streak, a suitable awe, an awareness of the unchanging and permanent, of something in its way more reliable and real, more secure and unthreatened, than mountains and pyramids, than the flight of days and nights. This kind of balance and proportion, of precision and elegance, can characterize a classificatory schematism, a conceptual structure. Such a structure, with its necessary interpretation, its tie to the world we know and in which we live, is surely a far cry from a succinct proof, a deftly axiomatized system, a geometry or such, but it does have an abstract structure with its own aesthetic elements. Of two classificatory schemas, each equivalent in organizing a domain of data, it is permissible to adopt the more beautiful.

Why not?

3.4 A Normative World Vision: (A Desiderated *Weltanschauung*)

On the "descriptivist" approach to these matters, one can, in theory, be axiologically neutral, rather as in measuring boards and weighing rocks. I think this naivety or arrogance is misplaced and founded on an illusion. "Scratch a descriptivist, find a prescriptivist." On the other hand, the prescriptivist, as opposed to the descriptivist, explicitly acknowledges the serious element of selection, and thus value judgment, involved in organizing data, and thus, in his way, its role in creating the human world.. Surely we have, from time to time, noted the essentiality of value judgment in the philosophical *engagement* or project. We have also noted the levels of such judgment, and the possibility of axiological diversity. Whereas much of what we have done requires no more than a conditional or instrumental, or relativistic, cognitivity, for value judgment, a truth or falsity relative to the attainment of particular ends, aims, and goals, it is almost inevitable that a very natural question should arise, one pertaining to the possibility of a final criterion, specifying a final end, aim, or goal, a *summum bonum*, or highest good, in terms of which axiological disagreement might be resolved.

The first thing to note is that such a criterion is not only logically possible, but empirically possible. We will consider three sorts of possibilities, all of which, one supposes, would be unpleasant, or, at least, objectionable. These three possibilities might be entitled iron-heel totalitarianism, velvet totalitarianism, and divine totalitarianism, all of which have in common the imposition of a *summum bonum* by force, of one sort or another. In "iron-heel totalitarianism," we have a collectivism which has the military and police power to articulate and enforce a particular vision of the world, with its attendant restrictions and coercions.. Collectivists, as is well known, feel it incumbent upon themselves to impose their own views and values on others, who may be too uninformed, stupid, or perverse to embrace them voluntarily. Most folks are to be oppressed for their own good, and liberated into a state of political correctness. Where propaganda fails the recalcitrant or dissatisfied may be rehabilitated by the application of suitable techniques in concentration camps, prisons, dungeons, *gulags*, and such. "Velvet totalitarianism," on the other hand, relies on social engineering, conditioning regimens, education, media manipulation, and such, contrived to produce a cooperative, docile populace, most of which has no inkling as to what has been done to them, why they feel as they do, and why they think as they do. They do not even know they are in prison. In this manner, collectivist ends are achieved subtly, gracefully, efficiently, and without too many hard feelings. Where this sort of thing proves ineffective, the recalcitrant or dissatisfied may be rehabilitated by the application of suitable techniques in concentration camps, prisons, dungeons, *gulags*, and such. In divine totalitarianism, we need a divine entity who is interested in these matters, for some reason, and takes it upon itself to impose its views of right and wrong on others. If we take a Calvinist view of these matters the divine entity does not impose certain views because they are right and wrong, but they are right and wrong because it imposes them. The divine entity is not to be subject to external conditions or outside influences. That would be a divine entity subsidiary to something else, not a "super divine entity," which seems to be desired, but a qualified, or conditioned divine entity, which is not desired. Such things, right and wrong, depend on its will. Its will makes things right and wrong. On this approach, for example, if the divine entity wished to make murder right and mercy wrong, then murder would be right and mercy wrong. We shall suppose that we ignore the middlemen here, who tell us what the divine entity thinks about these things, and make their living doing so. We shall suppose that the divine entity itself, say, Zeus, shows up with a handful of thunderbolts and makes the matter clear himself. A few skeptics might request the divine entity's credentials but a few thunderbolts later most folks would let the matter subside. And we, of course, to preclude various academic and epistemic subtleties, shall suppose that it is indeed the divine entity himself, in this case, Zeus. Now it seems that the ideal strategy for a divine entity at this point, one who is interested in these matters, at all, as one suspects Zeus might not be, would be to add a miracle, by means of which the particular values, views, world view, and such, it desires is happily accepted by all, who, one supposes, do not even realize they have been the victims of a miracle. Thus, we see, in a variety of ways, a normative world vision could be achieved.

To be sure, the preceding ways, sketched out above, in which a consensus with respect to a normative world vision might be achieved would most likely not be welcomed by most, unless it seemed likely that the totalitarianism involved would be one which they might manage, or, at least, find congenial.

Four remarks on utopias would seem to be in order.

First, most such proposals, aside from everyone being happy, and such, tend to be vague, if not silly, like Marx and Engels' notion in the "German Ideology," of a heavily urbanized, industrial civilization in which folks will hunt in the morning, fish in the afternoon, rear cattle in the evening, and have intellectual discussions after supper, which supper, presumably, has been prepared by someone, a sort of Arcadia amidst the factories and power stations. To be sure, this is perhaps being unfair, but it is hard to be either fair or unfair to what is, in effect, a blank page. Many Utopias are like Rorschach Blots, on which a diversity of fantasies might be imposed, a template or *tabula rasa* which the potential adherent is encouraged to fill in as he might please. In cat heaven there are many mice. In mouse heaven, there are no cats.

Second, most utopias seem to presuppose a collectivism, of one sort or another, a substantially homogenized population, a colony creature of sorts, in which "we," as defined by those in power, is accorded precedence over "you," who are essentially powerless. This may appeal to entities, one dares not say individuals, who enjoy being cells in a social organism, but it is likely to be less appealing to entities who prize liberty, who prefer independence, who respect difference, who would wish to be able to pursue their own ends and frame their own destiny to the extent possible. This is not an objection, of course, but an observation. Paradise for some may be a place on the galley bench for others.

Third, in the light of the second point, above, one notes that utopias seem usually to presuppose a one-size-fits-all policy. There is no reason to suppose that even every individual who might desire to be a cell in a social organism would desire to be a cell in the same sort of social organism. For example, religious, fascist, communistic, urban, agrarian, and so on, social organisms would be quite different. As normally understood, one does not get to choose one's utopia; one finds oneself within it, stuck with it, with nowhere else to go.

Fourth, the general notion of a utopia, as commonly conceived, presupposes an unrealistic view of human nature, a failure to recognize its complexity, and its blend, so to speak, of centrifugal and centripetal elements. Perhaps the best way to understand the human being is to inquire into what it has done. The record of deeds speaks louder than the speculations of utopians. History gives us a good idea of what the human being has been, is, and is likely to be. It suggests not only the implausibility of utopias, but the likelihood that, if achieved, they would be soon, by an aggressive elite, intent on its own interests, twisted into oppressive, exploitative tyrannies. The many roads to paradise have one thing in common; they all lead to the gates of hell.

The existence of a human nature, despite millennia of trying to turn it into something different, a product, an artifact, or such, may be taken as a given, one not likely to go away, though one supposes it might be bombed out of existence, or perhaps genetically engineered out of existence.

As long, however, as it is in existence, one must put up with it, and make do as one can.

Where there is a nature, whether of a dog or a cat, or a human, or an elm tree, there will be conditions under which it will languish, and conditions under which it will thrive, indeed, under which it might flourish. And, as noted earlier, if a particular form of life, given that it is at least capable of simple arithmetic, should be inveterately cruel, selfish, self-seeking, and exploitative, presumably conditions might be discovered in terms of which a consistent, compromising civil order might be maintained, even amongst such creatures, in which a rational prudentiality, if nothing else, aside from indigenous compassion and altruism, which, after all, are as real as anything else, would recommend at least a wary, harmonious coexistence. The problem here, then, is not analogous to designing a society for lions and lambs, whose interests are clearly antithetical, but one for lions, whose interests, given some common sense, are theoretically compatible.

The sought-for normative world vision, then, one in terms of which normative disagreement need not be surrendered to the arbitrations of weaponry, is obvious, if not practical. It is a vision of an open, tolerant society, particularly of a global nature. It is a macroutopia, so to speak, a utopia consisting of diverse utopias, or micro-utopias, of different sorts, an "over-utopia," so to speak, sheltering an ecology of sub-utopias, a live-and-let-live world of worlds.[51]

Such a world of worlds, of course, may not be practical. It would, however, at least in theory, have worlds for those who wish to be subsumed in wholes and those who wish to climb mountains, sail seas, paint pictures, found businesses, and go their own way. It would have worlds for the secular, of various sorts, and for the religious; of various sorts, for the passive and active, of various sorts, and so on.

One of the attractions of the smaller utopias, which might be expected to change, and develop, generation by generation, and which might interpenetrate, rather as commune members might work in a noncommunal society, is that the conflicts of ways of life and values might be minimized, as each enclave would be left largely alone to pursue its own ends. In this sense the *summum bonum* becomes a noncompetitive set of diverse *summum bonums*, each tolerant of the other. Too, naturally, it is anticipated that an individual might change

---

51. Whereas the concept of an open, tolerant society is surely familiar, and for generations, my thinking along these lines has been much influenced by some of the work of Robert Nozick. I commend the reader to his *Anarchy, State, and Utopia*. Nozick's word is 'meta-utopia'. His analysis is brilliant and detailed; it is an intellectual *tour de force* by one of the most imaginative and finest minds to characterize recent philosophy, too, that of a friend, who is much missed.

utopias freely, trying, say, to find that which would be best, on the whole, for him.

The major *caveat* involved here, as suggested earlier, is practicality.

First, whereas there might be a diversity of social and cultural worlds there is only one world of resources. Pending the manufacturing and populating of space colonies or putting camps and cities on alien worlds, there is a single world, ours, in which resources are limited, and might be put to diverse uses. There is no reason to suppose that these will be benevolently and peacefully accessed and distributed. Might not banditry and wars take place amongst these small worlds, as amongst larger, more uniform nations? Indeed, a sense of differences seems to underlie most conflicts, and each world might seem quite different, and alien, to several others, whereas in a larger nation there is, at least, likely to be a geographical scattering and diffusion of differences, in which many such differences might then be less obvious, even submerged, in a larger whole. Too, in a larger world there might be a transcendent allegiance, or patriotism, to which such differences might be subordinated.

Second, many individuals, given freedom of movement, might gravitate to the more successful worlds, those better endowed with resources, more salubrious in climate, more prosperous, or such, and this would presumably tend to produce social and economic hardships. And if individuals were denied freedom of movement, this, too, might produce bitterness, and militate against the basic concept of choice.

Third, there is a clear problem with how this transition into diverse utopias might be brought about. Are current nation states to disappear, or relinquish their sovereignty, or divide up their territories somehow into various enclaves, or what? If not, what would be the relationship of the utopias to the nation state? Might they be substates, counties, organizations, corporations, or what? How are these substates, or microworlds, to come into being? Would their members pay taxes to the national state, have civic duties, be able to vote, be subject to drafts, or such? Would the nation state have the right of eminent domain, be able to run roads through their territories, build bridges across their rivers, and such? And would the smaller "worlds" be able to afford water, sewerage, infrastructure, and such? There might be clear answers to such questions, but only after the political, economic, and social relationships of the utopias to the housing territory

were clarified. Similar questions would arise with respect to the
police, the military, the courts, currency, trade, and such.

Fourth, in the face of change, and in the absence of force, and
often in its presence, there is a tendency for human aggregates
to disagree, divide, fragment, secede from one another, and such.
Religions, for example, afford numerous examples of schisms,
revolts, reformations, and such. It is frequently the case that
today's coreligionist is tomorrow's infidel or heretic. Civil wars
and internecine conflict of one sort or another might be expected
to occur from time to time within our microworlds, conflicts which
might or might not be resoluble within the worlds in question.
Once again the question arises of an apparatus of adjudication,
in this case for suppressing or resolving such differences.

Fifth, some human beings enjoy prestige, attention, wealth,
power, and such, and, in one way or another, or under one
pretense or another, will seek the same. If the track record
of history is any indication, many a border will be crossed.
Where disproportions of territory or resources are perceived,
or enlargements of territory and resources are thought to be
desirable, even imperative, expansionism and imperialism may
be anticipated. A plenitude of utopias may not preclude war but,
rather, atomize and multiply it.

Sixth, depending on how these things are worked out, it might
be the case that a radical decentralization of population might
be produced, each group becoming, for most practical purposes,
self-sufficient. This could eventually produce a neofeudalism,
with its accompanying manorial economies. Historically, such a
development results in isolation, ignorance, stagnation, poverty,
barbarism, and war. Civilization becomes a rumor, wondered
about, by those who can read.

Seventh, the human being, statistically, is not prone to tolerance.
Xenophobia and ethnocentricity may have been socially selected
for, through millennia in which kinder, gentler, less suspicious,
more peaceful, more welcoming groups were overcome and
supplanted by, even extirpated by, more aggressive, predatory,
warlike groups. Hate, as well as love, may have its role in group
survival.

Eighth, many human beings are morally driven, for better or
for worse. Much depends on the morality. Some microworlds
might feel obliged, for better or for worse, to interfere in the
internal affairs of other microworlds. One cannot just assume

that all possible beliefs, rituals, policies, and practices of one microworld will be found just different, and acceptable, by the inhabitants of another microworld. For example, it is one thing for the inhabitants of microworld$_1$ to tolerate, say, polygamy in microworld$_2$, and quite another to tolerate burning widows, mass crucifixions, sacrificing virgins, casting infants to crocodiles, setting fire to heretics, putting out minority religionists for lions in public spectacles, and such, particularly if they are of your religion, and so on.

Some grounds for hope, of course, are found in the recognition, anthropologically acknowledged, that overwhelming moral agreement exists crossculturally. Even a group of Amazonian head hunters, apparently, shares most of its moral views with the rest of the world. Beyond that, it seems quite probable that most moral differences which remain might be screened out as due to differences in the environmental situation and differences in background beliefs. For example, in a situation of severely limited resources, it might make sense to kill the old, however regrettably, to conserve resources for the young, the future of the village or tribe. Similarly, if one really believed that the welfare of the village or tribe, even its survival, was consequent on blood sacrifice, that would be, within the belief system in question, a comprehensible justification for blood sacrifice. Along these lines, nations routinely send thousands of young people into situations where it is clear that lives, and perhaps a great many lives, will be lost. The major difference here seems to be in the nature of the belief system involved. On the other hand, acknowledging enormous crosscultural moral agreement, and noting that much of the residual disagreement might be discounted in the light of environmental differences and differences in belief systems, it seems probable, if only as a logical possibility, that some moral disagreement might remain. The degree and nature of the disagreement might be crucial. For example, if the disagreements were trivial or peripheral, they might not interfere, or much, in attaining a normative world vision; they might not, at least seriously, obviate or subvert the achievement of a shared normative world view, which might provide an ultimate criterion in the light of which moral discrepancies might be rationally adjudicated.

But here, of course, one leaves philosophy and speculates on the nature of reality, empirical and human.

It is quite possible that no final criterion in these matters, no final normative world vision, will ever become the object of a general rational and emotional agreement.

For example, there may be no objective way to resolve the differences between those who favor natural liberty and those who prefer societal conformity, between those who favor private property and those who abhor it, between those who respect a free market and those who desire the public appropriation and management of all major economic resources, between those who fear the government and wish to limit its power and those who favor an all-powerful, omnipotent state, which will enforce their will on the community at large. One supposes that there may be incommensurable moralities. For example, we do not expect Saint Anthony, the desert father, to see eye to eye on moral matters with Petronius Arbiter, the master of ceremonies in the court of Nero, nor Saint Francis with Richard Plantagenet, the Lion Hearted; nor Nietzsche with Marx.

In any event, life is apparently going to go on, at least for a time, but, if so, it is clear, given what science and technology have done, are doing, and are likely to do, it will not be going on in the same way.

And so, as noted earlier, philosophy has new things to do, and new places to go.

We have done some new things, and gone to some new places.

The future looms.

This book is finished.

Philosophy isn't. It has just begun.

ACKNOWLEDGMENTS

IT IS DIFFICULT, AND, TO SOME EXTENT, IMPOSSIBLE, to acknowledge the many influences and sources which have contributed, to one extent or another, to this effort. I should like, however, to thank my colleagues and my students. I have had the good fortune to work in a department of philosophy without a party line, where intellectual diversity is not only tolerated, but welcomed. How rare that is in contemporary academia. Truth is complex, abundant, and not easy to come by, and no particular individual or group, despite its convictions in the matter, owns it. Let all paths be explored, and new ones found. In this book, borders have been crossed, and philosophy has carried her flags into a new country, the future. In how many departments would that be possible? There is a new world out there. So thanks to a department which is a genuine philosophy department, and not another pea in an establishment's pod. In particular, I would like to thank my colleague, Professor James N. Jordan who, years ago, as department chairman, not only acceded to my request to teach an earlier version of this course, but even provided a name for it, which I have retained, both for the course, and, now, for this book, *Philosophy and the Challenge of the Future.* Secondly, I would very much like to thank Richard Curtis, and his editorial and production staff, at E-Rights/E-Reads, Ltd., for support and encouragement. Here we have a publisher who is committed to an open market and a free press, where more than one voice may be heard, a press which, too, is not another pea in an establishment's pod.

www.ingramcontent.com/pod-product-compliance
Lightning Source LLC
Chambersburg PA
CBHW032145080426
42735CB00008B/589